AF678847

New Insights into Glomerulonephritis

Contributions to Nephrology

Vol. 181

Series Editor

Claudio Ronco Vicenza

New Insights into Glomerulonephritis

Pathogenesis and Treatment

Volume Editor

Nan Chen Shanghai

29 figures, 13 in color, and 26 tables, 2013

KARGER Basel · Freiburg · Paris · London · New York · New Delhi · Bangkok · Beijing · Tokyo · Kuala Lumpur · Singapore · Sydney

Contributions to Nephrology
(Founded 1975 by Geoffrey M. Berlyne)

Prof. Nan Chen, MD
Department of Nephrology,
Ruijin Hospital, School of Medicine
Shanghai Jiao Tong University
No. 197 Ruijin Er Road
Shanghai 200025
PR China

Library of Congress Cataloging-in-Publication Data

New insights into glomerulonephritis : pathogenesis and treatment / volume editor, Nan Chen.
p. ; cm. -- (Contributions to nephrology, ISSN 0302-5144 ; v. 181)
Includes bibliographical references and index.
ISBN 978-3-318-02389-3 (hbk. : alk. paper) -- ISBN 978-3-318-02390-9 (e-ISBN)
I. Chen, Nan, 1948- II. Series: Contributions to nephrology ; v. 181. 0302-5144
[DNLM: 1. Glomerulonephritis--etiology. 2. Glomerulonephritis--therapy. W1 CO778UN v. 181 2013 / WJ 353]
RC918.G55
616.6'12--dc23
2013008448

Bibliographic Indices. This publication is listed in bibliographic services, including Current Contents® and Index Medicus.

Disclaimer. The statements, opinions and data contained in this publication are solely those of the individual authors and contributors and not of the publisher and the editor(s). The appearance of advertisements in the book is not a warranty, endorsement, or approval of the products or services advertised or of their effectiveness, quality or safety. The publisher and the editor(s) disclaim responsibility for any injury to persons or property resulting from any ideas, methods, instructions or products referred to in the content or advertisements.

Drug Dosage. The authors and the publisher have exerted every effort to ensure that drug selection and dosage set forth in this text are in accord with current recommendations and practice at the time of publication. However, in view of ongoing research, changes in government regulations, and the constant flow of information relating to drug therapy and drug reactions, the reader is urged to check the package insert for each drug for any change in indications and dosage and for added warnings and precautions. This is particularly important when the recommended agent is a new and/or infrequently employed drug.

All rights reserved. No part of this publication may be translated into other languages, reproduced or utilized in any form or by any means electronic or mechanical, including photocopying, recording, microcopying, or by any information storage and retrieval system, without permission in writing from the publisher.

© Copyright 2013 by S. Karger AG, P.O. Box, CH–4009 Basel (Switzerland)
www.karger.com
Printed in Germany on acid-free and non-aging paper (ISO 9706) by Bosch Druck, Ergolding
ISSN 0302–5144
e-ISSN 1662–2782
ISBN 978–3–318–02389–3
e-ISBN 978–3–318–02390–9

Contents

IX **Preface**
Chen, N. (Shanghai)

Overview

1 **Primary Glomerulonephritis in Mainland China: An Overview**
Xie, J.; Chen, N. (Shanghai)

12 **Advances in the Pathology of Glomerular Diseases**
Weening, J.J. (Hilversum); Ronco, P. (Paris); Remuzzi, G. (Bergamo)

22 **Changing Spectrum of Biopsy-Proven Primary Glomerular Diseases over the Past 15 Years: A Single-Center Study in China**
Pan, X.; Xu, J.; Ren, H.; Zhang, W.; Xu, Y.; Shen, P.; Li, X.; Wang, W.; Chen, X.; Wu, P.; Feng, X.; Hao, C.; Chen, N. (Shanghai)

31 **Evaluation of Anemia and Serum iPTH, Calcium, and Phosphorus in Patients with Primary Glomerulonephritis**
Li, Y.; Zhang, W.; Ren, H.; Wang, W.; Shi, H.; Li, X.; Chen, X.; Shen, P.; Wu, X.; Xie, J.; Chen, N. (Shanghai)

IgA Nephropathy

41 **Role of Podocyte Injury in IgA Nephropathy**
Menon, M.C.; Chuang, P.Y.; He, J.C. (New York, N.Y.)

52 **Genetic Studies of IgA Nephropathy: What Have We Learned from Genome-Wide Association Studies**
Xie, J. (Shanghai); Shapiro, S.; Gharavi, A. (New York, N.Y.)

65 **Immunopathological Predictors of Prognosis in IgA Nephropathy**
Tomino, Y. (Tokyo)

75 **Treatment of Progressive IgA Nephropathy: An Update**
Wang, W.; Chen, N. (Shanghai)

Focal Segmental Glomerulosclerosis/Minimal Change Disease

84 **Different Targets for Treating Focal Segmental Glomerular Sclerosis**
Coppo, R. (Turin)

91 **Screening of ACTN4 and TRPC6 Mutations in a Chinese Cohort of Patients with Adult-Onset Familial Focal Segmental Glomerulosclerosis**
Zhang, Q.; Ma, J.; Xie, J.; Wang, Z.; Zhu, B.; Hao, X.; Yang, L.; Ren, H.; Chen, N. (Shanghai)

101 **Increased Risk of Treatment Failure and End-Stage Renal Disease in Familial Focal Segmental Glomerular Sclerosis**
Hao, X.; Xie, J.; Ma, J.; Wang, Z.; Zhou, Q.; Yang, L.; Pan, X.; Ren, H.; Chen, N. (Shanghai)

109 **Treatment and Prognosis of Primary Focal Segmental Glomerulosclerosis**
Ren, H.; Shen, P.; Li, X.; Pan, X.; Zhang, Q.; Feng, X.; Zhang, W.; Chen, N. (Shanghai)

119 **New Insights into Treatment of Nephrotic Syndrome in Children**
Wang, X.; Xu, H. (Shanghai)

Membranous Nephropathy

131 **Pathogenesis of Membranous Nephropathy: A New Paradigm in Evolution**
Glassock, R.J. (Laguna Niguel, Calif.)

143 **Short-Term Complications of Membranous Nephropathy**
Barbour, S. (Vancouver, B.C.); Reich, H.; Cattran, D. (Toronto, Ont.)

152 **Tacrolimus Combined with Corticosteroids in Idiopathic Membranous Nephropathy: A Randomized, Prospective, Controlled Trial**
Xu, J.; Zhang, W.; Xu, Y.; Shen, P.; Ren, H.; Wang, W.; Li, X.; Pan, X.; Chen, N. (Shanghai)

Membranoproliferative Glomerulonephritis

163 **Membranoprolferative Glomerulonephritis – Mechanisms and Treatment**
Appel, G.B. (New York, N.Y.)

175 **Re-Evaluation of the Classification System for Membranoproliferative Glomerulonephritis**
Lu, Y.; Shen, P.; Li, X.; Xu, Y.; Pan, X.; Wang, W.; Chen, X.; Zhang, W.; Ren, H.; Chen, N. (Shanghai)

185 **C3 Glomerulopathy**
Servais, A.; Noël, L.-H.; Frémeaux-Bacchi, V.; Lesavre, P. (Paris)

194 **Hepatitis C Virus-Associated Glomerulonephritis**
Tang, S.C.W.; Lai, K.N. (Hong Kong SAR)

Crescentic Glomerulonephritis

207 **Pathogenesis of Rapidly Progressive Glomerulonephritis: What Do We Learn?**
Chen, Y.-X.; Chen, N. (Shanghai)

216 **ANCA-Associated Renal Vasculitis – An Update**
Tesar, V.; Hruskova, Z. (Prague)

229 **Management of Crescentic Glomerulonephritis: What Are the Recent Advances?**
Li, X.; Chen, N. (Shanghai)

240 **Plasma Exchange in the Treatment of Rapidly Progressive Glomerulonephritis**
Chen, X.; Chen, N. (Shanghai)

248 **Author Index**

249 **Subject Index**

Preface

Chronic kidney disease (CKD) is a major health issue not only in China but also in other parts of the world. Approximately 10% of the global population has CKD and the progression of the disease will lead to end-stage renal disease (ESRD) – a devastating condition requiring costly renal replacement therapy. Given this background, extensive understanding of the pathogenesis of the disease and exploring novel therapeutical targets will help us to alleviate the disease progression, improve prognosis and reduce its impact on the global economic burden.

Currently, primary glomerulonephritis is the leading cause of CKD and ESRD in China whilst in Western countries it is diabetic nephropathy and hypertensive nephropathy. Much progress regarding pathogenesis as well as treatment of primary glomerulonephritis has been achieved recently. As a result of active involvement in global nephrology, clinical data from Chinese studies on primary glomerulonephritis have significantly contributed to the making of international guidelines and histological classifications on glomerulonephritis such as the recently published *Kidney Disease: Improving Global Outcomes (KDIGO) Guidelines on Glomerulonephritis* and *Oxford Classification of IgA Nephropathy*. As not many books have been specifically published on primary glomerulonephritis, we decided to further research this topic.

As readers approach this book they will immediately find it is different from the classic textbooks which contain comprehensive information on every aspect of nephrology. It focuses on the cutting-edge knowledge and delivers up-to-date information on primary glomerulonephritis: in the *Overview* chapter, the epidemiological data of primary glomerulonephritis, changing spectrum of biopsy-proven primary glomerulonephritis in China and pathological advances of glomerulonephritis are reviewed and discussed; in the chapter on *IgA Nephropathy*, state-of-the-art renal science regarding the update of the genetic basis of the disease, role of podocytes injury, treatment advances and prognosis predictors are well presented; in the chapters on *Focal Segmental Glomerulosclerosis/*

Minimal Change Disease and *Membranous Nephropathy*, recent clinical evidence of treatment as well as new insights of pathogenesis are introduced; in the chapter on *Crescentic Glomerulonephritis*, evidence from recent clinical trials as well as promising data from animal models are revisited and bench-to-bedside knowledge is shared; in the chapter on *Membranoproliferative Glomerulonephritis*, two types of membranoproliferative glomerulonephritis are introduced and its classification is re-evaluated, and there are many others from which I am confident readers will find the intellectual and practical value of this book.

Being the editor of this edition of *Contributions to Nephrology*, there are so many people that I want to thank for their efforts in producing this project. I am particularly indebted to our renowned and devoted authors whose contributions constitute the essence of this project. All the authors have done a superb job by utilizing vast amounts of data and ensuring the accuracy of the information they provide. This book could not have been accomplished without the extraordinary efforts by the staff of S. Karger AG.

'The road ahead is afar, and I will search long and hard' – this is an ancient Chinese proverb that addresses the difficulty and persistency in our pursuing the knowledge to understand our world and improve human health. However, I do believe a useful book will help us to keep pace with this ever-changing and rapidly growing world of science. I hope the perspective from a team of well-respected global contributors as well as our local prominent nephrologists will keep readers at the forefront of nephrology. May this book serve as a bridge to connect Chinese nephrology to the world and may the knowledge introduced contribute to improve the outcome of patients with primary glomerulonephritis to the maximum extent possible.

Nan Chen, MD
Department of Nephrology, Ruijin Hospital
School of Medicine, Shanghai Jiao Tong University
Shanghai, PR China

Chen N (ed): New Insights into Glomerulonephritis.
Contrib Nephrol. Basel, Karger, 2013, vol 181, pp 1–11 (DOI: 10.1159/000348642)

Primary Glomerulonephritis in Mainland China: An Overview

Jingyuan Xie · Nan Chen

Nephrology Department, Ruijin Hospital, Shanghai Jiao Tong University, School of Medicine, Shanghai, PR China

Abstract

Chronic kidney disease (CKD) has received increased attention as a global health problem because of its increasing incidence and prevalence, poor outcome and high associated costs. Primary glomerulonephritis (GN) is the leading cause of CKD and end-stage renal disease in China. Immunoglobulin A nephropathy (IgAN) is the most common type of primary GN and accounts for 40–50% of the adult patients undergoing renal biopsy. Interestingly, the reported prevalence of focal segmental glomerulosclerosis (FSGS) in China varies by geographical region, and the prevalence of membranous nephropathy is increasing. Familial FSGS and IgAN are common in China, suggesting that genetic factors are involved in their pathogenesis. The evidence to guide the treatment of these glomerular diseases is limited. Most Chinese nephrologists follow the Kidney Disease: Improving Global Outcomes (KDIGO) guidelines during their practice. Here, we review the current prevalence, etiology, clinicopathological features and treatment of primary GN in China.

Copyright © 2013 S. Karger AG, Basel

Chronic kidney disease (CKD) has received increased attention as a global health problem due to its increasing incidence and prevalence, poor outcome and high associated costs. The prevalence of CKD in China is 10.8% [1]. A similar rate was found in developed countries such as the USA (13.1%) [2], but the rate of awareness is only 8.2% according to our previous community-based study [3]. The leading causes of CKD and end-stage renal disease (ESRD) in China are different from those in the USA. The main causes of ESRD in the USA are diabetic nephropathy (46.7%) and hypertensive nephropathy (23.8%), whereas only 9.4% of the ESRD cases in the USA are due to glomerulonephritis

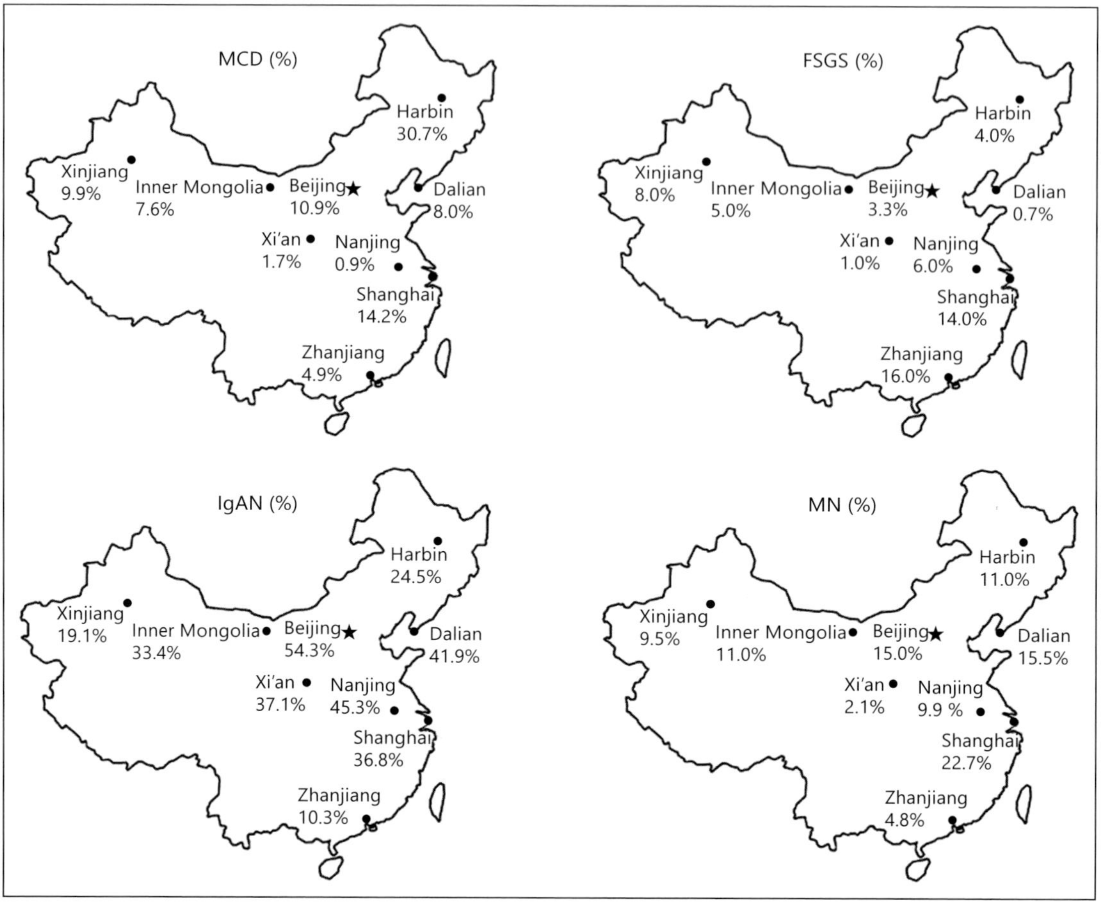

Fig. 1. Proportions of the pathological types of primary GN in different regions of China [5, 6, 10–14].

(GN) (2010 USRDS). According to the data from the China Pilot Dialysis Outcome and Practice Patterns Study (DOPPS), which was completed in 2011, the top three causes of ESRD in Chinese patients are GN (50%), diabetic nephropathy (20%) and hypertensive nephropathy (14%). Although the incidence of diabetic nephropathy and hypertensive nephropathy is increasing, GN remains the leading cause of ESRD in China. We have performed an epidemiological study that examined hospitalized patients in major hospitals in Shanghai, and we found that 55.3% of the patients had primary GN [4]. Data from other studies also showed that primary GN was the leading cause of CKD patients to undergo renal biopsies (61.7–68.6%) [5, 6]. The proportions of each pathological type of primary GN differ in various regions of China (fig. 1; table 1) and are

Table 1. Proportions (%) of pathological types of primary GN by geographical region

	USA [7]	Italy [8]	Japan [9]	Beijing [5]	Nanjing [6]	Shanghai[1]
MCD	13.0	9.3	10.5	10.9	0.9	14.2
FSGS	24.3	13.2	6.3	3.3	6.0	14.0
IgAN	18.5	34.5	54.2	54.3	45.3	36.8
MN	27.0	23.4	12.7	15.0	9.9	22.7
MPGN	9.6	6.6	2.6	1.1	3.4	0.8

[1] Data from Shanghai Ruijin Hospital.

also different from the proportions found in other regions of the world (table 1). In the past 20–30 years, the spectrum of primary GN has changed substantially. With improved living standards, better public health care and the early and effective antibiotic treatment of pharyngeal infections, the prevalence of poststreptococcal GN has declined sharply in China. Nevertheless, immunoglobulin A nephropathy (IgAN) remains the most common form of GN worldwide.

This review summarizes the prevalence, clinicopathological characteristics and treatment of primary GN in China, including minimal change disease (MCD), focal segmental glomerulosclerosis (FSGS), IgAN, membranous nephropathy (MN) and membranoproliferative glomerulonephritis (MPGN).

Minimal Change Disease

MCD is the most common cause of primary nephrotic syndrome (NS) in children and is also an important cause of primary NS in adults of all ages. MCD accounts for 46.9% of the pediatric and 25.3% of the adult cases of primary NS in China [5]. The first-line therapy for MCD is steroids. Although remission can be achieved in the majority of adult MCD patients using steroid therapy, 27.3% of them are resistant to steroid therapy. Steroid-dependent and steroid-resistant MCD are treated with a combination of prednisone and immunosuppressant agents such as cyclosporin A (CsA), cyclophosphamide (CTX), mycophenolate mofetil (MMF) and tacrolimus. A clinical study showed a 93% response to CsA therapy in children with steroid-resistant MCD [15]. Few studies have focused on adult MCD in China. Our unpublished data suggest that 84.8% of adult MCD patients achieved complete remission (CR) after steroid therapy alone and that 29.3% of the patients experienced at least one relapse during a median 20-month follow-up. A multicenter cohort trial reported that tacrolimus had a similar effect as intravenous pulse CTX but it resulted in a faster resolution of proteinuria [16].

Focal Segmental Glomerulosclerosis

Since the first clinicopathological studies of FSGS were published in the 1970s [17], there has been renewed interest because of the increasing incidence of this disease during the past 20 years [7]. The reported prevalence of FSGS in China varies by geographical region. For unknown reasons, there is an increasing prevalence gradient of FSGS from northern China to southern China (3.3% in Beijing, 6.0% in Nanjing, 14.0% in Shanghai, and 16% in Zhanjiang) [5, 6, 11]. FSGS and MCD are both podocyte diseases that are related to immunologic abnormality and podocyte injury. Although only a minority of FSGS patients have a family history of FSGS, genetic studies of familial FSGS have largely improved our understanding of this disease. During recent decades, a series of genes encoding podocyte-related molecules have been linked to FSGS; these include *NPHS1, NPHS2, ACTN4, TRPC6, INF2* and *MYO1E* [18]. Additionally, *NYH9* and *APOL1* are the genes most recently found to be correlated with FSGS and ESRD in African-American patients [19].

In the USA, approximately 50–60% of adults with FSGS have NS at presentation [20]. In our department in China, 97.7% of the FSGS patients present with proteinuria at initial onset, 27.7% with NS, 31.4% with hypertension and 38.2% with impaired renal function (eGFR <60 ml/min) [unpubl. data]. The KDIGO guidelines recommend that corticosteroid and immunosuppressive therapy be considered only in idiopathic FSGS associated with the clinical features of NS [21]. In one study, 32–47% of FSGS patients achieved CR during a 12-week steroid therapy, and 19–29% achieved partial remission [22]. However, few studies in China have discussed the treatment of adult idiopathic FSGS with NS. The KDIGO guidelines also suggest that an initial high dose of corticosteroids should be given for a minimum of 4 weeks, then a consequent high-dose corticosteroid therapy should continue for up to a maximum of 16 weeks, as tolerated or until CR has been achieved, whichever is earlier. However, from our experience, extending the steroid therapy from 8 to 12 weeks did not produce a higher rate of remission but may potentially increase the risk of side effects due to the steroid therapy. There are no guidelines for steroid therapy in Chinese adult FSGS patients. Because the benefits and risks of prolonged oral full-dose steroid treatment are unclear, future clinical studies should determine the best course of steroid therapy in Chinese FSGS patients.

Immunosuppressive therapy, including CTX, CsA and tacrolimus, is helpful in steroid-dependent and steroid-resistant FSGS patients. We reported that intravenous pulse CTX plus prednisone therapy achieved satisfactory therapeutic effects in steroid-dependent and steroid-resistant FSGS (55.6% achieved remission at 6 months and 66.7% at 12 months) [23]. Therefore, CTX therapy is still an appropriate choice for patients with primary FSGS, even in some refractory FSGS cases, and may be considered the first-line immunosuppressive agent due to its high effect-to-cost ratio. The KDIGO guidelines suggest that calcineurin inhibitors (CNIs) may be considered as a first-line therapy for patients with relative contraindications to high-dose steroids. A prospective multicenter Chinese trial showed that CR was achieved in 58.3% of steroid-resistant NS patients after 6 months of tacrolimus plus prednisone therapy [24]. However, in patients with steroid- and CTX-resistant NS, CR was achieved in 64.7% of them. Unfortunately, all of these previous studies were limited by a small sample size and a relatively short follow-up period, so better-designed clinical trials are required to confirm whether tacrolimus is more effective than other agents.

Familial FSGS is not rare in China, but it is often overlooked in clinical practice. We have found that approximately 10.7% of FSGS patients in Ruijin Hospital have a family history of FSGS. Compared with sporadic FSGS, familial FSGS is characterized by more severe pathological lesions, a lower response rate to therapy and a lower renal survival rate [25]. Causal variations in podocyte-related proteins are less frequently reported in familial FSGS patients in the Chinese population than in the Caucasian population [18]. This finding indicates that novel causal variations may underlie Chinese FSGS. New techniques such as genome-wide association studies (GWAS) and next-generation sequencing may help to explore novel genes of FSGS in the Chinese population.

IgA Nephropathy

IgAN is the most common form of GN worldwide and is the leading cause of ESRD in China. As shown in table 1, there is a marked regional difference in the frequency of IgAN amongst primary GN patients. IgAN occurs with greatest frequency in Asian countries and accounts for 36.8–54.2% of primary GN, but it occurs with modest frequency in the USA (18.5%) [26].

Various clinical manifestations have been documented in patients with IgAN. Our data show that 20.2% of IgAN patients had gross hematuria, 20.7% had NS and 46.9% had hypertension [27]. The extreme variability of the clinical course of IgAN has also been reported, ranging from a benign condition to progressive renal failure. In all, 15–40% of IgAN patients will develop the impaired renal function that eventually progresses to ESRD in 10–20 years [28–30]. Several semiquantitative pathological grading systems of IgAN have been developed. The Oxford classification, which is the most recently published IgAN grading system, is considered the most promising one to date. Several centers in China have performed validation studies of the Oxford classification [31, 32]. These studies confirmed the association of mesangial hypercellularity, segmental glomerulosclerosis and tubular atrophy-interstitial fibrosis with renal prognosis in patients with IgAN. Due to the high variability of clinical features and prognosis of IgAN patients, it remains a challenge to precisely predict the prognosis of IgAN patients. Several studies indicate that higher proteinuria, hypertension, impaired renal function, hypoproteinemia, hyperuricemia and histological grade are independent predictors of an unfavorable renal outcome [33, 34]. We have developed a four-variable equation (eGFR, serum albumin, hemoglobin and systolic blood pressure) to calculate a clinical risk score for disease progression based on data from 619 Han Chinese with IgAN [27]. Our risk score explained nearly 22% of the total variance in predicting ESRD in our patient population. The new progression risk score calculator based on these four baseline variables is now available on a website [http://www.columbiamedicine.org/divisions/gharavi/calc_progression.php] developed by our collaborator, and this new risk score offers a simple clinical tool for the risk stratification of IgAN.

Familial aggregation is widely reported in IgAN and accounts for 12.3% of IgAN patients undergoing renal biopsy at Ruijin Hospital. Familial IgAN has a worse prognosis than sporadic IgAN, so more patients with a family history progress to ESRD [35]. However, a recent study showed that patients with familial IgAN had similar outcomes to those with sporadic IgAN [36]. Three GWAS have been performed on IgAN patients, and these studies have identified seven independent susceptibility loci [37–39]; however, causal variations remain undefined, and further studies should be conducted to clarify the pathogenesis of IgAN. The pathogenesis of IgAN remains uncertain, but genetic, environmental and immunological factors [40] contribute to its development.

Although there is no specific therapy for IgAN, individual therapy should be conducted according to the patient's condition. We recommend angiotensin-converting-enzyme inhibitor (ACEI) or angiotensin II receptor blocker (ARB) treatment in IgAN patients with proteinuria levels <1 g/day. One recent study showed that ACEI plus calcitriol resulted in a safe decrease of proteinuria in patients with IgAN [41]. The KDIGO guidelines suggest that IgAN patients with persistent proteinuria over 2 g/day, despite 3–6 months of optimal supportive care, and with GFR >50 ml/min should receive a 6-month course of corticosteroid therapy. A meta-analysis suggested that steroid use is associated with a decrease of proteinuria and with a significant reduction in ESRD risk. Prednisolone plus CTX or azathioprine may be effective therapies in IgAN if the patients' serum creatinine levels are >1.5 mg/dl [42]. Another study reported that tacrolimus caused a rapid proteinuria remission in refractory IgAN patients. A 3-year follow-up study showed that MMF alone did not help to reduce proteinuria or delay the progression of renal function in IgAN patients [43]. Based on this evidence, the KDIGO guidelines do not suggest the use of MMF alone for IgAN. However, MMF alone may be helpful for reducing proteinuria in Asian IgAN patients, although more clinical studies need to be conducted to confirm this [44].

As another type of mesangial proliferative glomerulonephritis besides IgAN, non-IgA mesangial proliferative glomerulonephritis (MsPGN) has few investigations. The prevalence of MsPGN was reported approximately 10–20% of primary GN, however less occurrence was found based on data from Ruijin Hospital (<5%) [see X.X. Pan et al., pp. 22–30]. Because there is insufficient evidence on pathogenesis, prognosis and treatment in this group of patients, we will not describe this subtype in detail in the current review.

Idiopathic Membranous Nephropathy

Idiopathic membranous nephropathy (IMN) is a common pattern of primary GN. Our data suggest an increasing trend of primary GN amongst renal biopsies, from 7.1% in 2000 to 22.7% in 2009–2011 [see X.X. Pan et al., pp. 22–30]; this increase may be associated with changes in the environment, drug use, the spectrum of kidney disease and the age of the patients undergoing renal biopsy. The pathogenesis of IMN remains unclear. PLA2R is present in the normal podocytes and in the immune deposits of patients with IMN, indicating that PLA2R is a major antigen in this disease [45]. At Ruijin Hospital, NS is present in 78.5% of IMN cases and hypertension in 30.9%, and less than 5% of the IMN patients progress to ESRD [unpubl. data]. Approximately 10% of MN patients in Japan progress to ESRD, which is significantly less than those in Italy (33.3%) [9]. Individualized therapy for IMN should be given according to the patient's renal function, proteinuria, serum albumin level and severity of thromboembolism. From our experience, we suggest that IMN patients with normal renal function, proteinuria levels <4 g/day and serum albumin levels >25 g/l should be treated with ACEI or ARB combined with prophylactic anticoagulant and/or antiplatelet therapy, while patients with proteinuria levels >4 g/day and serum albumin levels <25 g/l should be treated with at least 6 months of steroids combined with immunosuppressants, preferably CTX. From our experience, for patients who have contraindications to CTX, we suggest that CNIs may be used.

Chinese studies reported that 68.8–88.0% of nephrotic IMN patients achieved remission after steroid plus immunosuppressive therapy [46, 47]. Furthermore, a Chinese intention-to-treat analysis suggested that a significantly larger decrease in proteinuria occurred in tacrolimus-treated patients than in CTX-treated patients, but no difference in remission rate between the two groups was found at the end of the 12-month follow-up [47]. We suggest that CTX is the best immunosuppressive therapy for nephrotic MN patients. Tacrolimus is also helpful in refractory nephrotic MN; in one study, 35.7% of patients with refractory nephrotic MN achieved CR after treatment with tacrolimus plus prednisone, and 42.9% of the patients achieved partial remission [48].

Membranoproliferative Glomerulonephritis

MPGN accounts for approximately 1–3% of adult cases of biopsy-confirmed GN [49]. With diverse clinical presentations, MPGN can occur at any age. The pathogenesis of MPGN remains uncertain, but it may be related to infection, autoimmune disease, complement dysfunction and monoclonal gammopathy. On the basis of electron-microscopic findings, MPGN is traditionally classified as primary (idiopathic) MPGN type I, type II, type III or secondary MPGN. In China, MPGN type I ranks first of the three primary types. A new classification system for MPGN based on immunofluorescence microscopy was developed recently which is helpful for understanding the pathogenesis of MPGN [49], however its relationship to the clinical outcome remains to be determined. The KDIGO guidelines suggest that adults and children with presumed idiopathic MPGN accompanied by NS and a progressive decline of kidney function should receive oral CTX or MMF plus low-dose corticosteroids, with the initial therapy limited to less than 6 months. However, the related clinical research is insufficient in China.

C3 glomerulopathy is a group of GN characterized by glomerular deposits of C3, with no or only limited glomerular deposits of immunoglobulin [50]. C3 glomerulopathy, associated with the dysregulation of the alternative complement pathway, can be classified as idiopathic MPGN type I, dense deposit disease, CFHR5 nephropathy, C3 GN and familial MPGN III. Unfortunately, no effective therapies have been reported.

In conclusion, primary GN is the leading cause of CKD and ESRD in China. The prevalence of GN varies between countries and even within China. The underlying reason for the variability is unknown but may be due to the different genetic backgrounds and environment of the people living in the different regions. The pathogenesis of primary GN is still largely unknown, but some promising findings have been reported, particularly from genetic studies because of the rapidly evolving technologies in this field. The treatment of GN is still a major challenge for nephrologists. The KDIGO guidelines for GN have been published, and these guidelines will help Chinese nephrologists develop more effective treatment strategies and improve patient care. Although we have implemented the KDIGO guidelines in clinical practice, further studies are still needed to validate and adjust the recommendations in the guidelines for the Chinese population.

Acknowledgements

This work was supported by grants from the National Basic Research Program of China 973, Program No. 2012CB517600 (No. 2012CB517604), the National Key Technology R&D Program (12-5)2011BAI10B00 (2011BAI10B06), the National Natural Science Foundation of China (81070568 and 81000295), the Key Program of Shanghai Science and Technology Commission (08dz1900502), and the Specialized Project on Scientific Research within the Healthcare Circle (201002010).

References

1 Zhang L, Wang F, Wang L, Wang W, Liu B, Liu J, Chen M, He Q, Liao Y, Yu X, et al: Prevalence of chronic kidney disease in China: a cross-sectional survey. Lancet 2012;379:815–822.

2 Coresh J, Selvin E, Stevens LA, Manzi J, Kusek JW, Eggers P, Van Lente F, Levey AS: Prevalence of chronic kidney disease in the United States. JAMA 2007;298:2038–2047.

3 Chen N, Wang W, Huang Y, Shen P, Pei D, Yu H, Shi H, Zhang Q, Xu J, Lv Y, et al: Community-based study on CKD subjects and the associated risk factors. Nephrol Dial Transplant 2009;24:2117–2123.

4 Chen N, Ma XB, Zhang JY, Mao PJ, He LQ, Yu JP, Ding XQ, Jiang GR, Gu Y, Zhang W, Wang NS, Mei CL, Ni ZH: Multicenter epidemiology survey of chronic kidney disease in the departments of nephrology in eastern China. 48th ERA-EDTA Congress, Prague, June 23–26, 2011.

5 Zhou FD, Zhao MH, Zou WZ, Liu G, Wang H: The changing spectrum of primary glomerular diseases within 15 years: a survey of 3,331 patients in a single Chinese centre. Nephrol Dial Transplant 2009;24:870–876.

6 Li LS, Liu ZH: Epidemiologic data of renal diseases from a single unit in China: analysis based on 13,519 renal biopsies. Kidney Int 2004;66:920–923.

7 Braden GL, Mulhern JG, O'Shea MH, Nash SV, Ucci AA Jr, Germain MJ: Changing incidence of glomerular diseases in adults. Am J Kidney Dis 2000;35:878–883.

8 Gesualdo L, Di Palma AM, Morrone LF, Strippoli GF, Schena FP: The Italian experience of the national registry of renal biopsies. Kidney Int 2004;66:890–894.

9 Sugiyama H, Yokoyama H, Sato H, Saito T, Kohda Y, Nishi S, Tsuruya K, Kiyomoto H, Iida H, Sasaki T, et al: Japan Renal Biopsy Registry: the first nationwide, web-based, and prospective registry system of renal biopsies in Japan. Clin Exp Nephrol 2011;15:493–503.

10 Li J, Sun L, Sun ZG, Wang B, Bai LL, Zhao ZY: Analysis of 1,042 cases renal biopsy in Dalian area. J Dalian Med Univ 2011;33:492.

11 Wu YQ, Wang Z, Xu HF, Jin XM, Zhou HZ: Frequency of primary glomerular disease in northeastern China. Braz J Med Biol Res 2011;44:810–813.

12 Yue H, Xu SR, Lu C, Liu WL, Jiang H, Song X, Qiao LP, AL: Analysis of the epidemiology and pathologic types of 1,323 patients undergoing renal biopsy. China Clin Pract Med 2008;2:59.
13 Zhao XY, Jia YF: Analysis of the Epidemiology and Pathologic Types of 616 Primary Glomerular Disease in Inner Mongolia, Chifeng Region. Hohhot, Inner Mongolia Medical College, 2010.
14 Lin HP, Liu HF: Analysis of 1,627 Renal Clinical and Biopsy Data. Guangdong, Guangdong Medical College, 2008.
15 Cao Q, Huang WY, Xu H, Zhou LJ: Long-term effects of cyclosporin A in children with steroid-resistant idiopathic nephrotic syndrome and outcomes of the patients (in Chinese). Zhonghua er ke za zhi (Chin J Pediatr) 2008;46:698–701.
16 Li H, Shi X, Shen H, Li X, Wang H, Xu G, Chen J: Tacrolimus versus intravenous pulse cyclophosphamide therapy in Chinese adults with steroid-resistant idiopathic minimal change nephropathy: a multicenter, open-label, nonrandomized cohort trial. Clin Ther 2012;34:1112–1120.
17 Report of the International Study of Kidney Disease in Children: Nephrotic syndrome in children: prediction of histopathology from clinical and laboratory characteristics at time of diagnosis. Kidney Int 1978;13:159–165.
18 Zhu B, Chen N, Wang ZH, Pan XX, Ren H, Zhang W, Wang WM: Identification and functional analysis of a novel TRPC6 mutation associated with late onset familial focal segmental glomerulosclerosis in Chinese patients. Mutat Res 2009;664:84–90.
19 Genovese G, Friedman DJ, Ross MD, Lecordier L, Uzureau P, Freedman BI, Bowden DW, Langefeld CD, Oleksyk TK, Uscinski Knob AL, et al: Association of trypanolytic ApoL1 variants with kidney disease in African-Americans. Science 2010;329:841–845.
20 D'Agati VD, Kaskel FJ, Falk RJ: Focal segmental glomerulosclerosis. N Engl J Med 2011;365:2398–2411.
21 KDIGO Clinical Practice Guideline for Glomerulonephritis. Kidney Int Suppl 2012; 2:181.
22 Cattran DC, Rao P: Long-term outcome in children and adults with classic focal segmental glomerulosclerosis. Am J Kidney Dis 1998; 32:72–79.
23 Ren H, Shen P, Li X, Pan XX, Zhang W, Chen N: Tacrolimus versus cyclophosphamide in steroid-dependent or steroid-resistant focal segmental glomerulosclerosis: a randomized controlled trial. Am J Nephrol 2013;37:84–90.
24 Fan L, Liu Q, Liao Y, Li Z, Ji Y, Yang Z, Chen J, Fu J, Zhang J, Kong Y, et al: Tacrolimus is an alternative therapy option for the treatment of adult steroid-resistant nephrotic syndrome: a prospective, multicenter clinical trial. Int Urol Nephrol 2012 (E-pub ahead of print).
25 Hao X, Xie J, Ma J, Zhou Q, Xue W, Ren H, Chen N: Increased risk of treatment failure and end-stage renal disease in familial FSGS. J Am Soc Nephrol 2012;23:535A.
26 Kiryluk K, Li Y, Sanna-Cherchi S, Rohanizadegan M, Suzuki H, Eitner F, Snyder HJ, Choi M, Hou P, Scolari F, et al: Geographic differences in genetic susceptibility to IgA nephropathy: GWAS replication study and geospatial risk analysis. PLoS Genet 2012; 8:e1002765.
27 Xie J, Kiryluk K, Wang W, Wang Z, Guo S, Shen P, Ren H, Pan X, Chen X, Zhang W, et al: Predicting progression of IgA nephropathy: new clinical progression risk score. PLoS One 2012;7:e38904.
28 Donadio JV, Grande JP: IgA nephropathy. N Engl J Med 2002;347:738–748.
29 Goto M, Wakai K, Kawamura T, Ando M, Endoh M, Tomino Y: A scoring system to predict renal outcome in IgA nephropathy: a nationwide 10-year prospective cohort study. Nephrol Dial Transplant 2009;24:3068–3074.
30 Berthoux F, Mohey H, Laurent B, Mariat C, Afiani A, Thibaudin L: Predicting the risk for dialysis or death in IgA nephropathy. J Am Soc Nephrol 2011;22:752–761.
31 Shi SF, Wang SX, Jiang L, Lv JC, Liu LJ, Chen YQ, Zhu SN, Liu G, Zou WZ, Zhang H, et al: Pathologic predictors of renal outcome and therapeutic efficacy in IgA nephropathy: validation of the oxford classification. Clin J Am Soc Nephrol 2011;6:2175–2184.
32 Zeng CH, Le W, Ni Z, Zhang M, Miao L, Luo P, Wang R, Lv Z, Chen J, Tian J, et al: A multicenter application and evaluation of the oxford classification of IgA nephropathy in adult Chinese patients. Am J Kidney Dis 2012;60:812–820.

33 Lv J, Zhang H, Zhou Y, Li G, Zou W, Wang H: Natural history of immunoglobulin A nephropathy and predictive factors of prognosis: a long-term follow up of 204 cases in China. Nephrology (Carlton) 2008;13:242–246.

34 Shi Y, Chen W, Jalal D, Li Z, Mao H, Yang Q, Johnson RJ, Yu X: Clinical outcome of hyperuricemia in IgA nephropathy: a retrospective cohort study and randomized controlled trial. Kidney Blood Press Res 2012;35:153–160.

35 Schena FP, Cerullo G, Rossini M, Lanzilotta SG, D'Altri C, Manno C: Increased risk of end-stage renal disease in familial IgA nephropathy. J Am Soc Nephrol 2002;13:453–460.

36 Izzi C, Ravani P, Torres D, Prati E, Viola BF, Guerini S, Foramitti M, Frasca G, Amoroso A, Ghiggeri GM, et al: IgA nephropathy: the presence of familial disease does not confer an increased risk for progression. Am J Kidney Dis 2006;47:761–769.

37 Feehally J, Farrall M, Boland A, Gale DP, Gut I, Heath S, Kumar A, Peden JF, Maxwell PH, Morris DL, et al: HLA has strongest association with IgA nephropathy in genome-wide analysis. J Am Soc Nephrol 2010;21:1791–1797.

38 Gharavi AG, Kiryluk K, Choi M, Li Y, Hou P, Xie J, Sanna-Cherchi S, Men CJ, Julian BA, Wyatt RJ, et al: Genome-wide association study identifies susceptibility loci for IgA nephropathy. Nat Gen 2011;43:321–327.

39 Yu XQ, Li M, Zhang H, Low HQ, Wei X, Wang JQ, Sun LD, Sim KS, Li Y, Foo JN, et al: A genome-wide association study in Han Chinese identifies multiple susceptibility loci for IgA nephropathy. Nat Genet 2012;44: 178–182.

40 Chen N, Nusbaum P, Halbwachs-Mecarelli L, Lesavre P: Light-chain composition of serum IgA1 and in vitro IgA1 production in IgA nephropathy. Nephrol Dial Transplant 1991; 6:846–850.

41 Liu LJ, Lv JC, Shi SF, Chen YQ, Zhang H, Wang HY: Oral calcitriol for reduction of proteinuria in patients with IgA nephropathy: a randomized controlled trial. Am J Kidney Dis 2012;59:67–74.

42 Goumenos DS, Davlouros P, El Nahas AM, Ahuja M, Shortland JR, Vlachojannis JG, Brown CB: Prednisolone and azathioprine in IgA nephropathy – a ten-year follow-up study. Nephron Clin Pract 2003;93:C58–C68.

43 Frisch G, Lin J, Rosenstock J, Markowitz G, D'Agati V, Radhakrishnan J, Preddie D, Crew J, Valeri A, Appel G: Mycophenolate mofetil vs. placebo in patients with moderately advanced IgA nephropathy: a double-blind randomized controlled trial. Nephrol Dial Transplant 2005;20:2139–2145.

44 Tang S, Leung JC, Chan LY, Lui YH, Tang CS, Kan CH, Ho YW, Lai KN: Mycophenolate mofetil alleviates persistent proteinuria in IgA nephropathy. Kidney Int 2005;68:802–812.

45 Beck LH Jr, Bonegio RG, Lambeau G, Beck DM, Powell DW, Cummins TD, Klein JB, Salant DJ: M-type phospholipase A2 receptor as target antigen in idiopathic membranous nephropathy. N Engl J Med 2009;361:11–21.

46 Li J, Zhang YM, Qu Z, Zhao MH, Liu G: Low-dose cyclosporine treatment in Chinese nephrotic patients with idiopathic membranous nephropathy: an uncontrolled study with prospective follow-up. Am J Med Sci 2010;339:532–536.

47 Chen M, Li H, Li XY, Lu FM, Ni ZH, Xu FF, Li XW, Chen JH, Wang HY: Tacrolimus combined with corticosteroids in treatment of nephrotic idiopathic membranous nephropathy: a multicenter randomized controlled trial. Am J Med Sci 2010;339:233–238.

48 Chen W, Liu Q, Liao Y, Yang Z, Chen J, Fu J, Zhang J, Kong Y, Fu P, Lou T, et al: Outcomes of tacrolimus therapy in adults with refractory membranous nephrotic syndrome: a prospective, multicenter clinical trial. Am J Med Sci 2013;345:81–87.

49 Sethi S, Fervenza FC: Membranoproliferative glomerulonephritis – a new look at an old entity. N Engl J Med 2012;366:1119–1131.

50 Fakhouri F, Fremeaux-Bacchi V, Noel LH, Cook HT, Pickering MC: C3 glomerulopathy: a new classification. Nat Rev Nephrol 2010;6: 494–499.

Prof. Nan Chen, MD
Department of Nephrology, Ruijin Hospital
School of Medicine, Shanghai Jiao Tong University
No. 197 Ruijin Er Road, Shanghai 200025 (PR China)
E-Mail chen-nan@medmail.com.cn

Overview

Chen N (ed): New Insights into Glomerulonephritis.
Contrib Nephrol. Basel, Karger, 2013, vol 181, pp 12–21 (DOI: 10.1159/000348639)

Advances in the Pathology of Glomerular Diseases

Jan J. Weening[a] · Pierre Ronco[b] · Giuseppe Remuzzi[c]

[a]Departments of Pathology Erasmus MC, Rotterdam & Tergooiziekenhuizen Blaricum, Hilversum, The Netherlands; [b]INSERM UMR-S 702, UPMC Université Paris 6, AP-HP, Tenon Hospital, Department of Nephrology, Hôpital Tenon, Paris, France; [c]Istituto Mario Negri Bergamo, Mario Negri Institute for Pharmacological Research, Centro Anna Maria Astori, Science and Technology Park Kilometro Rosso, Bergamo, Italy

Abstract

Glomerular injury can be caused by numerous insults including hemodynamics, infections and immunity, hereditary and metabolic diseases, and toxicity. Basic and translational experimental studies in combination with clinical research in patients with renal disease have advanced our understanding of the etiology and pathogenesis of many forms of glomerulonephritis. This new knowledge has facilitated classification and treatment and has contributed to a better outcome of patients with renal disease. Since renal disease almost without exception leads to systemic cardiovascular complications, these advances are also of general health interest. Here, we shall briefly review general principles in the pathology of glomerular injury and discuss recent developments in the study of podocytopathies; membranous glomerulopathy; ANCA-associated vasculitis; C3 glomerulopathies and the role of complement in endothelial injury; and the prognostic value of the renal biopsy in predicting long-term outcome in lupus nephritis, vasculitis and IgA nephropathy.
Copyright © 2013 S. Karger AG, Basel

During the 20th century, glomerular pathophysiology and pathology were advanced considerably by the combination of clinical and pathological observations documenting accurately the various glomerular patterns of injury in relation to clinical syndromes such as the acute nephritic syndrome, the nephrotic syndrome and chronic renal disease. Experimental studies provided a rational basis for understanding the etiology and pathogenesis of several forms of renal disease and their evolution over time [1]. The introduction of the renal biopsy made it possible to observe the patterns of renal disease in man by microscopic analysis, including electron microscopy and immunofluorescence techniques.

With these new tools, combined with major advances in the field of immunology and cell biology in the second half of the 20th century, renal pathology rapidly advanced. The molecular and cellular basis of the glomerular microcirculation and its permselectivity, and the myriad of tubular functions in health and disease were unraveled.

In diagnostic renal pathology, four patterns of glomerular injury can be distinguished based on compartmentalization of glomerular injury: epithelial, endothelial, mesangial injury and chronic glomerulosclerosis. The four patterns encompass most glomerular diseases and correlate with four distinct clinical syndromes respectively: the nephrotic syndrome, the nephritic syndrome, hematuria and asymptomatic proteinuria (which in addition to mesangial injury can also be due to structural basement membrane deficiencies), and chronic renal failure.

Epithelial Cell Injury – Podocytopathies and Membranous Nephropathy

New developments in the study of glomerular visceral epithelial cell (GVEC) injury have contributed to a better understanding of minimal change nephropathy (MCNS) and different forms of focal and segmental glomerulosclerosis (FSGS). Injury to the podocyte leads to a loss of interdigitating foot processes with detachment from the underlying basement membrane and disruption of the glomerular sieve and charge-selective barrier resulting in massive proteinuria. In MCNS, this injury can only be observed by electron microscopy and is usually sensitive to steroid treatment. MCNS is the most frequent form of nephrotic syndrome in children but can also occur in adults. Clinical and experimental studies have suggested MCNS to be an immune-mediated disease in which podocytes are affected by circulating cytokines with or without an association with hypersensitivity reactions or malignancies such as Hodgkin's lymphoma [2]. However, the molecular basis for MCNS is still uncertain and one of the last enigmas in renal pathology.

The idiopathic form of FSGS is considered to be part of the same spectrum as MCNS and is often, in the early phase, sensitive to steroid treatment [3]. In many cases, FSGS shows a relapsing pattern of nephrotic syndrome ultimately leading to an unresponsive form of glomerular injury and to end-stage renal failure.

Originally, FSGS was a pattern of injury described in a wide variety of renal diseases. Over the past three decades and in particular over the last 10 years, this descriptive diagnosis has been challenged and classified into a number of well-characterized podocytopathies, which can be divided into forms based on disruption of podocyte genes encoding cell-adhesion and cell-signaling proteins

Table 1. Two groups of podocytopathies identified as causes for FSGS

Gene disruption involving nephrin, podocin, CD2AP, α-actinin, integrin α3β1, CD151 tetraspanin, TRPC6, laminin B2, collagen XVII, Ilk, Rhoa, Myo1E, INFN2, MYH9, dystroglycan and heparan sulfate moieties
Extrinsic factors including intracapillary hypertension, viral infection, toxicity, complement activation, metabolic storage diseases

and into forms based on extrinsic or systemic factors (table 1). In addition, modifying genes such as APOL1 [4] may affect cellular injury of podocytes under stress, facilitating the development of FSGS.

The identification of the different causes of podocytopathies has facilitated the management of patients with secondary forms of FSGS and has improved their prognostic evaluation, including the prospect of transplantation. Studies on the biology of GVECs have revealed the pathophysiology of cell-cell and cell-matrix interaction, governing the integrity of the cell and the glomerular capillary filter and its malfunction in the nephrotic syndrome in MCNS and FSGS [3], including among others the significance of mechanosensing, heparan sulfate modulation, proton pump function and cathepsin L activation. A role of the soluble urokinase receptor (suPAR) as a permeability factor in FSGS has been proposed based on data that in mice suPAR-activated podocyte β3 integrin causes proteinuria and FSGS-like glomerulopathy and that significantly higher serum levels of suPAR were reported in FSGS patients than in patients with other glomerulopathies [5]. However, in a later study, serum suPAR levels were not different among idiopathic FSGS, secondary FSGS and MCNS, nor did they predict responsiveness to steroid therapy, which raised serious doubts on the potential role of suPAR in FSGS pathogenesis [6].

In idiopathic MCNS and FSGS currently available treatment includes steroids, cyclophosphamide, cyclosporine and anti-CD20 antibodies, but a rational approach is hampered by the lack of accurate understanding of their etiology.

Pathology of Membranous Glomerulopathy

Another cause of the nephrotic syndrome due to GVEC injury is membranous nephropathy (MN), a form of immune-mediated glomerular injury, in which immune complexes of immunoglobulins and complement accumulate in a granular pattern along the outer side of the GBM leading to GVEC detachment. The immunoglobulins can be directed against autoantigens, produced by the

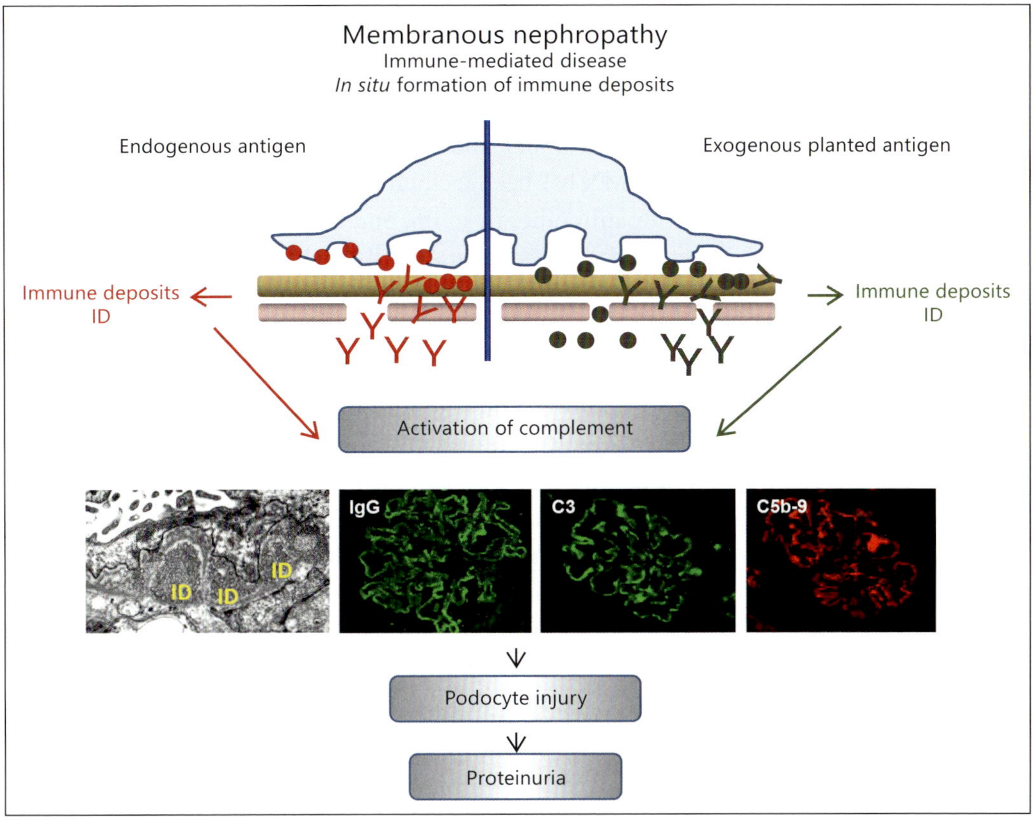

Fig. 1. Schematic illustration of the pathogenesis of proteinuria in MN.

GVEC or extrinsic antigens accumulating at the specific site due to charge interactions as was first shown by experimental studies and recently proven for a clinical condition involving cationic bovine serum albumin [7] (fig. 1). The podocyte in this case is merely an innocent bystander and likely is injured by the local formation of complement stimulated by the nearby immune complexes. This pattern is more reminiscent of secondary MN that occurs with hepatitis B antigenemia or class V lupus, in which it is hypothesized that positively charged viral proteins or nuclear histones may also target immune deposit formation to the subepithelial side of the GBM. Bovine serum albumin is the first dietary antigen to be unequivocally involved in immune-mediated glomerulopathies, thus pointing to a role for absorbed dietary antigens and more generally for environmental factors in the pathogenesis of these diseases.

In most patients with the idiopathic autoimmune form of MN, the autoantibodies are of the IgG4 and IgG1 type and are directed to the phospholipase A2 receptor (PLA2-R) [8]. These antibodies may affect the GVEC directly through interaction with PLA2-R and they can bind complement components which through different pathways ultimately lead to the formation of the membrane attack complex causing cell injury and death. Distinction between idiopathic and secondary forms of MN has been facilitated by the introduction of assays to determine PLA2-R autoantibodies in serum and PLA2-R antigen in renal biopsies [9, 10]. Further work is needed to establish whether circulating levels of anti-PLA2R1 antibodies can predict disease outcome or recurrence in the grafted kidney after transplantation [11, 12].

Genome-wide association studies in large cohorts of MN patients in Europe have revealed the autoimmune response in patients with idiopathic MN to be based on two single nucleotide polymorphisms – one for the gene encoding PLA2-R, and one for the gene encoding the HLA antigen-presenting peptide within the HLA-DQA1 complex [13]. This implies a highly restricted autoimmune response based on an inappropriate interaction between the pathogenetic peptide and the complementary antigen-presenting molecule. The cause of the induction of recognition, which often does not occur but at the adult age, is at this moment unknown and could be due to demasking of the epitope by infection, toxicity or degradation processes during cell or tissue renewal.

Further insight in the process underlying PLA2-R self-recognition may facilitate the design of epitope-blocking reagents, which may inhibit autoimmunity, and can be used to monitor and prevent recurrence of the disease after transplantation. After decades of aspecific cytotoxic therapeutic attempts, the anti-CD20 monoclonal antibody rituximab has been shown to be a promising therapy for idiopathic MN. In patients with persistent nephrotic syndrome despite optimized conservative therapy, rituximab was well tolerated and led to significant reduction of proteinuria and all patients with at least 4 years of follow-up achieved complete or partial remission [14]. This represents a paradigm shift in treatment, allowing fresh insights in pathophysiology.

Furthermore, treatment can be aimed at the level of tissue injury in particular targeting complement component activation and regulation (see below under C3 glomerulopathies).

Other antibodies directed against enolase, superoxide dismutase 2, and aldose reductase have also been identified in the serum [15] but their role in the initial development of MN remains controversial because they recognize cytoplasmic antigens. These antigens can however be routed to the membrane in the aggressed podocyte where they will possibly serve as targets for circulating antibodies.

Endothelial Cell Injury – ANCA-Associated Renal Pathology

At about the same time as genome-wide association studies revealed the genetic cause of idiopathic MN, a similar multicenter approach in a large cohort of patients, revealed similar highly restricted polymorphisms in genes encoding HLA-DP and proteinase 3 and SERPINA1 in patients with autoantibodies directed against proteinase 3 suffering from ANCA-associated systemic vasculitis and glomerulonephritis [16]. In this disease, the endothelial compartment of the glomerulus is injured leading to fibrinoid necrosis, capillary wall destruction, and extracapillary proliferation, resulting in permanent damage of the glomerular tuft.

Since the first description of the relevance of ANCA in systemic vasculitis in 1985 [17], the etiology and pathogenesis have been clarified to a major level based on seminal discoveries, such as the identification of proteinase 3 and myeloperoxidase as targets for ANCA, the pathogenesis of endothelial injury and vasculitis due to a direct interaction of ANCA with neutrophils and endothelium, and the role of infection with *Staphylococcus aureus* and the relevance of prevention by antibiotic treatment [18]. New developments will be aimed at epitope-guided treatment in an attempt to modify the genetic response of this serious form of autoimmune disease.

The role of complement in the ANCA-associated endothelial injury is less well understood. At immunofluorescence, only segmental low-level deposition of immunoglobulins and complement components can be observed which has led to the designating term *pauci-immune glomerulonephritis*, notwithstanding the undisputed pathogenetic action of the ANCA autoantibodies. Probably regulation of complement activation and breakdown is of importance [19] and treatment aimed at these mechanisms may prove to be successful in limiting tissue injury.

Complement System and Glomerular Pathology

The serine proteases of the complement system have an important role in opsonization and removal of microbial antigens and dead cell and tissue fragments. In this process, complement components can be activated by antibodies, by serum proteins of the collectin family and by microbial surface glycoproteins and extracellular matrix components, known as the three routes of activation: the classical, mannan-binding lectin, and alternative pathways (fig. 2) [20].

The complement system can induce removal of debris in a quiescent way through digestion by macrophages and B cells with induction of immune tolerance, but in cases of massive or persistent debris the system may initiate an in-

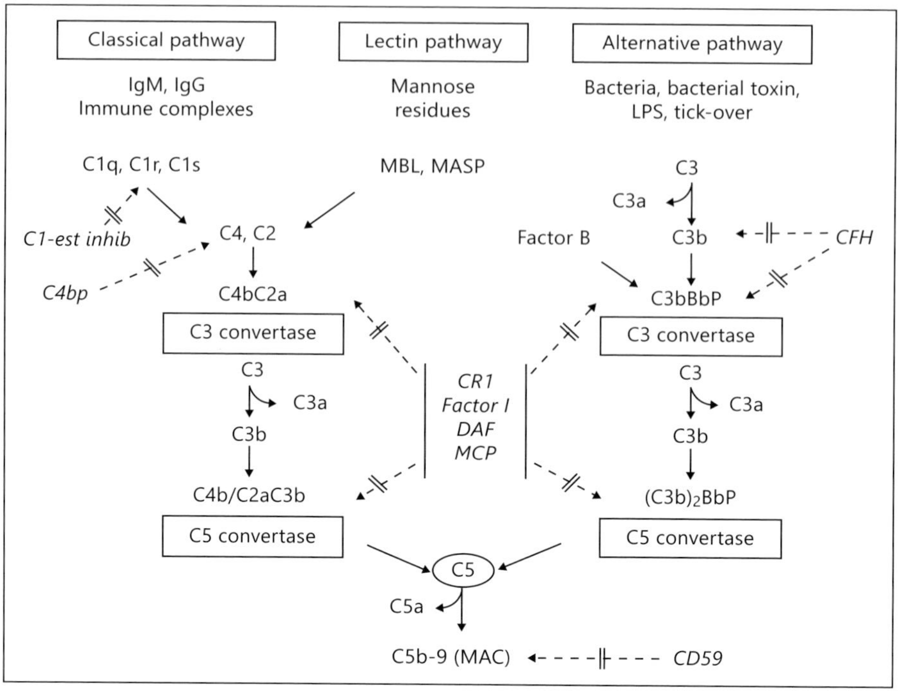

Fig. 2. The three pathways of the complement system.

flammatory pathway by recruitment of phagocytes through chemotaxis resulting in tissue inflammation and breakdown leading to immunization.

The components of the complement system are held in check by a large number of regulatory and inhibitory proteins in the circulation and at the surface of endothelial cells. Recent experimental and clinical studies have revealed how dysregulation of the complement system may cause induction of systemic lupus erythematosus by inadequate removal of apoptotic nuclear fragments leading to autoimmunity to DNA and phospholipids [21].

Endothelial Cell Injury – Complement and HUS

Inappropriate activation of the complement system due to ineffective inhibition may lead to prolonged tissue damage. This mechanism has recently been revealed to underlie the *Escherichia coli*-associated and atypical forms of hemolytic uremic syndrome (aHUS) [20] and the two idiopathic forms of membranoproliferative glomerulonephritis (MPGN) known previously as type I and type II MPGN (also called dense deposit disease – DDD) [22].

Sixty percent of aHUS cases are related to mutations in genes encoding complement regulatory proteins, factor H (CFH), membrane-cofactor protein (MCP), factor I (CFI) and thrombomodulin (THBD), or the alternative pathway C3 convertase components C3 and factor B (FB) or to inhibitory anti-CFH autoantibodies. Patients with combined mutations in two or three different complement genes have also been reported underlying the complexity of aHUS genetics.

A recent immunofluorescence-based classification approach distinguishes those forms of MPGN with isolated C3 deposits – known as C3 glomerulopathies and characterized by defective control of the alternative pathway of complement – from MPGN type I with deposits of immunoglobulin and complement and characterized by activation of the classical pathway by antigen-antibody immune complexes [23]. However, such distinction is not universally accepted and evidence is accumulating on involvement of the alternative pathway of complement in MPGN with C3 and immune deposits as well. C3NeF, an autoantibody stabilizing the AP C3 convertase, is found in 50–75% of patients with C3 glomerulopathies but also in about half of patients with MPGN I. Autoantibodies against CFH and factor B and complement gene mutations have also been reported.

Major therapeutic progress has been made recently in aHUS with the discovery of the efficacy of Eculizumab, an anti-C5 antibody [20, 24]. For C3 glomerulopathies, there is no specific treatment. Eculizumab reduced proteinuria in a few patients but others did not respond.

Renal Pathology and the Prognosis of Glomerular Disease

The three major glomerular diseases – lupus nephritis, ANCA-associated glomerulonephritis and IgA-nephropathy – have recently been reclassified [25–27] and for all three, clinicopathological observations have revealed a strong correlation of chronic glomerular injury (adhesions, scars) with poor outcome [28]. In lupus nephritis, more sophisticated quantification of chronic lesions in the glomerular, tubulointerstitial and vascular compartment was reported to have no additional value over the relatively simple analysis of the percentage of glomeruli with chronic lesions [28]. Recently, molecular and proteomic studies on renal tissue and urine have been initiated to correlate transcription and peptide profiles with outcome in order to allow a more precise intervention in signaling pathways involved in inflammation and repair.

References

1 Weening JJ, Jennette JC: Historical milestones in renal pathology. Virchows Archiv 2012;461:3–11.

2 Berg JG, Weening JJ: Role of the immune system in the pathogenesis of idiopathic nephrotic syndrome. Clin Sci 2004;107:125–136.

3 D'Agati VD, Kaskel FJ, Falk RJ: Focal segmental glomerulosclerosis. N Engl J Med 2011;365:2398–2411.

4 Genovese G, Friedman DJ, Ross MD, Lecordier L, Uzureau P, Freedman BI, Bowden DW, Langefeld CD, Oleksyk TK, Uscinski Knob AL, Bernhardy AJ, Hicks PJ, Nelson GW, Vanhollebeke B, Winkler CA, Kopp JB, Pays E, Pollak MR: Association of trypanolytic ApoL1 variants with kidney disease in African-Americans. Science 2010;329:841–845.

5 Wei C, El Hindi S, Li J, Fornoni A, Goes N, Sageshima J, Maiguel D, Karumanchi SA, Yap HK, Saleem M, Zhang Q, Nikolic B, Chaudhuri A, Daftarian P, Salido E, Torres A, Salifu M, Sarwal MM, Schaefer F, Morath C, Schwenger V, Zeier M, Gupta V, Roth D, Rastaldi MP, Burke G, Ruiz P, Reiser J: Circulating urokinase receptor as a cause of focal segmental glomerulosclerosis. Nat Med 2011; 17:952–960.

6 Maas RJ, Wetzels JF, Deegens JK: Serum-soluble urokinase receptor concentration in primary FGS. Kidney Int 2012;81:1043–1044.

7 Debiec H, Lefeu F, Kemper MJ, Niaudet P, Deschênes G, Remuzzi G, Ulinski T, Ronco P: Early-childhood membranous nephropathy due to cationic bovine serum albumin. N Engl J Med 2011;364:2101–2110.

8 Beck LH Jr, Bonegio RG, Lambeau G, Beck DM, Powell DW, Cummins TD, Klein JB, Salant DJ: M-type phospholipase A2 receptor as target antigen in idiopathic membranous nephropathy. N Engl J Med 2009;361:11–21.

9 Hofstra JM, Debiec H, Short CD, Pellé T, Kleta R, Mathieson PW, Ronco P, Brenchley PE, Wetzels JF: Antiphospholipase A2 receptor antibody titer and subclass in idiopathic membranous nephropathy. J Am Soc Nephrol 2012;23:1735–1743.

10 Debiec H, Ronco P: PLA2R autoantibodies and PLA2R glomerular deposits in membranous nephropathy. N Engl J Med 2011;364: 689–690.

11 Beck LH Jr, Fervenza FC, Beck DM, Bonegio RG, Malik FA, Erickson SB, Cosio FG, Cattran DC, Salant DJ: Rituximab-induced depletion of anti-PLA2R autoantibodies predicts response in membranous nephropathy. J Am Soc Nephrol 2011;22:1543–1550.

12 Debiec H, Martin L, Jouanneau C, Dautin G, Mesnard L, Rondeau E, Mousson C, Ronco P: Autoantibodies specific for the phospholipase A2 receptor in recurrent and de novo membranous nephropathy. Am J Transplant 2011; 11:2144–2152.

13 Stanescu HC, Arcos-Burgos M, Medlar A, Bockenhauer D, Kottgen A, Dragomirescu L, Voinescu C, Patel N, Pearce K, Hubank M, Stephens HA, Laundy V, Padmanabhan S, Zawadzka A, Hofstra JM, Coenen MJ, den Heijer M, Kiemeney LA, Bacq-Daian D, Stengel B, Powis SH, Brenchley P, Feehally J, Rees AJ, Debiec H, Wetzels JF, Ronco P, Mathieson PW, Kleta R: Risk HLA-DQA1 and PLA2R1 alleles in idiopathic membranous nephropathy. N Engl J Med 2011;364:616–626.

14 Ruggenenti P, Cravedi P, Chianca A, Perna A, Ruggiero B, Gaspari F, Rambaldi A, Marasà M, Remuzzi G: Rituximab in idiopathic membranous nephropathy. J Am Soc Nephrol 2012;23:1416–1425.

15 Murtas C, Bruschi M, Candiano G, Moroni G, Magistroni R, Magnano A, Bruno F, Radice A, Furci L, Argentiero L, Carnevali ML, Messa P, Scolari F, Sinico RA, Gesualdo L, Fervenza FC, Allegri L, Ravani P, Ghiggeri GM: Coexistence of different circulating antipodocyte antibodies in membranous nephropathy. Clin J Am Soc Nephrol 2012;7: 1394–1400.

16 Lyons PA, Rayner TF, Trivedi S, Holle JU, Watts RA, Jayne DR, Baslund B, Brenchley P, Bruchfeld A, Chaudhry AN, Cohen Tervaert JW, Deloukas P, Feighery C, Gross WL, Guillevin L, Gunnarsson I, Harper L, Hrušková Z, Little MA, Martorana D, Neumann T, Ohlsson S, Padmanabhan S, Pusey CD, Salama AD, Sanders JS, Savage CO, Segelmark M, Stegeman CA, Tesař V, Vaglio A, Wieczorek S, Wilde B, Zwerina J, Rees AJ, Clayton DG, Smith KG: Genetically distinct subsets within ANCA-associated vasculitis. N Engl J Med 2012;367:214–223.

17 Van der Woude FJ, Rasmussen N, Lobatto S, Wiik A, Permin H, van Es LA, van der Giessen M, van der Hem GK, The TH: Autoantibodies against neutrophils and monocytes: tool for diagnosis and marker of disease activity in Wegener's granulomatosis. Lancet 1985;1:425–429.
18 Kallenberg CG: Pathogenesis of ANCA-associated vasculitis, an update. Clin Rev Allergy Immunol 2011;41:224–231.
19 Gou SJ, Yuan J, Chen M, Yu F, Zhao MH: Circulating complement activation in patients with antineutrophil cytoplasmic antibody-associated vasculitis. Kidney Int 2013; 83:129–137.
20 Noris M, Mescia F, Remuzzi G: STEC-HUS, atypical HUS and TTP are all diseases of complement activation. Nat Rev Nephrol 2012;8:622–633.
21 Chen M, Daha MR, Kallenberg CG: The complement system in systemic autoimmune disease. J Autoimmun 2010;34:J276–J286.
22 Sethi S, Fervenza FC: Membranoproliferative glomerulonephritis, a new look at an old entity. N Engl J Med 2012;366:1119–1131.
23 D'Agati VD, Bomback AS: C3 glomerulopathy: what's in a name? Kidney Int 2012;82: 379–381.
24 Zuber J, Fakhouri F, Roumenina LT, Loirat C, Frémeaux-Bacchi V; French Study Group for aHUS/C3G: Use of eculizumab for atypical haemolytic uraemic syndrome and C3 glomerulopathies. Nat Rev Nephrol 2012;8: 643–657.
25 Weening JJ, D'Agati VD, Schwartz MM, Seshan SV, Alpers CE, Appel GB, Balow JE, Bruijn JA, Cook T, Ferrario F, Fogo AB, Ginzler EM, Hebert L, Hill G, Hill P, Jennette JC, Kong NC, Lesavre P, Lockshin M, Looi LM, Makino H, Moura LA, Nagata M; International Society of Nephrology Working Group on the Classification of Lupus Nephritis; Renal Pathology Society Working Group on the Classification of Lupus Nephritis: The classification of glomerulonephritis in systemic lupus erythematosus revisited. Kidney Int 2004;65:521–530.
26 Berden AE, Ferrario F, Hagen EC, Jayne DR, Jennette JC, Joh K, Neumann I, Noël LH, Pusey CD, Waldherr R, Bruijn JA, Bajema IM: Histopathologic classification of ANCA-associated glomerulonephritis. J Am Soc Nephrol 2010;21:1628–1636.
27 Working Group of the International IgA Nephropathy Network and the Renal Pathology Society, Cattran DC, Coppo R, Cook HT, Feehally J, Roberts IS, Troyanov S, Alpers CE, Amore A, Barratt J, Berthoux F, Bonsib S, Bruijn JA, D'Agati V, D'Amico G, Emancipator S, Emma F, Ferrario F, Fervenza FC, Florquin S, Fogo A, Geddes CC, Groene HJ, Haas M, Herzenberg AM, Hill PA, Hogg RJ, Hsu SI, Jennette JC, Joh K, Julian BA, Kawamura T, Lai FM, Leung CB, Li LS, Li PK, Liu ZH, Mackinnon B, Mezzano S, Schena FP, Tomino Y, Walker PD, Wang H, Weening JJ, Yoshikawa N, Zhang H: The Oxford classification of IgA nephropathy: pathology definitions, correlations and classification. Kidney Int 2009;76:546–556.
28 Hiramatsu N, Kuroiwa T, Ikeuchi H, Maeshima A, Kaneko Y, Hiromura K, Ueki K, Nojima Y: Revised classification of lupus nephritis is valuable in predicting renal outcome with an indication of the proportion of glomeruli affected by chronic lesions. Rheumatology 2008;47:702–707.

Jan J. Weening, MD
Departments of Pathology, Erasmus MC
Rotterdam & Tergooiziekenhuizen Blaricum
POB 10016, NL–1201 DA Hilversum (The Netherlands)
E-Mail jweening@tergooiziekenhuizen.nl

Chen N (ed): New Insights into Glomerulonephritis.
Contrib Nephrol. Basel, Karger, 2013, vol 181, pp 22–30 (DOI: 10.1159/000348638)

Changing Spectrum of Biopsy-Proven Primary Glomerular Diseases over the Past 15 Years: A Single-Center Study in China

Xiaoxia Pan · Jing Xu · Hong Ren · Wen Zhang · Yaowen Xu · Pingyan Shen · Xiao Li · Weiming Wang · Xiaonong Chen · Pei Wu · Xiaobei Feng · Cuilan Hao · Nan Chen

Department of Nephrology, Ruijin Hospital, School of Medicine, Shanghai Jiao Tong University, Shanghai, PR China

Abstract

The prevalence of chronic kidney disease (CKD) is reported to be 10.8–11.8% of the Chinese population. With economic development and longer life expectancy, the spectrum of CKD etiology has kept changing. Primary glomerular diseases (PGD) are still the most common renal diseases in China. To investigate the changing pattern of PGD in China, we retrospectively analyzed consecutive native renal biopsies performed in our hospital from 1997 to 2011. The patients were grouped according to a 3-year interval, 1997–1999 (period 1), 2000–2002 (period 2), 2003–2005 (period 3), 2006–2008 (period 4), 2009–2011 (period 5), and divided into three age groups (<20, 20–59, and ≥60 years old). 8,909 qualified cases were enrolled in this study. Among 8,909 specimens, 6,337 (71.13%) were diagnosed as PGD, while this prevalence decreased significantly from 77.61% in 1997–1999 to 66.73% in 2006–2008. IgA nephropathy (IgAN) was the most common PGD (36.66%), without any significant difference in the 5 periods ($p = 0.185$). IgAN was the most common PGD both in patients between the 20- to 59-year-old group (45.58%) and <20-year-old group (19.29%) as well. Membranous nephropathy (MN) was the most frequently found PGD in patients at age ≥60 years (39.64%). The frequency of MN was increased significantly from 6.48% in 1997–1999 to 22.79% in 2009–2011 ($p < 0.001$). The proportion of elderly patients increased significantly from 3.18% in 1997–1999 to 15.21% in 2009–2011 ($p < 0.001$). The prevalence of endocapillary proliferative glomerulonephritis (EnPGN) has decreased since

1997. PGD has remained the most common renal disease in China, although with a descending trend. The spectrum of PGD is different in different age groups. The frequency of EnPGN has decreased significantly, while that of MN has increased significantly.
Copyright © 2013 S. Karger AG, Basel

Chronic kidney disease (CKD) has become a major public health challenge in the world, including China. It is reported that the prevalence of CKD is 11.8% in Shanghai citizens [1] and 10.8% in the populations of 13 Chinese provinces [2]. In China, primary glomerular diseases (PGD) are still the most common CKD in patients where renal biopsies are performed and the most common cause of ESRD. According to the Shanghai Dialysis Registration in 2011, PGD accounted for 60.0 and 46.2% of all patients receiving hemodialysis and peritoneal dialysis in 1998 and 2011, respectively (data from the Shanghai Dialysis Registry Report, 2011). The prevalence of PGD in renal biopsy specimens is reported to range from 57.7 to 68.6% [3, 4].

The pattern of PGD is different and changes with time in different ethnic groups and geographical areas. IgA nephropathy (IgAN) is the most common PGD in most European and Asian areas [3–12]. In the USA, IgAN was the most frequent PGD in young adult Caucasians, while it was rare in African-Americans in whom focal segmental glomerulosclerosis (FSGS) remained more common [13, 14]. FSGS is the most common PGD in Brazil and the Middle East as well [15, 16]. Several reports have demonstrated that the prevalence of FSGS [14, 17–19] and IgAN [4, 10, 14, 17, 20] is increasing, while the prevalence of postinfectious glomerulonephritis is decreasing in most industrialized countries but remains high in some developing communities [21]. With economic development and changing lifestyles, aging of the population, the pattern of PGD in China is also changing. To investigate the changing pattern of PGD in China, we retrospectively analyzed the data of all the patients who received renal biopsies in our department over the last 15 years.

Patients and Methods

The records of 8,965 patients with native renal biopsy performed in Ruijin Hospital from January 1997 to December 2011 were retrospectively analyzed. Indications for renal biopsy were as follows: (1) nephrotic syndrome or nephritic range proteinuria, (2) nephritic syndrome, (3) rapidly progressive glomerulonephritis syndrome, (4) asymptomatic hematuria with proteinuria, (5) acute renal failure with glomerular injury with/without vascular damage, and (6) chronic renal failure with a kidney of relatively normal size. The indication of renal biopsy remained unchanged during the whole observation period. Unclassified diagnosis due to inadequate sampling (<5 glomeruli in the specimen for light microscopy when there were no typical findings in immunofluorescence or electron microscopy or absence of a glomerulus in immunofluorescence), repeat biopsies and incomplete records were excluded. Altogether, 8,909 qualified cases were enrolled in this study, of whom 6,337 were diagnosed as PGD. A diagnosis of PGD was considered if there was, at the time of biopsy, (1) no history of any known associated systemic or autoimmune disease, (2) negative serology for hepatitis virus, HIV, *Treponema pallidum*, antinuclear factor, and (3) no familial history of renal disease.

Renal biopsy specimens were processed and stained for light microscopy with PAS, Masson, HE, and Jones. Immunohistochemistry with polyclonal antisera was used against human IgG, IgM, IgA, C3c, C4, C1q, fibrinogen, κ, and λ light chains. Electron microscopy was performed on 6,039 (95.30%) biopsy specimens from PGD patients. The biopsy specimens were examined by the same group of pathologists during the years 1997–2008 and changed to other pathologists after 2009. The pathologic diagnostic criteria have remained consistent over the last 15 years. When new entities were recognized or new diagnostic criteria appeared, we updated the previous diagnosis one by one according to the new criteria.

PGD was classified as follows: minimal change disease (MCD), FSGS, IgAN, non-IgA mesangioproliferative glomerulonephritis (MsPGN), membranous nephropathy (MN), endocapillary proliferative glomerulonephritis (EnPGN), crescentic glomerulonephritis (CreGN), membranoproliferative glomerulonephritis (MPGN) and others; the last category includes rare diseases such as fibrillary glomerulonephritis, immunotacoid glomerulopathy, fibronectin glomerulopathy and collagen III glomerulopathy, etc. Patients with the following renal diseases were excluded: secondary glomerular diseases, hereditary glomerular diseases such as Alport's syndrome which fallibly confound with FSGS in LM [22], vascular disease-associated nephropathy such as benign or malignant nephroangiosclerosis, acute or chronic tubulointerstitial nephritis, acute tubular necrosis and unclassified nephropathies.

The patients were grouped into 5 periods according to a 3-year interval: 1997–1999 (period 1), 2000–2002 (period 2), 2003–2005 (period 3), 2006–2008 (period 4) and 2009–2011 (period 5). Further analysis was performed by dividing patients into three groups according to different ages at the time of renal biopsy (<20, 20–59, and ≥60 years, respectively).

Statistical Analysis

Data were stored in a database file. All data were analyzed by SPSS version 11.0 (SPSS, Chicago, Ill., USA). Numerical variables in descriptive analysis were reported as mean ± SD, or median (interquartile range) as appropriate to their distribution. Statistical analysis was performed using the χ^2 test and Fisher's exact test for categorical variables, and the Student's t test for continuous variables. $p < 0.05$ was considered statistically significant.

Results

General Information of Patients with PGD

From January 1997 to December 2011, 8,909 qualified renal biopsies were identified, of which 6,337 (71.13%) patients were diagnosed as having PGD. As shown in figure 1, PGD was still the leading kidney disease in patients receiving

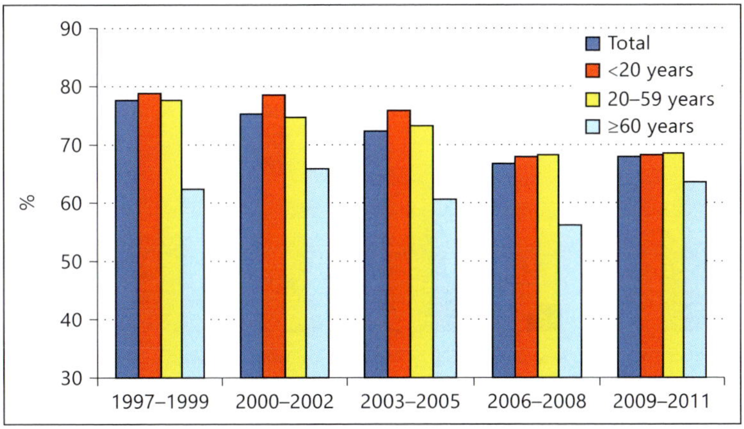

Fig. 1. Prevalence of PGD according to period and age.

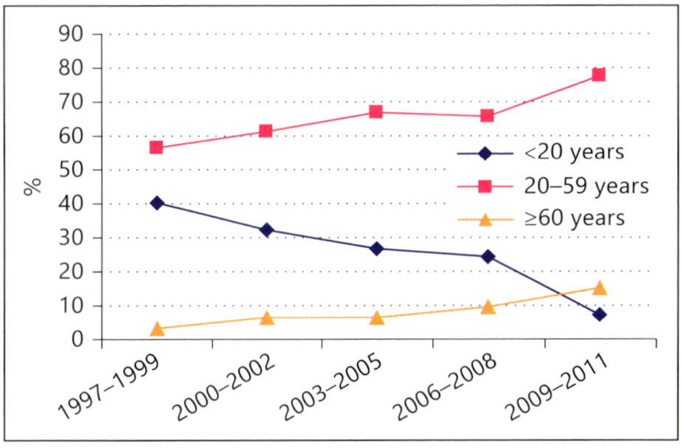

Fig. 2. Proportion of patients with PGD at different ages in different periods.

renal biopsies in our center over the last 15 years. The prevalence of PGD significantly decreased from 77.61% in the 1997–1999 period to 66.73% in the 2006–2008 period and 67.8% in the 2009–2011 period ($p < 0.05$).

Among 6,337 PGD patients, 3,307 were male and 3,030 female (male:female ratio 1.09:1). Mean age was 33.64 ± 17.98 years. There was a significant increase in the mean age, which was 26.50 ± 16.32, 30.50 ± 17.93, 32.80 ± 17.61, 34.02 ± 18.55 and 40.65 ± 16.19 years in periods 1, 2, 3, 4 and 5, respectively ($p < 0.001$). As shown in figure 2, the proportion of elderly patients (≥60 years old) increased significantly, from 3.18% in the 1997–1999 period to 15.21% in the 2009–2011 period ($p < 0.001$) and the younger patients (<20 years old) decreased signifi-

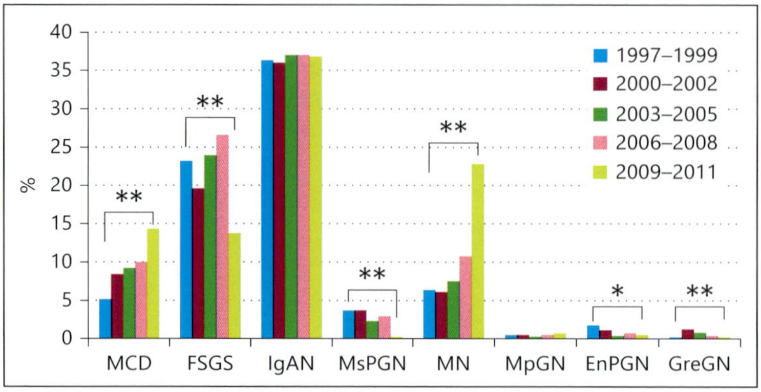

Fig. 3. Prevalence of various PGD according to period. * p < 0.01, ** p < 0.001.

cantly in our study ($p < 0.001$). The prevalence of PGD in patients <60 years old was higher than that in elderly patients (≥60 years), while it gradually decreased from 78.91% at age <20 years and 77.76% at age 20–59 years in the 1997–1999 period to 68.21% at age <20 years and 68.63% at age 20–59 years in the 2009–2011 period ($p < 0.001$ and $p < 0.001$, respectively).

Spectrum of PGD in Total and Different Periods

In a total of 6,337 patients with PGD, the leading cause was IgAN (36.66%), followed by FSGS (21.52%), MN (11.36%), MCD (9.97%), MsPGN (2.49%), EnPGN (0.85%), CreGN (0.60%), and MPGN (0.57%).

As indicated in figure 3, IgAN was the most common PGD, accounting for 36.47, 35.95, 36.97, 37.02 and 36.90% in periods 1, 2, 3, 4 and 5, respectively, without any significant difference ($p = 0.185$). The second PGD in periods 1–4 was FSGS (23.38% in period 1, 19.61% in period 2, 23.98% in period 3, and 26.61% in period 4) and changed to MN in period 5. The proportion of MN increased significantly from 6.48% in period 1 to 22.79% in period 5 ($p < 0.001$), the same as MCD (5.08% in period 1 to 14.25% in period 5, $p < 0.001$). The proportion of non-IgA MsPGN and EnPGN decreased significantly from 3.68 and 1.78% in period 1 to 0.28 and 0.49% in period 5 ($p < 0.001$ and $p = 0.002$, respectively). There was no significant difference in the proportion of MPGN within 5 periods ($p > 0.05$). The proportion of CreGN decreased from 1.31% in period 2 to 0.14% in period 5.

Spectrum of PGD after Age Stratification

When stratified by age, there were 1,550, 4,232 and 555 patients with renal biopsy performed in the <20-, 20- to 59-, and ≥60-year-old groups, respectively.

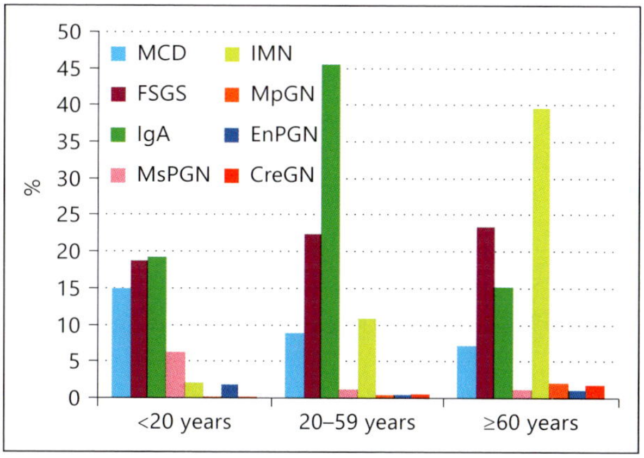

Fig. 4. Prevalence of various PGD according to age.

In patients <20 years old, the most common PGD was IgAN (19.29%), followed by FSGS (18.84%), MCD (14.90%) and MsPGN (6.26%). In the 20- to 59-year-old group, IgAN was still the leading type of PGD, accounting for 45.58%, followed by FSGS (22.26%), MN (11.03%) and MCD (8.79%). In elderly patients aged ≥60 years, the most common PGD was MN (39.64%), followed by FSGS (23.42%), IgAN (15.13%) and MCD (7.21%) (fig. 4). IgAN had the highest prevalence in middle-aged patients and the lowest in elderly patients. Along with growing age, the proportion of MPGN was increasing, accounting for 0.13% in age <20 years, 0.52% in age 20–59 years and 2.16% in age >60 years ($p < 0.001$), as well as MN and CreGN. On the contrary, the proportion of MCD went in the opposite direction. EnPGN had a higher prevalence in younger patients (<20 years, 1.87%) and elderly patients (≥60 years, 1.08%).

Discussion

In the current study, we analyzed the pathological data of 6,337 cases of PGD, diagnosed between 1997 and 2011, in a single renal center in Eastern China. It was shown that PGD remains the most common disease among patients receiving renal biopsy in China. The proportion of PGD decreased significantly from 77.61% in the 1997–1999 period to 66.73% in the 2006–2008 period and remained consistent in the 2009–2011 period. This finding is similar to the reports from Nanjing and Beijing [3, 4]. With age stratification the PGD predominated in patients <60 years old, whereas in the elderly the frequency of PGD decreased. The decreasing trend in years was mainly observed in younger and middle-aged patients.

The results confirmed that IgAN is the most common PGD in Shanghai, although compared to the reports from Nanjing and Beijing, the prevalence of IgAN in our group is the lowest (36.66% in 1997–2011 vs. Nanjing 45.23% in 1979–2002, Beijing 54.3% in 1993–2007) [3, 4]. This is consistent with other Asian countries and some European countries and Australia [5–12, 23]. It remained stable for the last 15 years, while varying between age groups. The highest prevalence was observed in the 20- to 59-year-old group, accounting for 45.58%. Only 15.13% of the elderly patients were diagnosed as IgAN.

With the economic improvement of Chinese society, people's life expectancy has increased. More and more elderly patients have undergone a renal biopsy. The mean age of patients undergoing biopsies in our group has been rising and the proportion of MN has increased significantly. In patients over 60 years of age, MN instead of IgAN became first place of PGD in our center. An increase in frequency of MN [8] has also been reported from other centers in China [4] and other countries [6, 7]. However, Hanko et al. [10] reported that the proportion of PGD due to MN fell significantly in the UK biopsy population. The reason for this is unclear.

Our current study hereby confirmed that there was a significant decline in EnPGN. The proportion of EnPGN descended from 1.78% in 1997–1999 to 0.49% in 2009–2011 ($p < 0.001$). Given to ameliorative living conditions, advanced public healthcare and the effective antibiotic treatment of infections in the modern world, it has been reported that the prevalence of EnPGN broadly declined in other centers in China. A decrease in frequency of EnPGN has also been reported by most industrialized countries and some developing countries, for instance Brazil and Tunisia [15, 24].

It has been reported that the proportion of MPGN decreased in some countries [4, 7, 8, 25], however this phenomenon was not observed in our group. With rising ages, the proportion of PGD due to MPGN increased from 0.13% in patients <20 years of age to 2.16% in patients over 60 years of age. This is consistent with biopsy data from Beijing and Spain [4, 6].

The prevalence of FSGS, MCD and MsPGN in the Chinese population is variable in different reports [3, 4, 26]. The renal pathology pattern can be characteristic among different age groups and ethnics, as well as the influence of different regions and environments. Regional differences were reflected in the prevalence of several glomerular diseases. Consequently, the factors inherent to the stereotype by a single pathologist or technical improvement can influence both the analysis and results.

Acknowledgements

This work was supported by grants from the National Basic Research Program of China 973, grant No. 2012CB517600 (No. 2012CB517604), National Key Technology R&D Program (12-5)2011BAI10B00 (2011BAI10B06), National Natural Science Foundation of China (30871001), and Key Program of Shanghai Science and Technology Commission (08dz1900502).

References

1 Chen N, Wang W, Huang Y, et al: Community-based study on CKD subjects and the associated risk factors. Nephrol Dial Transplant 2009;24:2117–2123.
2 Zhang L, Wang F, Wang L, et al: Prevalence of chronic kidney disease in China: a cross-sectional survey. Lancet 2012;379:815–822.
3 Li LS, Liu ZH: Epidemiologic data of renal diseases from a single unit in China: analysis based on 13,519 renal biopsies. Kidney Int 2004;66:920–923.
4 Zhou FD, Zhao MH, Zou WZ, Liu G, Wang HY: The changing spectrum of primary glomerular diseases within 15 years: a survey of 3,331 patients in a single Chinese centre. Nephrol Dial Transplant 2009;24:870–876.
5 Schena FP: Survey of the Italian Registry of Renal Biopsies. Frequency of the renal diseases for 7 consecutive years. The Italian Group of Renal Immunopathology. Nephrol Dial Transplant 1997;12:418–426.
6 Rivera F, López-Gómez JM, Pérez-García R, et al: Frequency of renal pathology in Spain 1994–1999. Nephrol Dial Transplant 2002;17:1594–1602.
7 Rychlík I, Jancová E, Tesar V, et al: The Czech Registry of Renal Biopsies. Occurrence of renal diseases in the years 1994–2000. Nephrol Dial Transplant 2004;19:3040–3049.
8 Simon P, Ramee MP, Boulahrouz R, et al: Epidemiologic data of primary glomerular diseases in western France. Kidney Int 2004;66:905–908.
9 Sipiczki T, Ondrik Z, Abrahám G, et al: The incidence of renal diseases as diagnosed by biopsy in Hungary. Orv Hetil 2004;145:1373–1379.
10 Hanko JB, Mullan RN, O'Rourke DM, et al: The changing pattern of adult primary glomerular disease. Nephrol Dial Transplant 2010;25:490–496.
11 Research Group on Progressive Chronic Renal Disease: Nationwide and long-term survey of primary glomerulonephritis in Japan as observed in 1,850 biopsied cases. Nephron 1999;82:205–213.
12 Chang JH, Kim DK, Kim HW, et al: Changing prevalence of glomerular diseases in Korean adults: a review of 20 years of experience. Nephrol Dial Transplant 2009;24:2406–2410.
13 Nair R, Walker PD: Is IgA nephropathy the commonest primary glomerulopathy among young adults in the USA? Kidney Int 2006;69:1455–1458.
14 Swaminathan S, Leung N, Lager DJ, et al: Changing incidence of glomerular disease in Olmsted County, Minnesota: a 30-year renal biopsy study. Clin J Am Soc Nephrol 2006;1:483–487.
15 Polito MG, de Moura LA, Kirsztajn GM: An overview on frequency of renal biopsy diagnosis in Brazil: clinical and pathological patterns based on 9,617 native kidney biopsies. Nephrol Dial Transplant 2009;24:3050–3054.
16 Huraib S, Al Khader A, Shaheen FA, et al: The spectrum of glomerulonephritis in Saudi Arabia: the results of the Saudi Registry. Saudi J Kidney Dis Transpl 2000;11:434–441.
17 Hass M, Meechan SM, Karrison TG, et al: Changing etiologies of unexplained adult nephritic syndrome: a comparison of renal biopsy findings from 1976–1979 and 1995–1997. Am J Kidney Dis 1997;30:621–631.
18 Narasimhan B, Chacko B, John GT, et al: Characterization of kidney lesions in Indian adults: towards a renal biopsy registry. J Nephrol 2006;19:205–210.

19 Dragovic D, Rosenstock JL, Wahl SJ, et al: Increasing incidence of total segmental glomerulosclerosis and examination of demographic patterns. Clin Nephrol 2005;63: 1–7.

20 Woo KT, Chiang GS, Pall A, et al: The changing pattern of glomerulonephritis in Singapore over the past two decades. Clin Nephrol 1999;52:96–102.

21 Kanjanabuch T, Kittikowit W, Eiam-Ong S: An update on acute postinfectious glomerulonephritis worldwide. Nat Rev Nephrol 2009;5:259–269.

22 Pan XX, Yan JY, Chen N, et al: Detection of COL4A5 gene mutations in Chinese patients with Alport syndrome. Nephrol Dial Transplant 2004;19:1123–1128.

23 Briganti EM, Dowling J, Finlay M, et al: The incidence of biopsy-proven glomerulonephritis in Australia. Nephrol Dial Transplant 2001;16:1364–1367.

24 Ben Maïz H, Abderrahim E, Ben Moussa F, et al: Epidemiology of glomerular diseases in Tunisia from 1975 to 2005. Influence of changes in healthcare and society. Bull Acad Natl Med 2006;190:403–418.

25 Covic A, Schiller A, Volovat C, et al: Epidemiology of renal disease in Romania: a 10-year review of two regional renal biopsy databases. Nephrol Dial Transplant 2006;21: 419–424.

26 Wu YQ, Wang Z, Xu HF, et al: Frequency of primary glomerular disease in northeastern China. Braz J Med Biol Res 2011;44:810–813.

Prof. Nan Chen, MD
Department of Nephrology, Ruijin Hospital
School of Medicine, Shanghai Jiao Tong University
No. 197 Ruijin Er Road, Shanghai 200025 (PR China)
E-Mail chen-nan@medmail.com.cn

Chen N (ed): New Insights into Glomerulonephritis.
Contrib Nephrol. Basel, Karger, 2013, vol 181, pp 31–40 (DOI: 10.1159/000348636)

Evaluation of Anemia and Serum iPTH, Calcium, and Phosphorus in Patients with Primary Glomerulonephritis

Ya Li · Wen Zhang · Hong Ren · Weiming Wang · Hao Shi · Xiao Li · Xiaonong Chen · Pingyan Shen · Xiaojing Wu · Jingyuan Xie · Nan Chen

Department of Nephrology, Ruijin Hospital, School of Medicine, Shanghai Jiao Tong University, Shanghai, PR China

Abstract

Glomerulonephritis (GN) remains a major cause of morbidity and mortality in chronic kidney disease (CKD). Our study aimed to investigate the prevalence of anemia, abnormal serum intact parathyroid hormone (iPTH), calcium, and phosphorus in a Chinese patient population with primary GN. Medical histories and laboratory test results were collected from 2,924 patients with primary GN hospitalized in Ruijin Hospital of Shanghai between January 2003 and August 2009. The leading cause of CKD was primary glomerular diseases, which were responsible for up to 53.5% of all cases. IgA nephropathy was the most common cause, accounting for 38.7%, followed by focal segmental glomerulosclerosis (FSGS). The anemia rate of GN patients in early stages of CKD (stages 1–2 and 3) was 16–36%, and rapidly accelerated to 65.8 and 80.2% in advanced CKD stage 4 and stage 5, respectively. There was no significant decline observed in the level of serum calcium in patients with CKD stages 1–4 ($p > 0.05$). However, in patients with CKD stage 5 the prevalence of hypocalcaemia increased significantly (13.7%, $p = 0.000$). The prevalence of hyperphosphatemia did not significantly increase in patients with CKD stages 1–3 ($p < 0.05$), but was much higher in patients with CKD stages 4 and 5 ($p = 0.001$ and $p = 0.021$, respectively) and showed a negative correlation with renal function. Serum iPTH levels did not increase significantly in GN patients with CKD stages 1–2. The median iPTH levels were 54.7, 88.6, and 289.2 pg/ml ($p = 0.000$) for CKD stages 3–5, respectively, all of which showed negative correlation with renal function. The proportion of vitamin D insufficiency and deficiency increased to 29.3 and 11.2%, respectively, as the glomerular filtration rate fell below 15 ml/min/1.73 m^2. Primary glomerular disease remains the major cause of CKD in China, and complications such as anemia and metabolic bone disease are frequently present in GN patients.

Copyright © 2013 S. Karger AG, Basel

Chronic kidney disease (CKD) is becoming a major public health concern worldwide, as its prevalence continues to rise [1]. It is estimated that one in ten people worldwide suffer from CKD [2]. The International Society of Nephrology (ISN) has predicted that approximately 36 million people will die of cardiovascular disease (CVD) secondary to CKD by the year 2015 [3]. In China, the prevalence of CKD is ~10.8%, which is similar to that reported for developed countries [4–6]. If not treated properly, patients with CKD will suffer serious complications, such as cardiovascular and bone diseases, and can eventually progress to end-stage renal disease (ESRD) [7].

Glomerulonephritis (GN) remains a major cause of morbidity and mortality from CKD in many regions of the world, particularly Asian countries. GN patients account for 46.2% of dialysis recipients in Shanghai and GN is implicated as the cause of ESRD in 16% of all dialysis patients in the United States. GN variation in incidence among different geographical areas is associated with socioeconomic conditions, ethnicity, genetic variability, and environmental factors. In developing countries, most GN cases are related to exposure to microbial, parasitic and viral infections. In contrast, in more developed countries, antigenic exposure may be related to diet, allergens, lifestyle, and ethnicity. An effective GN treatment profile should include accurate diagnosis and specific treatment of the disease, including dialysis and transplantation when kidney failure occurs. Further, effective treatment should also include intervention therapy at early stages, as well as a combination of evaluation and treatment with the goals of delaying deterioration of renal function and reducing the risk of CVD [8]. In addition to CVD, the main complications of GN are anemia, hypertension, mineral and bone disorders, neuropathy, and malnutrition [9].

Anemia, a major complication of CKD, is closely related to survival rate, CVD complications, and disease progression in GN patients. CKD-induced anemia increases morbidity and mortality of cardiovascular complications, potentially leading to further deterioration of renal function and the establishment of a vicious cycle known as the 'cardiorenal anemia syndrome' [10]. The presence of left ventricular hypertrophy is also associated with decreased survival of patients on dialysis [11]. In addition, anemia has been identified as an independent predictor of death in stable coronary artery disease patients with CKD [12]. CKD-associated mineral bone disorders significantly increase mortality in CKD patients. In fact, hyperphosphatemia is one of the most important risk factors associated with CVD in patients on dialysis [13]. Therefore, detailed understanding of GN complications and their mechanisms will be extremely helpful for clinicians to manage GN patients.

Currently, there are limited data available regarding CKD-associated anemia and abnormal calcium and phosphorus metabolism in patients with GN in China. This study included newly diagnosed GN patients admitted to the Nephrology Department of Ruijin Hospital (Shanghai, China) over the past 6 years. To evaluate the complications in different stages of CKD, the relationship between renal function and anemia, abnormal calcium, and abnormal phosphorus metabolism in GN patients was investigated.

Materials and Methods

Patients and Study Design

Patients with GN complications including anemia, calcium and phosphorus metabolism, were retrospectively analyzed. Data were collected using an electronic database exclusively designed for this study. The data were obtained by review of the hospital, departmental and computerized laboratory databases of the central laboratory. A total of 5,466 patients with CKD admitted to the Ruijin Hospital between January 2003 and August 2009 were screened. Patients with primary GN, ranging in age from 14 to 80 years, were included in the study. However, presence of acute kidney injury, congenital renal dysplasia, history of primary parathyroid disease, or uncontrolled digestive diseases and hematological diseases were cause for study exclusion. Finally, 2,924 patients with primary GN were enrolled. Medical history, medications, and blood samples were collected.

According to the 'Clinical Practice Guidelines for All Stages of Disease' in the National Kidney Foundation Disease Outcomes Quality Initiative (NKF-KDOQI), the estimated glomerular filtration rate (eGFR) was calculated by the Chronic Kidney Disease Epidemiology Collaboration (CKD-EPI) formula: $141 \times \min(\text{serum creatinine (Scr)}/\kappa, 1)^{\alpha} \times \max(\text{Scr}/\kappa, 1)^{-1.209} \times 0.993^{\text{Age}} \times 1.018$ [if female] $\times$ 1.159 [if Black] [14, 15]. κ is 0.7 for females and 0.9 for males; α is –0.329 for females and –0.411 for males; min indicates the minimum of Scr/κ or 1; max indicates the maximum of Scr/κ or 1. The Modification of Diet in Renal Disease (MDRD) equation was also used to calculate eGFR: $186 \times (\text{Scr})^{-1.154} \times \text{Age}^{-0.203} \times$ [0.742 if female] $\times$ [1.212 if Black] [14].

Anemia was defined as hemoglobin (Hb) <120 g/l in adult female patients and <130 g/l in adult male patients, according to Clinical Practice Guidelines for anemia of CKD published by The Kidney Disease: Improving Global Outcomes (KDIGO) Foundation [16]. For calcium-phosphorus metabolism disorder, the corrected serum calcium (g/l) and the corrected total calcium was calculated by the formula: [40 – serum albumin (g/l)]/40 + actual serum calcium (g/l) and total calcium (mg/dl) + 0.8 × [4 – serum albumin (g/dl)]. Abnormal calcium-phosphorus metabolism was defined as a calcium <2.1 mmol/l (8.5 mg/dl) and p >1.45 mmol/l (4.5 mg/dl). In patients with CKD stages 3–5 who were not on dialysis, the optimal parathyroid hormone (PTH) level is not known. In patients with CKD stage 5, KDIGO suggests maintaining intact PTH levels in the range of approximately two to nine times the upper normal limit for the assay. According to NKF-KDOQI guidelines, 25(OH)D >30 ng/ml indicates the normal reserves of vitamin D, while 16 ng/ml <25(OH)D ≤30 ng/ml defines inadequacy of 25(OH)D, and 5 ng/ml ≤25(OH)D ≤15 ng/ml implies deficiency of 25(OH)D [17].

The laboratory data included values for Scr, Hb, hematocrit, serum iron, ferritin, transferrin saturation, triglyceride, cholesterol, calcium, phosphorus, intact PTH (iPTH), and 25(OH)D. The etiologies of CKD, age distribution, CKD stages were assessed.

Statistical Analyses

Descriptive statistics of continuous variables are presented as means with standard deviations or medians, depending on the distribution of the variable. Categorical variables are presented as distributions (i.e. frequencies and percentages). eGFR was used as both a continuous and categorical variable, with intervals of 30 ml/min forming five categories. The statistical significance of differences in the measured variables between subgroups was tested by the Pearson's χ^2 test for categorical analysis, and the t test or the Wilcoxon's rank sum test, where appropriate, for continuous variables.

Results

Etiologies of CKD and Clinical Characteristics of CKD Patients

General Characteristics of the Study Population

From January 2003 to August 2009, 5,466 patients were screened. There were 2,921 males and 2,545 females at a ratio of approximately 1:0.87. The ages ranged between 18 and 79 years (median 44.9), with patients of <20, 20–40, 40–60, and ≥60 years accounting for 7.6% (n = 416), 33.5% (n = 1,831), 36.2% (n = 1,979), and 22.7% (n = 1,240), respectively. The distribution of CKD stages was as follows: 25.6% for stage 1, 19.8% for stage 2, 19.6% for stage 3, 12.9% for stage 4, and 22.1% for stage 5.

Causes of CKD

The leading cause of CKD was primary glomerular diseases, which were responsible for up to 53.5% of the cases (n = 2,924). Of these 2,924 patients, 2,360 underwent kidney biopsy examination. IgA nephropathy was the most common cause of primary glomerular disease, accounting for 38.7%, followed by focal segmental glomerulosclerosis (FSGS), which accounted for 25.6%. Secondary kidney disease was the other predominant cause of CKD, accounting for 27.9% of cases (n = 1,523). The prevalence of autoimmune kidney injury and connective tissue diseases was 14.2%, and was mainly represented by systemic lupus erythematous nephritis and systemic vasculitis.

Association between Anemia and eGFR in GN Patients

Anemia was observed in 2,497 cases, accounting for 45.7% of the total number of CKD patients. With the progression of CKD, the prevalence of anemia gradually increased. The anemia rate of GN patients in early stages of CKD (stages 1–2 and 3) was 16–36%, and rapidly reached 65.8 and 80.2% when eGFR decreased to 15–29 and <15 ml/min/1.73 m^2, respectively (fig. 1). Hb levels decreased with reductions in GFR level. In CKD stages 4 and 5, Hb levels were 113.00 and 85.00 g/l, respectively, and these levels significantly decreased ($p < 0.05$) when the eGFR fell to 30 ml/min/1.73 m^2. The rate of anemia in GN pa-

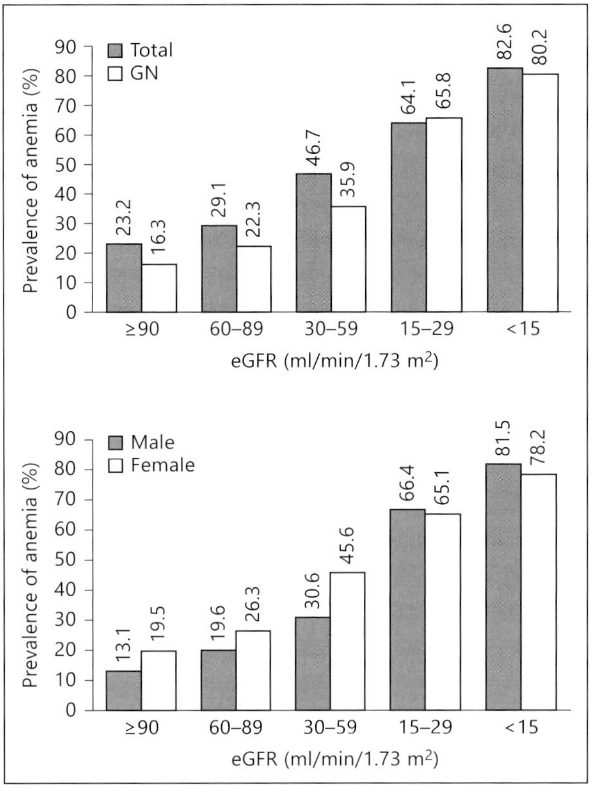

Fig. 1. Prevalence of anemia in GN patients.

tients was significantly lower than in total CKD patients at stages 1–3 ($p < 0.05$). No significant difference was observed between women and men at stages 4–5 ($p > 0.05$). The prevalence of anemia increased when eGFR levels decreased in both the male group and the female group (fig. 1).

There was a significant decline of serum iron and transferrin saturation in GN patients with CKD stage 5, compared to that in GN patients with CKD stages 1–4 ($p < 0.05$). No significant differences were observed in the ferritin levels ($p > 0.05$) (table 1).

Association of eGFR, Calcium, and Phosphorus in GN Patients

We did not observe a significant decline in serum calcium levels in GN patients with CKD stages 1–4 ($p > 0.05$). However, in CKD stage 5, the rate of hypocalcaemia increased significantly (13.7%, $p = 0.000$) (fig. 2).

The prevalence of hyperphosphatemia did not significantly increase in patients with CKD stages 1–3 ($p > 0.05$), but was much higher in the patients with

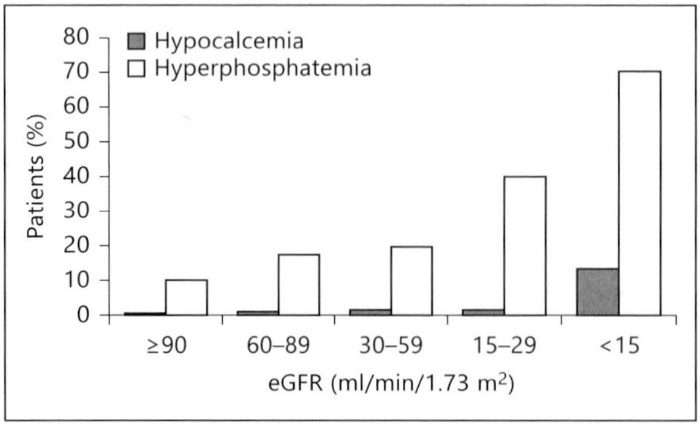

Fig. 2. Prevalence of abnormal serum calcium and phosphorus by eGFR in GN patients.

Table 1. Laboratory findings of GN patients (median)

	GFR, ml/min/1.73 m^2				
	≥90	60–89	30–59	15–29	<15
Hb, g/l	137.0	136.0	128.0	113.0*	85.0*
Iron, µmol/l	13.8	12.6	12.9	13.3	11.9*
Ferritin, ng/ml	93.8	184.8	182.8	133.3	195.5
Transferrin saturation, %	32.2	26.4	28.1	27.8	22.0*
Alb, g/l	33.0	33.0	34.0	33.0	31.0
Pre-Alb, g/l	271.0	277.0	296.0	306.0	295.0
Tch, mmol/l	5.7	5.8	5.5	5.2	4.7
TG, mmol/l	1.9	2.3	2.2	2.0	1.6
LDL-C, mmol/l	3.4	3.4	3.2	3.1	2.7
Ca, mmol/l	2.4	2.4	2.4	2.4	2.1*
P, mmol/l	1.34	1.25	1.28	1.44*	1.87*
iPTH, pg/ml	41.2	40.9	54.7	88.6*	289.2*
Vitamin D, nmol/l	36.0	47.0	38.7	34.0	32.7*

* $p < 0.05$ vs. GN patients with CKD stage 1.

CKD stages 4 and 5 ($p = 0.001$ and $p = 0.021$, respectively), showing a negative correlation with kidney function. The median serum phosphorus level increased from 1.34 to 1.87 mmol/l.

The serum levels of iPTH showed no significant increase in the GN patients with CKD stages 1–2 (41.2 and 40.9 pg/ml, respectively), but were much higher in the GN patients with stage 4–5. There was also an increase in the level of iPTH across declining eGFR levels. Median iPTH levels were 54.7, 88.6, and 289.2 pg/ml ($p = 0.000$) with stages 3–5, respectively, and showed negative correlation with renal function. The medians of calcium and phosphorus values were within normal ranges, and increases in iPTH began to occur at stage 4. There was a significant decline in serum levels of vitamin D in GN patients with CKD stage 5 ($p > 0.05$). In GN patients with early stage disease, the serum levels of active vitamin D were normal. The proportions of vitamin D insufficiency and deficiency increased to 29.3 and 11.2%, respectively, when eGFR was <15 ml/min/1.73 m^2.

Discussion

A decrease in GFR correlates with the prevalence of CKD-associated complications, however most of these complications can be detected and treated early [18]. Here, we describe an increase in the prevalence of anemia and abnormalities of vitamin D metabolites, iPTH, calcium, and phosphorus in a large-scale, observational study of an unselected cohort of GN patients with varying levels of kidney dysfunction.

In our study, the median age of CKD patients was 44.9 years, with middle-aged patients (20–60 years old) accounting for a large proportion (~70%). In developed countries, the largest proportion of patients with CKD are the elderly (>60 years old) [19]. Our data suggest that young and middle-aged patients should undergo annual routine examinations. In our study, 22.1% of patients were at stage 5 of CKD, indicating that kidney disease had progressed significantly before the patient first sought medical attention. Therefore, it is imperative to emphasize education in the community to promote early screening, diagnosis, and treatment of CKD.

The United States Renal Data System (USRDS) indicates diabetic nephropathy, hypertension nephrosclerosis, and GN are the leading causes of CKD [20]. Epidemiological research in patients on dialysis in Shanghai demonstrated that chronic glomerular disease, hypertension, and diabetes were the major causes of ESRD [21]. Although the number of patients with hypertension and diabetes increased annually, the present screening of 5,466 CKD patients revealed that primary glomerular disease accounted for 53.5% of causes resulting in CKD, among which IgA nephropathy was the most common disease, followed by FSGS. In secondary renal disease, lupus nephritis was the most common, accounting for 24.3%.

Anemia is often observed in GN patients, and the prevalence of CKD-related anemia is ~50% [22]. Our study found a similar prevalence rate of 45.7%. Although anemia occurs in all stages of CKD patients, evidence indicates that the prevalence of anemia is closely related to the severity of CKD. The Third National Health and Nutrition Examination Survey (NHANES III, 1998–1994) noted that the prevalence rate of anemia in patients with stage 1 and stage 2 CKD was 1.8 and 1.3%, respectively [23]. While results from McClellan et al. [24] show the prevalence of anemia in patients with stage 1 and 2 CKD was 26.7%, and increased to 41.6, 53.6, and 75.5% in patients with stage 3, 4, and 5, respectively, a difference that may be related to the different study populations. Our study revealed that the rates were 16.3 and 22.3% in GN patients with stage 1 and 2 CKD, respectively, implying that anemia might occur in early stages of GN. The prevalence of anemia in GN patients is lower compared with the total CKD population, and perhaps the cause plays an important role. Finally, secondary glomerular nephritis patients were more susceptible to anemia.

Calcium and phosphorus metabolism dysregulation and bone disease are important complications in CKD patients. These are especially prevalent in patients with ESRD, which often results in decreased quality of life, bone calcification, and increased mortality from CVD [25–27]. During the deterioration of renal function, the prevalence of hypocalcaemia and hyperphosphatemia increases. Epidemiological evidence suggests that hyperphosphatemia is an important risk factor, not only for secondary hyperparathyroidism but also for CVD. Levin et al. [26] demonstrated that calcium and phosphorus values did not become abnormal until GFR fell below 40 ml/min/1.73 m^2, and were relatively stable until GFR fell below 20 ml/min/1.73 m^2. A recent Dialysis Outcomes and Practice Patterns Study (DOPPS) [28] indicated that 12.4% of dialysis patients with corrected calcium had >8.4 mg/dl and 34.2% had phosphorus below 5.5 mg/dl. In our study, both hyperphosphatemia and hypocalcaemia were uncommon in patients with CKD stages 1–3. The prevalence of hyperphosphatemia increased to 40 and 70% when GFR was <30 ml/min/1.73 m^2. The percentage of patients with hyperphosphatemia in our study was higher than in the DOPPS. It is necessary to lower phosphorus in GN patients with hyperphosphatemia using phosphate binders, limiting dietary phosphate intake, and/or increasing the frequency or duration of dialysis.

Because of sample collection and assay variability, KDIGO raises significant concerns with regard to the validity of absolute levels of PTH and their strict use as a clinically relevant biomarker for targeting specific values. Here, we also demonstrate elevated iPTH levels, which increased significantly at eGFR values below 30 ml/min/1.73 m^2. The median values significantly increased to 289.2 pg/l when GFR levels were <15 ml/min/1.73 m^2. There was not a significant increase in iPTH from stage 1 to stage 2 CKD. The optimal iPTH level is not known for patients with stages 3–5 CKD who are not on dialysis. However, KDIGO suggests that patients with levels of iPTH above the upper normal limit of the assay should first be evaluated for hyperphosphatemia, hypocalcaemia, and vitamin D deficiency.

Vitamin D metabolism disorder plays an important role in the secondary hyperparathyroidism in CKD patients. Levin et al. [26] revealed that the prevalence of 25(OH)D deficiency remained relatively stable until eGFR fell below 30 ml/min/1.73 m^2. Our results confirmed that the level of vitamin D decreased significantly when the eGFR level was <15 ml/min/1.73 m^2, while remaining normal in patients with stages 1–4 CKD. The proportion of vitamin D insufficiency and shortage increased significantly when patients progressed to stage 5.

In conclusion, the present study shows that primary glomerular disease remains a major cause of CKD. Many complications, such as anemia, renal bone disease and CVD, often occur in combination. The prevalence of anemia, hyperphosphatemia, hypocalcaemia, elevated PTH, and decreased 25(OH)D are associated with the decline of eGFR. Future research should be directed toward attempting to further describe the relationship between anemia, abnormalities of mineral metabolism, and clinically meaningful outcomes by testing the impact of early interventions.

Acknowledgments

This work was supported by grants from the National Basic Research Program of China, 973 Program (Nos. 2012CB517600, 2012CB517604, and 2012CB517700), the National Key Technology R&D Program (12-5) (Nos. 2011BAI10B00 and 2011BAI10B06), and the National Nature Science Foundation of China (No. 30600291).

References

1 Zhang QL, Rothenbacher D: Prevalence of chronic kidney disease in population-based studies: systematic review. BMC Public Health 2008;8:117–130.

2 Coresh J, Astor BC, Greene T, Eknoyan G, Levey AS: Prevalence of chronic kidney disease and decreased kidney function in the adult US population: Third National Health and Nutrition Examination Survey. Am J Kidney Dis 2003;41:1–12.

3 Levey AS, Andreoli SP, DuBose T, Provenzano R, Collins AJ: Chronic kidney disease: common, harmful, and treatable – World Kidney Day 2007. Clin J Am Soc Nephrol 2007;2:401–405.

4 Chen N, Wang W, Huang Y, Shen P, Pei D, Yu H, Shi H, Zhang Q, Xu J, Lv Y, Fan Q: Community-based study on CKD subjects and the associated risk factors. Nephrol Dial Transplant 2009;24:2117–2123.

5 Chen N, Pan XX, Gu Y, Mao PJ, Ni ZH, Ding XQ, Zhang JH, Wang NS, Yuan WJ, Zhang JY: Analysis of early kidney damage in hospitalized patients with chronic kidney disease: a multicenter study. Ren Fail 2012;34:329–333.

6 Zhang L, Zhang P, Wang F, Zuo L, Zhou Y, Shi Y, Li G, Jiao S, Liu Z, Liang W, Wang H: Prevalence and factors associated with CKD: a population study from Beijing. Am J Kidney Dis 2008;51:373–384.

7 Ati AS, Remuzzi G: Chronic renal diseases as a public health problem: epidemiology, social and economic implications. Kidney Int 2005;68:7–10.

8 Levey AS, Coresh J, Balk E, Kausz AT, Levin A, Steffes MW, Hogg RJ, Perrone RD, Lau J, Eknoyan G: National Kidney Foundation. National Kidney Foundation practice guidelines for chronic kidney disease: evaluation, classification, and stratification. Ann Intern Med 2003;139:137–147.

9 Stevens LA, Li S, Wang C, Huang C, Becker BN, Bomback AS, Brown WW, Burrows NR, Jurkovitz CT, McFarlane SI, Norris KC, Shlipak M, Whaley-Connell AT, Chen SC, Bakris GL, McCullough PA: Prevalence of CKD and comorbid illness in elderly patients in the United States: results from the Kidney Early Evaluation Program (KEEP). Am J Kidney Dis 2010;55:S23–S33.

10 Besarab A, Levin A: Defining a renal anemia management period. Am J Kidney Dis 2000; 36:S13–S23.

11 Tarng DC: Cardiorenal anemia syndrome in chronic kidney disease. J Chin Med Assoc 2007;70:424–429.

12 Eckardt KU, Scherhag A, Macdougall IC, Tsakiris D, Clyne N, Locatelli F, Zaug MF, Burger HU, Drueke TB: Left ventricular geometry predicts cardiovascular outcomes associated with anemia correction in CKD. J Am Soc Nephrol 2009;20:2651–2660.

13 Lee GH, Benner D, Regidor DL, et al: Impact of kidney bone disease and its management on survival of patients on dialysis. J Ren Nutr 2007;17:38–44.

14 National Kidney Foundation: K/DOQI clinical practice guidelines for chronic kidney disease: evaluation, classification, and stratification. Am J Kidney Dis 2002;39:S1–S266.

15 Sabanayagam C, Wong TY, Tai ES: The CKD-EPI equation and MDRD study equation find similar prevalence of chronic kidney disease in Asian populations. Ann Intern Med 2009;151:892–893.

16 KDOQI: KDOQI Clinical Practice Guideline and Clinical Practice Recommendations for anemia in chronic kidney disease: 2007 update of hemoglobin target. Am J Kidney Dis 2007;50:471–530.

17 Wald R, Tentori F, Tighiouart H, Zager PG, Miskulin DC: Impact of the Kidney Disease Outcomes Quality Initiative (KDOQI) Clinical Practice Guidelines for Bone Metabolism and Disease in a large dialysis network. Am J Kidney Dis 2007;49:257–266.

18 Chen XN, Pan XX, Yu HJ, Shen PY, Zhang QY, Zhang W, Ren H, Qian Y, Zhu P, Chen N: Analysis of cardiovascular disease in Chinese inpatients with chronic kidney disease. Intern Med 2011;50:1797–1801.

19 Coresh J, Selvin E, Stevens LA, Manzi J, Kusek JW, Eggers P, Van Lente F, Levey AS: Prevalence of chronic kidney disease in the United States. JAMA 2007;298:2038–2047.

20 Foley RN, Collins AJ: End-stage renal disease in the United States: an update from the United States Renal Data System. J Am Soc Nephrol 2007;18:2644–2648.

21 Yao Q, Zhang W, Qian J: Dialysis status in China: a report from the Shanghai Dialysis Registry (2000–2005). Ethn Dis 2009;19(suppl 1):23–26.

22 Robinson BE: Epidemiology of chronic kidney disease and anemia. J Am Med Dir Assoc 2006;7:S3–S6.

23 Astor BC, Muntner P, Levin A, Eustace JA, Coresh J: Association of kidney function with anemia: the Third National Health and Nutrition Examination Survey (1988–1994). Arch Intern Med 2002;162:1401–1408.

24 McClellan W, Aronoff SL, Bolton WK, Hood S, Lorber DL, Tang KL, Tse TF, Wasserman B, Leiserowitz M: The prevalence of anemia in patients with chronic kidney disease. Curr Med Res Opin 2004;20:1501–1510.

25 Kovesdy CP, Kalantar-Zadeh K: Bone and mineral disorders in pre-dialysis CKD. Int Urol Nephrol 2008;40:427–440.

26 Levin A, Bakris GL, Molitch M, Smulders M, Tian J, Williams LA, Andress DL: Prevalence of abnormal serum vitamin D, PTH, calcium, and phosphorus in patients with chronic kidney disease: results of the study to evaluate early kidney disease. Kidney Int 2007;71:31–38.

27 Vassalotti JA, Uribarri J, Chen SC, et al: Trends in mineral metabolism: Kidney Early Evaluation Program (KEEP) and the National Health and Nutrition Examination Survey (NHANES) 1999–2004. Am J Kidney Dis 2008;51:S56–S68.

28 Dialysis Outcomes and Practice Patterns Study. DOPPS Annual Report 2010. http//www.dopps.org/annual report/index.htm.

Prof. Nan Chen, MD
Department of Nephrology, Ruijin Hospital
School of Medicine, Shanghai Jiao Tong University
No. 197 Ruijin Er Road, Shanghai 200025 (PR China)
E-Mail chen-nan@medmail.com.cn

Chen N (ed): New Insights into Glomerulonephritis.
Contrib Nephrol. Basel, Karger, 2013, vol 181, pp 41–51 (DOI: 10.1159/000348461)

Role of Podocyte Injury in IgA Nephropathy

Madhav C. Menon · Peter Y. Chuang · John Cijiang He

Division of Nephrology, Department of Medicine, Mount Sinai School of Medicine, New York, N.Y., USA

Abstract

IgA nephropathy (IgAN) encompasses different pathological entities, all of which are characterized by the mesangial deposition of IgA. Aberrantly glycosylated IgA molecules appear to play a major role in the pathogenesis of IgAN. Both genetic and environmental factors contribute to the formation of IgG antibodies that binds to aberrantly glycosylated IgA molecules and the glomerular deposition of circulating polymeric IgA complexes. Mesangial cells serve as primary glomerular cell type injured in IgAN with a variety of pathological changes leading to progressive renal dysfunction. Recent evidence suggests that indirect podocyte injury also contributes to glomerular damage observed in IgAN. These mechanisms were demonstrated in cultured podocytes exposed to mesangial cell-conditioned medium of human IgAN. Large-scale biopsy series of human kidneys also suggest that podocyte injury plays a role in IgAN. In this review, we discuss the recent advances that have helped elucidate the mechanisms of podocyte injury in IgAN and how these mechanisms correlate with clinically important outcomes in human studies.
Copyright © 2013 S. Karger AG, Basel

IgA nephropathy (IgAN) was first described in the late 1960s by Berger and Hinglais [1]. A considerable amount has been learned about the pathogenesis, clinical course and treatment of IgAN since then [for reviews, see 2–6]. It remains the commonest primary glomerulonephritides in Asia and accounts for 10–20% of all renal biopsies in Europe and North America [7–9]. The geographic distribution and genetic factors predisposing to IgAN have been discussed elsewhere in this issue. IgA1-predominant mesangial deposits serve as the pathologic hallmark of IgAN. Mesangial cell proliferation and extracellular

Table 1. Pathological lesions in IgAN and their frequency. Single-center consecutive biopsies of IgAN [13]

Pathological lesions encountered in IgAN	% of patients (n = 668)
No lesions by light microscopy	4
Mesangial proliferation only	13
Focal proliferative glomerulonephritis	37
Diffuse proliferative glomerulonephritis	28
Crescentic glomerulonephritis (>50% glomeruli with crescents)	4
Focal sclerosis (without proliferation)	6*
Chronic sclerosing glomerulonephritis	6
Lesions not otherwise specified	2

* Newer series have found a higher prevalence of FSGS lesions [19, 20].

matrix expansion are present in the earliest stages of the disease. Progression to glomerulosclerosis with tubular atrophy and interstitial fibrosis with the development of end-stage renal disease (ESRD) is estimated to occur in 20–40% of patients within 20 years [10–12]. The spectrum of pathological lesions and frequency with which these were encountered in a large single-center biopsy registry (668 patients) of IgAN are listed in table 1 [13]. The clinical spectrum of IgAN ranges from asymptomatic mesangial IgA1 deposits in healthy subjects to rapidly progressive renal failure and ESRD [2, 14, 15]. The unique clinical course manifesting in an individual patient has a clear bearing on outcomes [12]. For instance, persistent proteinuria (>1 g/day) and hypertension have been shown to be clinical predictors for progression to ESRD in IgAN [12, 16, 17]. A multinational working group developed the Oxford classification of IgAN according to the degree of mesangial hypercellularity, segmental glomerulosclerosis, endocapillary hypercellularity, and tubular atrophy/interstitial fibrosis as an overall predictor of prognosis [18]. Although mesangial cells have been considered as the major cell type injured in IgAN, recent studies suggest that podocyte injury also occurs in IgAN. Podocyte injury in IgAN presents phenotypically as significant proteinuria, ultrastructurally as foot-process effacement and histologically as glomerular capsular adhesions and/or focal segmental sclerosis (FSGS). Podocyte injury has gained acceptance as a key mechanism leading to the disease progression in IgAN [19–21]. In the following review, we discuss the recent advances in unraveling the pathomechanisms of podocyte injury in IgAN and how they have correlated with observed clinical consequences.

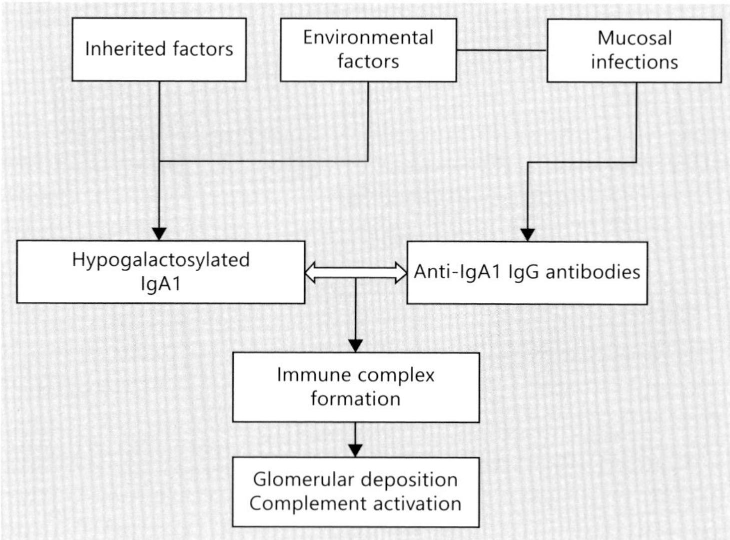

Fig. 1. Pathogenesis of IgAN – formation and deposition of IgA complexes. Pathogenic IgA complexes are produced as a consequence of cumulative inherited (aberrant enzymatic glycosylation) and acquired (mucosal infections) insults. These lead to increased production, enhanced polymerization, and decreased clearance of aberrantly glycosylated IgA molecules. Anti-glycan IgG antibodies that target the hypogalactosylated hinge region of the IgA molecules are present in a high proportion of IgAN patients. Glomerular deposition of these complexes initiates a cascade of events leading to glomerular injury.

Pathogenesis

Polymeric Hypogalactosoylated IgA1 Complexes, Glomerular Deposition and Local Complement Activation

The pathogenesis of IgAN has been extensively reviewed from both molecular and clinicopathologic perspectives [2–6]. An overview of pathogenesis is provided in figures 1 and 2. The formation of circulating IgA1 immune complexes containing aberrantly galactosylated IgA (subclass IgA1) is an essential element of IgAN. These IgA1 molecules are characterized by reduced terminal galactosylation of O-linked N-acetylgalactosamine residues at the hinge region of their heavy chains. The high recurrence of IgAN posttransplant strongly implies extrarenal triggers for IgA1 production [22]. Increased circulating IgA1-secreting B cells, relative IL-4 and IL-5 excess in T cells – a milieu favoring IgA production – and increases in IgA1-producing memory B cells in the bone marrow have been identified in IgAN [3]. A 'second hit' characterized by the formation of IgG antibodies against the hinge region of IgA1 molecules which have specificity for the hypogalactosylated N-acetylgalactosamine residues occurs in a high propor-

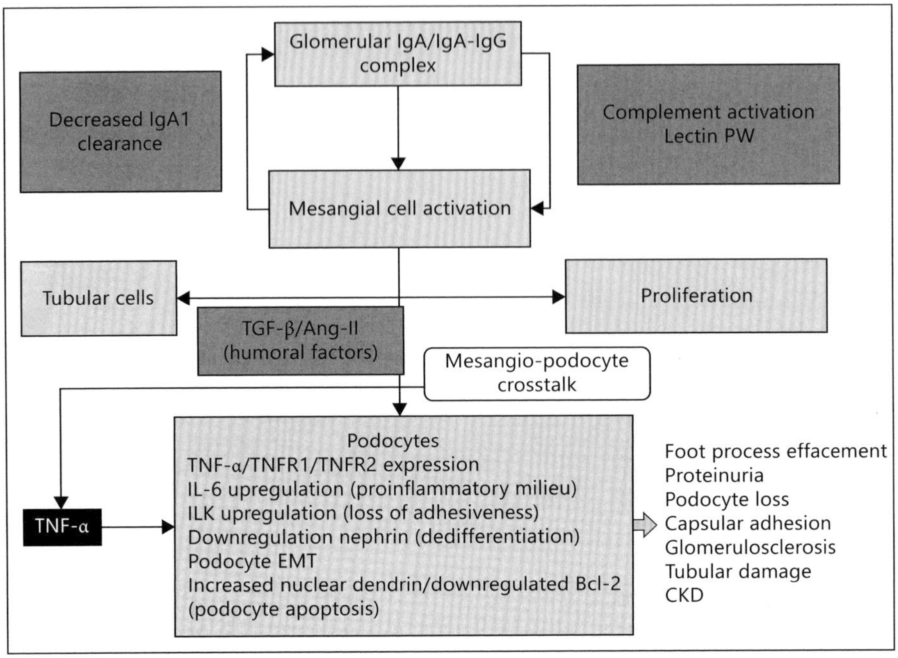

Fig. 2. Pathogenesis of IgAN – mesangial cell activation, mesangio-podocyte crosstalk and podocyte injury. pIgA complexes directly activate complement (lectin pathway) and mesangial cells with autocrine and paracrine effects. Mesangial cell proliferation occurs. Major mediators for mesangio-podocyte crosstalk are TGF-β and Ang-II, which lead to podocyte dedifferentiation with cytoskeletal disorganization and loss of adhesiveness. Production of TNF-α along with activation of TNFR1/TNFR2 induces proinflammatory response (IL-6 upregulation) and promotes podocyte apoptosis. The clinicopathologic correlations of these changes are manifested by proteinuria, foot-process effacement, podocyte loss with capsular adhesions, and progressive glomerulosclerosis with subsequent tubulointerstitial fibrosis.

tion of IgAN patients [23–25]. Premature and/or increased sialation of IgA1 may also facilitate hypogalactosylation of IgA1 in IgAN [26, 27]. IgAN-IgA1 shows increased tendency to self-aggregate (polymerized or p-IgA1) and diminished clearance by hepatocytes [3, 28]. Glomerular deposition (mesangial/paramesangial) of predominant or co-dominant IgA1 is the hallmark of IgAN. Mesangial IgA1 complexes activate complement locally via the mannose-binding lectin pathway [29–31]. The co-deposition of complement C3 without C1q encountered in IgAN deposits testifies against the classical pathway activation. Direct activation of mesangial cells (discussed below), secretion of cytokines and recruitment of inflammatory cells also occurs leading to progressive injury in IgAN [32–34].

Podocyte Injury: Mesangio-Podocyte Crosstalk

Podocytes are specialized, terminally differentiated cells of mesenchymal origin characterized by a complexly regulated actin cytoskeleton and interdigitating cellular extensions (foot processes). Processes interfering with podocyte's structural or functional integrity lead to disruption of the glomerular filtration barrier [35, 36]. It is apparent now that mesangio-podocyte crosstalk is fundamental to podocyte injury in IgAN. Studies suggest that cultured human podocytes are not able to bind polymeric hypogalactosylated IgA1 from IgAN (IgAN-pIgA1) patients while mesangial cells are able to bind it with tenfold higher affinity [37]. The mesangium expresses high levels of CD71 (transferrin receptor), which can bind IgA1 and co-localizes with pathogenic deposits in IgAN patients [38]. However, CD71 is reported to be expressed ubiquitously in all kidney cell types [37]. A candidate receptor for IgA in the mesangial cell is the asialoglycoprotein receptor, which is responsible for hepatic clearance of IgA and has been shown indirectly to bind hypogalactosylated IgAN-pIgA1 [39]. Another putative IgA1 receptor on mesangial cells is CD89 (Fc-α receptor) and its overexpression transgenic mice developed IgAN-like renal disease [40]. Other studies have implied the absence of both these receptors on mesangial cells [37, 41]. However, numerous in vitro studies show that mesangial cells can be activated by IgAN-pIgA1 [37, 42, 43]. Since PIgA1 had no direct effect on podocytes, Lai and colleagues [37, 44, 45] designed a series of elegant in vitro experiments to treat mesangial cells with pIgA1 isolated from IgAN sera. This mesangial cell-conditioned medium was then used to treat podocytes and tubular cells to study phenotypic and secretory changes in these cells.

Using this system, they demonstrated that mesangial cells can crosstalk with podocytes. A schema for mesangio-podocyte crosstalk with downstream effects culminating in podocyte injury is represented in figure 2. Initial observations demonstrated that IgAN-pIgA1 (not monomeric IgA1) upregulated TGF-β synthesis in cultured mesangial cells [43, 46]. TGF-β induced extracellular matrix marker (fibronectin-1) by mesangial cells in an autocrine fashion [46]. Suppression of human podocyte differentiation markers (nephrin, ezrin, podocin) by conditioned medium from mesangial cells was observed suggesting a paracrine effect [45]. Both autocrine and paracrine effects could be diminished by anti-TGF-β antibody [45, 46]. Furthermore, the podocyte dedifferentiation observed above was reproduced by direct stimulation of TGF-β alone at the comparable concentrations [45].

Mesangial cell-derived TNF-α had also been shown to be important mediator for glomerulotubular crosstalk [44]. In a prior study, Lai et al. [37] observed a dose- and time-dependent increase in TNF-α production by human podocytes exposed to IgAN-pIgA1-conditioned medium. This was not seen when serum from healthy controls was used to stimulate human MCs. It is noteworthy that TNF-α is predominantly of podocyte origin and not from mesangial cells themselves. Furthermore, podocyte-derived TNF-α upregulated the expression of TNF receptors 1 and 2 and IL-6 in an autocrine fashion. Anti-TNF-α antibody was synergistic with anti-TGF-β in inhibiting the podocyte dedifferentiation induced by IgAN-pIgA1-conditioned media, suggesting both mediators contribute to mesangial cell-dependent podocyte effects [45]. As controls, podocytes treated with IgAN-pIgA1 medium from other cell types (tubular cells, podocytes, endothelial cells) did not show dedifferentiation [37, 45].

Increased synthesis of angiotensin-II by human and mouse mesangial cells in response to IgAN-pIgA1 compared to controls has been reported by independent groups [42, 43]. Wang et al. [42] demonstrated that downregulation of nephrin in response to PIgA1-conditioned medium could be partly reversed by pretreatment with enalaprilat or valsartan. Mesangial TGF-β synthesis was also significantly inhibited by captopril and losartan [43]. Integrin-linked kinase is recruited in podocytes upon activation of α3-β1 integrin which is essential for podocyte-basement membrane interaction. Dysregulation of integrin-linked kinase causes podocyte injury manifested by podocyte epithelial-mesenchymal transition and FSGS in mice [47, 48]. Podocyte epithelial-mesenchymal transition associated with increased permeability to albumin as a manifestation of injury has also recently been observed with IgAN-pIgA1 medium [49]. Ye et al. [50] showed upregulation of integrin-linked kinase and decreased podocyte adhesiveness in vitro upon treatment with pIgA1-conditioned medium. These changes correlated with angiotensin-II levels in the medium and were partially reversed with valsartan. Impaired podocyte adhesiveness in association with dedifferentiation was observed with pIgA1 medium treatment [42]. Taken together, these studies suggest a major role for angiotensin-II in the mesangio-podocyte crosstalk in IgAN.

Podocyte apoptosis is a major injury leading to glomerular disease. Upregulation of TNF, TNF-R1 and TNF-R2 has been observed in podocytes with pIgA1 medium in vitro and in IgAN tissue in vivo [37]. Activation of TNF-R1 promotes apoptotic signals through casapase-8 activation [51]. Lai et al. [37] reported increased podocyte apoptosis as a consequence of upregulated TNF-α-TNF-R1 interaction using an in vitro model. TGF-β increases podocyte apoptosis through Smad-7 and p-38 MAP kinase activation in a dose-dependent manner [52, 53]. A study by Wang et al. [54] has also shown direct induction of apoptosis in podocytes exposed to pIgA1. Dendrin is a podocyte protein located normally at the slit diaphragm whose translocation to the nucleus has been linked to podocyte apoptosis in different injury models [55, 56]. Recently, dendrin-positive nuclei were observed to be significantly increased in acute glomerular injury in IgAN. In addition, urinary podocytes stained positively with both dendrin and annexin V (apoptosis marker) were increased in patients with IgAN versus patients with minimal change disease [55]. These data suggest that podocyte apoptosis occurs in IgAN, likely through mesangio-podocyte communication.

Podocyte Injury: Clinicopathologic Correlation in IgAN

Podocyte dedifferentiation (loss of podocyte-specific markers – nephrin, podocin, ezrin, GLEPP-1, synaptopodin) with progression of clinical disease has been observed in IgAN biopsies by several groups [19, 45, 57]. The mesenchymal markers (FSP-1 and smooth muscle actin) were proportionally increased in podocytes along with the amount of proteinuria [49]. Podocytopenia (loss of podocytes), measured histologically after immunostaining, has been found to correlate with the amount of proteinuria and renal outcomes in IgAN patients [58, 59]. A surrogate for podocyte adhesiveness and apoptosis is net urinary podocyte loss. Cumulative urinary podocyte shedding and urinary podocalyxin expression are correlated with the extent of histological and clinical injury in IgAN patients [60, 61]. Downregulation of anti-apoptotic Bcl-2 in podocytes is strongly associated with elevated serum creatinine, decline in creatinine clearance, and magnitude of proteinuria at the time of diagnostic by renal biopsy [62]. The efficacy of inhibitors of the renin-angiotensin system in the treatment of IgAN highlights the importance of angiotensin-II-mediated mesangio-podocyte and glomerulotubular communication in progressive renal decline from a clinical perspective [10, 63, 64]. Recent studies have provided new insight on the clinical significance of podocyte injury in IgAN [19, 20]. Firstly, capsular adhesions representing areas of podocyte loss (injury) were identified in 41% of 128 biopsies (while only seen in 8% of lupus biopsies). Secondly, FSGS lesions were encountered in 80% of all biopsies. Importantly, FSGS in IgAN recapitulated all the different morphologic variants described for primary FSGS, not simply for the perihilar (secondary) variant [21]. Finally, when FSGS was identified, it correlated with tubulointerstitial fibrosis and was associated with significantly worse patient outcomes. Together, these studies suggest that podocyte injury is singularly important in the pathogenesis of IgAN, more so than previously recognized.

In conclusion, multiple heritable and environmental insults converge for the production and glomerular deposition of pathogenic IgA complexes in IgAN. Mesangial cell activation is an initial consequence of IgA deposition. Pathways of mesangio-podocyte crosstalk have been increasingly elucidated. This crosstalk leads to indirect podocyte injury which has cytoskeletal and signaling consequences and correlates to clinical and histological findings in IgAN. Podocyte injury and apoptosis are more frequently encountered in IgAN than previously recognized and portend adverse outcomes.

Acknowledgements

J.C.H. is supported by NIH 1R01DK088541, NIH 1R01DK078897, Chinese 973 Fund 2012CB517601, and a VA Merit Award, P.Y.C. is supported by NIH 5K08DK082760, and M.C.M. is supported by an ASN Research Fellowship.

References

1 Berger J, Hinglais N: Intercapillary deposits of IgA-IgG (in French). J Urol Nephrol 1968; 74:694–695.
2 Donadio JV, Grande JP: IgA nephropathy. N Engl J Med 2002;347:738–748.
3 Lai KN: Pathogenesis of IgA nephropathy. Nat Rev Nephrol 2012;8:275–283.
4 Moura IC, Benhamou M, Launay P, Vrtovsnik F, Blank U, Monteiro RC: The glomerular response to IgA deposition in IgA nephropathy. Semin Nephrol 2008;28:88–95.
5 Barratt J, Feehally J, Smith AC: Pathogenesis of IgA nephropathy. Semin Nephrol 2004;24: 197–217.
6 Floege J: The pathogenesis of IgA nephropathy: what is new and how does it change therapeutic approaches? Am J Kidney Dis 2011; 58:992–1004.
7 D'Amico G: The commonest glomerulonephritis in the world: IgA nephropathy. Q J Med 1987;64:709–727.
8 Li LS, Liu ZH: Epidemiologic data of renal diseases from a single unit in China: analysis based on 13,519 renal biopsies. Kidney Int 2004;66:920–923.
9 Simon P, Ramee MP, Boulahrouz R, Stanescu C, Charasse C, Ang KS, Leonetti F, Cam G, Laruelle E, Autuly V, Rioux N: Epidemiologic data of primary glomerular diseases in western France. Kidney Int 2004;66:905–908.
10 Julian BA, Novak J: IgA nephropathy: an update. Curr Opin Nephrol Hypertens 2004;13: 171–179.
11 D'Amico G: Natural history of idiopathic IgA nephropathy: role of clinical and histological prognostic factors. Am J Kidney Dis 2000;36: 227–237.
12 Philibert D, Cattran D, Cook T: Clinicopathologic correlation in IgA nephropathy. Semin Nephrol 2008;28:10–17.
13 Nachman PH, Jennette JC, Falk RJ: Brenner and Rector's the Kidney, ed 8. Philadelphia, Saunders, 2008.
14 Suzuki K, Honda K, Tanabe K, Toma H, Nihei H, Yamaguchi Y: Incidence of latent mesangial IgA deposition in renal allograft donors in japan. Kidney Int 2003;63:2286–2294.
15 Waldherr R, Rambausek M, Duncker WD, Ritz E: Frequency of mesangial IgA deposits in a non-selected autopsy series. Nephrol Dial Transplant 1989;4:943–946.
16 Alamartine E, Sabatier JC, Guerin C, Berliet JM, Berthoux F: Prognostic factors in mesangial IgA glomerulonephritis: an extensive study with univariate and multivariate analyses. Am J Kidney Dis 1991;18:12–19.
17 Rekola S, Bergstrand A, Bucht H: Deterioration of GFR in IgA nephropathy as measured by ^{51}Cr-EDTA clearance. Kidney Int 1991;40: 1050–1054.

18 Working Group of the International IgA Nephropathy Network and the Renal Pathology Society: Cattran DC, Coppo R, Cook HT, Feehally J, Roberts IS, Troyanov S, Alpers CE, Amore A, Barratt J, Berthoux F, Bonsib S, Bruijn JA, D'Agati V, D'Amico G, Emancipator S, Emma F, Ferrario F, Fervenza FC, Florquin S, Fogo A, Geddes CC, Groene HJ, Haas M, Herzenberg AM, Hill PA, Hogg RJ, Hsu SI, Jennette JC, Joh K, Julian BA, Kawamura T, Lai FM, Leung CB, Li LS, Li PK, Liu ZH, Mackinnon B, Mezzano S, Schena FP, Tomino Y, Walker PD, Wang H, Weening JJ, Yoshikawa N, Zhang H: The Oxford classification of IgA nephropathy: rationale, clinicopathological correlations, and classification. Kidney Int 2009;76:534–545.

19 Hill GS, Karoui KE, Karras A, Mandet C, Duong Van Huyen JP, Nochy D, Bruneval P: Focal segmental glomerulosclerosis plays a major role in the progression of IgA nephropathy. I. Immunohistochemical studies. Kidney Int 2011;79:635–642.

20 El Karoui K, Hill GS, Karras A, Moulonguet L, Caudwell V, Loupy A, Bruneval P, Jacquot C, Nochy D: Focal segmental glomerulosclerosis plays a major role in the progression of IgA nephropathy. II. Light microscopic and clinical studies. Kidney Int 2011;79:643–654.

21 Cook HT: Focal segmental glomerulosclerosis in IgA nephropathy: a result of primary podocyte injury? Kidney Int 2011;79:581–583.

22 Berger J: Recurrence of IgA nephropathy in renal allografts. Am J Kidney Dis 1988;12: 371–372.

23 Suzuki H, Fan R, Zhang Z, Brown R, Hall S, Julian BA, Chatham WW, Suzuki Y, Wyatt RJ, Moldoveanu Z, Lee JY, Robinson J, Tomana M, Tomino Y, Mestecky J, Novak J: Aberrantly glycosylated IgA1 in IgA nephropathy patients is recognized by IgG antibodies with restricted heterogeneity. J Clin Invest 2009;119:1668–1677.

24 Tomana M, Novak J, Julian BA, Matousovic K, Konecny K, Mestecky J: Circulating immune complexes in IgA nephropathy consist of IgA1 with galactose-deficient hinge region and antiglycan antibodies. J Clin Invest 1999; 104:73–81.

25 Schena FP, Pastore A, Ludovico N, Sinico RA, Benuzzi S, Montinaro V: Increased serum levels of IgA1-IgG immune complexes and anti-F(ab')$_2$ antibodies in patients with primary IgA nephropathy. Clin Exp Immunol 1989;77:15–20.

26 Leung JC, Poon PY, Lai KN: Increased sialylation of polymeric immunoglobulin A1: mechanism of selective glomerular deposition in immunoglobulin A nephropathy? J Lab Clin Med 1999;133:152–160.

27 Suzuki H, Moldoveanu Z, Hall S, Brown R, Vu HL, Novak L, Julian BA, Tomana M, Wyatt RJ, Edberg JC, Alarcon GS, Kimberly RP, Tomino Y, Mestecky J, Novak J: IgA1-secreting cell lines from patients with IgA nephropathy produce aberrantly glycosylated IgA1. J Clin Invest 2008;118:629–639.

28 Narita I, Gejyo F: Pathogenetic significance of aberrant glycosylation of IgA1 in IgA nephropathy. Clin Exp Nephrol 2008;12:332–338.

29 Hisano S, Matsushita M, Fujita T, Endo Y, Takebayashi S: Mesangial IgA2 deposits and lectin pathway-mediated complement activation in IgA glomerulonephritis. Am J Kidney Dis 2001;38:1082–1088.

30 Roos A, Rastaldi MP, Calvaresi N, Oortwijn BD, Schlagwein N, van Gijlswijk-Janssen DJ, Stahl GL, Matsushita M, Fujita T, van Kooten C, Daha MR: Glomerular activation of the lectin pathway of complement in IgA nephropathy is associated with more severe renal disease. J Am Soc Nephrol 2006;17:1724–1734.

31 Oortwijn BD, Eijgenraam JW, Rastaldi MP, Roos A, Daha MR, van Kooten C: The role of secretory IgA and complement in IgA nephropathy. Semin Nephrol 2008;28:58–65.

32 Arima S, Nakayama M, Naito M, Sato T, Takahashi K: Significance of mononuclear phagocytes in IgA nephropathy. Kidney Int 1991;39:684–692.

33 Falk MC, Ng G, Zhang GY, Fanning GC, Roy LP, Bannister KM, Thomas AC, Clarkson AR, Woodroffe AJ, Knight JF: Infiltration of the kidney by αβ and γδ T cells: effect on progression in IgA nephropathy. Kidney Int 1995;47:177–185.

34 Kanamaru Y, Arcos-Fajardo M, Moura IC, Tsuge T, Cohen H, Essig M, Vrtovsnik F, Loirat C, Peuchmaur M, Beaudoin L, Launay P, Lehuen A, Blank U, Monteiro RC: Fc-α receptor I activation induces leukocyte recruitment and promotes aggravation of glomerulonephritis through the Fcr-γ adaptor. Eur J Immunol 2007;37:1116–1128.

35 Pavenstadt H, Kriz W, Kretzler M: Cell biology of the glomerular podocyte. Physiol Rev 2003;83:253–307.

36 Menon MC, Chuang PY, He CJ: The glomerular filtration barrier: components and crosstalk. Int J Nephrol 2012;2012:749010.

37 Lai KN, Leung JC, Chan LY, Saleem MA, Mathieson PW, Lai FM, Tang SC: Activation of podocytes by mesangial-derived TNF-α: Glomerulo-podocytic communication in IgA nephropathy. Am J Physiol Renal Physiol 2008;294:F945–F955.

38 Haddad E, Moura IC, Arcos-Fajardo M, Macher MA, Baudouin V, Alberti C, Loirat C, Monteiro RC, Peuchmaur M: Enhanced expression of the CD71 mesangial IgA1 receptor in Berger disease and Henoch-Schönlein nephritis: association between CD71 expression and IgA deposits. J Am Soc Nephrol 2003;14:327–337.

39 Gomez-Guerrero C, Duque N, Egido J: Mesangial cells possess an asialoglycoprotein receptor with affinity for human immunoglobulin A. J Am Soc Nephrol 1998;9:568–576.

40 Launay P, Grossetete B, Arcos-Fajardo M, Gaudin E, Torres SP, Beaudoin L, Patey-Mariaud de Serre N, Lehuen A, Monteiro RC: Fc-α receptor (CD89) mediates the development of immunoglobulin A (IgA) nephropathy (Berger's disease). Evidence for pathogenic soluble receptor-IgA complexes in patients and CD89 transgenic mice. J Exp Med 2000; 191:1999–2009.

41 Leung JC, Tsang AW, Chan DT, Lai KN: Absence of CD89, polymeric immunoglobulin receptor, and asialoglycoprotein receptor on human mesangial cells. J Am Soc Nephrol 2000;11:241–249.

42 Wang C, Liu X, Ye Z, Zhang J, Tang H, Chen Z, Zhang H, Lou T: Mesangial medium with IgA1 from IgA nephropathy inhibits nephrin expression in mouse podocytes. Eur J Clin Invest 2009;39:561–567.

43 Lai KN, Tang SC, Guh JY, Chuang TD, Lam MF, Chan LY, Tsang AW, Leung JC: Polymeric IgA1 from patients with IgA nephropathy upregulates transforming growth factor-β synthesis and signal transduction in human mesangial cells via the renin-angiotensin system. J Am Soc Nephrol 2003;14:3127–3137.

44 Chan LY, Leung JC, Tsang AW, Tang SC, Lai KN: Activation of tubular epithelial cells by mesangial-derived TNF-α: glomerulotubular communication in IgA nephropathy. Kidney Int 2005;67:602–612.

45 Lai KN, Leung JC, Chan LY, Saleem MA, Mathieson PW, Tam KY, Xiao J, Lai FM, Tang SC: Podocyte injury induced by mesangial-derived cytokines in IgA nephropathy. Nephrol Dial Transplant 2009;24:62–72.

46 Lopez-Armada MJ, Gomez-Guerrero C, Egido J: Receptors for immune complexes activate gene expression and synthesis of matrix proteins in cultured rat and human mesangial cells: Role of TGF-β. J Immunol 1996; 157:2136–2142.

47 Kang YS, Li Y, Dai C, Kiss LP, Wu C, Liu Y: Inhibition of integrin-linked kinase blocks podocyte epithelial-mesenchymal transition and ameliorates proteinuria. Kidney Int 2010; 78:363–373.

48 El-Aouni C, Herbach N, Blattner SM, Henger A, Rastaldi MP, Jarad G, Miner JH, Moeller MJ, St-Arnaud R, Dedhar S, Holzman LB, Wanke R, Kretzler M: Podocyte-specific deletion of integrin-linked kinase results in severe glomerular basement membrane alterations and progressive glomerulosclerosis. J Am Soc Nephrol 2006;17:1334–1344.

49 Wang C, Liu X, Ke Z, Tang Y, Li CC, Li CM, Ye Z, Zhang J, Lou T: Mesangial medium from IgA nephropathy patients induces podocyte epithelial-to-mesenchymal transition through activation of the phosphatidyl inositol-3-kinase/Akt signaling pathway. Cell Physiol Biochem 2012;29:743–752.

50 Ye ZC, Wang C, Tang Y, Liu X, Peng H, Zhang H, Lou TQ: Serum IgA1 from patients with IgA nephropathy upregulates integrin-linked kinase synthesis and inhibits adhesive capacity in podocytes through indirect pathways. Clin Invest Med 2009;32:E20–E27.

51 Wajant H: Death receptors. Essays Biochem 2003;39:53–71.

52 Schiffer M, Bitzer M, Roberts IS, Kopp JB, ten Dijke P, Mundel P, Bottinger EP: Apoptosis in podocytes induced by TGF-β and Smad7. J Clin Invest 2001;108:807–816.
53 Wu DT, Bitzer M, Ju W, Mundel P, Bottinger EP: TGF-β concentration specifies differential signaling profiles of growth arrest/differentiation and apoptosis in podocytes. J Am Soc Nephrol 2005;16:3211–3221.
54 Wang C, Peng H, Tang H, Liu X, Chen Z, Lou T, Yu X: Serum IgA1 from IgA nephropathy patients induces apoptosis in podocytes through direct and indirect pathways. Clin Invest Med 2007;30:E240–E249.
55 Asanuma K, Akiba-Takagi M, Kodama F, Asao R, Nagai Y, Lydia A, Fukuda H, Tanaka E, Shibata T, Takahara H, Hidaka T, Asanuma E, Kominami E, Ueno T, Tomino Y: Dendrin location in podocytes is associated with disease progression in animal and human glomerulopathy. Am J Nephrol 2011;33:537–549.
56 Asanuma K, Campbell KN, Kim K, Faul C, Mundel P: Nuclear relocation of the nephrin and CD2AP-binding protein dendrin promotes apoptosis of podocytes. Proc Natl Acad Sci USA 2007;104:10134–10139.
57 Tian J, Wang HP, Mao YY, Jin J, Chen JH: Reduced glomerular epithelial protein 1 expression and podocyte injury in immunoglobulin A nephropathy. J Int Med Res 2007; 35:338–345.
58 Lemley KV, Lafayette RA, Safai M, Derby G, Blouch K, Squarer A, Myers BD: Podocytopenia and disease severity in IgA nephropathy. Kidney Int 2002;61:1475–1485.
59 Xu L, Yang HC, Hao CM, Lin ST, Gu Y, Ma J: Podocyte number predicts progression of proteinuria in IgA nephropathy. Mod Pathol 2010;23:1241–1250.
60 Hara M, Yanagihara T, Kihara I: Cumulative excretion of urinary podocytes reflects disease progression in IgA nephropathy and Schönlein-Henoch purpura nephritis. Clin J Am Soc Nephrol 2007;2:231–238.
61 Asao R, Asanuma K, Kodama F, Akiba-Takagi M, Nagai-Hosoe Y, Seki T, Takeda Y, Ohsawa I, Mano S, Matsuoka K, Kurosawa H, Ogasawara S, Hirayama Y, Sekine S, Horikoshi S, Hara M, Tomino Y: Relationships between levels of urinary podocalyxin, number of urinary podocytes, and histologic injury in adult patients with IgA nephropathy. Clin J Am Soc Nephrol 2012;7:1385–1393.
62 Qiu LQ, Sinniah R, I-Hong Hsu S: Downregulation of Bcl-2 by podocytes is associated with progressive glomerular injury and clinical indices of poor renal prognosis in human IgA nephropathy. J Am Soc Nephrol 2004;15: 79–90.
63 Praga M, Gutierrez E, Gonzalez E, Morales E, Hernandez E: Treatment of IgA nephropathy with ACE inhibitors: a randomized and controlled trial. J Am Soc Nephrol 2003;14:1578–1583.
64 Campbell R, Sangalli F, Perticucci E, Aros C, Viscarra C, Perna A, Remuzzi A, Bertocchi F, Fagiani L, Remuzzi G, Ruggenenti P: Effects of combined ACE inhibitor and angiotensin II antagonist treatment in human chronic nephropathies. Kidney Int 2003;63:1094–1103.

John Cijiang He, MD, PhD
Division of Nephrology, Box 1243
Mount Sinai School of Medicine
One Gustave L. Levy Place, New York, NY 10029 (USA)
E-Mail cijiang.he@mssm.edu

Chen N (ed): New Insights into Glomerulonephritis.
Contrib Nephrol. Basel, Karger, 2013, vol 181, pp 52–64 (DOI: 10.1159/000348652)

Genetic Studies of IgA Nephropathy: What Have We Learned from Genome-Wide Association Studies

Jingyuan Xie[a] · Samantha Shapiro[b] · Ali Gharavi[b]

[a]Department of Nephrology, Ruijin Hospital, Shanghai Jiao Tong University, School of Medicine, Shanghai, PR China, and [b]Department of Medicine, College of Physicians and Surgeons, Columbia University, New York, N.Y., USA

Abstract

Primary IgA nephropathy (IgAN) is the most common glomerulonephritis in the world. It is most common in Asian populations, followed by Caucasians, yet relatively infrequent amongst African populations. The striking difference in the prevalence of IgAN between world populations, together with the known familial aggregation of disease, suggests an inherited mechanism. Recently three genome-wide association studies (GWAS) of IgAN have identified seven susceptibility loci, providing initial insight into the genetic architecture of this trait. While genetic studies of complex traits are challenging, applying new techniques and methods of analysis, especially Next-Generation Sequencing, will push the genetic studies of IgAN forward. Copyright © 2013 S. Karger AG, Basel

Primary IgA nephropathy (IgAN) is the most common glomerulonephritis in the world, and is especially prevalent in Asia, where it accounts for up to 45% of primary glomerulonephritis. Additionally, IgAN is one of the most common causes of end-stage renal disease, with 15–40% of patients progressing to end-stage renal disease [1]. There is strong epidemiological and now molecular evidence for involvement of genetic factors in the pathogenesis of the disease [2, 3]. In an effort to identify the susceptibility loci involved in sporadic cases of IgAN, genome-wide association studies (GWAS) have been undertaken. GWAS have heralded a breakthrough in understanding the genetic basis of many complex diseases, enabling the discovery of common risk alleles that have a small effect on disease phenotype.

Pathogenesis

IgAN manifests as recurrent episodes of hematuria and is defined on renal biopsy by the presence of mesangial proliferation with the deposition of IgA1 containing immune-complexes. Humans have two IgA subtypes, IgA1 and IgA2. In evolutionary terms, IgA1 is relatively new, found only in humans and hominid primates. It presumably arose from a duplication of the IgA heavy-chain constant-region genes. The main structural difference between the IgA1 and IgA2 molecules is at the hinge region between the two Fab fragments. The hinge region of IgA1 contains multiple serine and threonine residues, each of which can be O-glycosylated by the addition of N-acetyl-galactosamine (GalNAc); these GalNAc moieties are subsequently galactosylated and sialylated. In a majority of IgAN patients, the GalNAc moiety on IgA1 is galactose-deficient. This exposed GalNAc moiety can be recognized as an antigen, stimulating production of anti-glycan antibodies and ultimately leading to the formation of immune complexes. Mesangial deposition of immune-complexes containing galactose-deficient IgA1 (Gd-IgA1) and subsequent complement activation are likely responsible for the glomerular injury seen in IgAN.

It is now known that serum Gd-IgA1 levels are elevated in most patients with both familial and sporadic IgAN regardless of ethnicity or age [4]. Recent studies suggest that an inherited defect in IgA1-producing cells leads to the preferential production of Gd-IgA1 [3]. More recent studies, in Caucasian, Asian, and African-American cohorts, have also demonstrated that 25–33% of first-degree relatives of patients with IgAN also have elevated levels of serum Gd-IgA1, despite having no clinical manifestation of renal injury [4]. Serum Gd-IgA1 levels appear to be heritable in a dominant pattern and appear to co-segregate with IgAN susceptibility alleles [4]. Of note, a recent study demonstrated that pediatric patients with Henoch-Schönlein purpura nephritis (HSPN) and a large fraction of their first-degree relatives also have significantly higher Gd-IgA1 levels compared with age- and ethnicity-matched controls. This provides evidence of a shared pathogenesis between IgAN and HSPN [5].

Genetics of IgAN: Before the GWAS Era

Prior to the GWAS era, candidate-gene association studies and linkage analysis were the methods used to identify susceptibility genes or loci for IgAN. Genome-wide linkage scans identified significant and suggestive susceptibility loci, but no causative genes have been identified [6–8]. Linkage analysis did demonstrate significant genetic heterogeneity, such that different genes account for disease in different families [6–8]. Additionally, there have been over 100 candidate-gene studies for IgAN, the majority of which focus on genes involved in adaptive immunity (HLA, IgAFc receptor), cytokine signaling (TGF-β1, TNF), and the glycosylation pathways (C1GALT1 and ST6GALNAC2, cosmc) [9]. While many of these studies had positive results, virtually none of these findings have been replicated. In general these older studies suffered from inadequate sample sizes and general methodological problems.

Genetics of IgAN: GWAS Era

Recent advances in high-density genotyping technology allow simultaneous analysis of several million SNPs. By using a large number of common SNPs to screen the entire genome, GWAS enable the discovery of common alleles that confer low risk of developing complex polygenic diseases. Since the first GWAS was published in 2005, identifying *complement factor H* (*CFH*) as a susceptibility gene for age-related macular degeneration [10], numerous investigators have used GWAS to study complex diseases in populations worldwide. At this point, over 1,100 susceptibility loci have been identified and confirmed in diseases such as type 2 diabetes, schizophrenia, and IgAN.

To date, GWAS have identified seven susceptibility loci for IgAN in European and Asian populations [11–13]. The first IgAN GWAS was conducted by Feehally et al. [11] in a British cohort composed of 431 cases and 4,980 controls. This study used 286,200 SNPs for analysis and identified association signals at the major histocompatibility (MHC) locus, specifically across the HLA-B, DRB1, DQA, and DQB loci. The second IgAN GWAS was performed in a Chinese cohort that consisted of 1,194 cases and 902 controls. It was subsequently replicated in two additional independent cohorts, a Chinese cohort of 712 cases and 748 controls and an Italian cohort of 1,238 cases and 1,172 controls [12]. This GWAS identified five susceptibility loci for IgAN: three loci on chromosome 6p21 in the MHC region, one on chromosome 22q12 called the *HORMAD2* locus, and one on chromosome 1q32 containing the *CFH* gene cluster. The most recent GWAS, with discovery in an independent Chinese cohort of 1,434 cases and 4,270 controls, replicated four of the previously identified IgAN susceptibility loci and identified two novel loci, on chromosome 17p13 and 8q23. Table 1 summarizes the seven IgAN susceptibility loci identified in the three GWAS.

The strongest signal in all three IgAN GWAS was at the MHC locus on Chr. 6p21, although there is still uncertainty about the precise origin of the signal. The HLA locus has been the subject of multiple candidate-gene and linkage studies, but these studies have yielded contradictory results. In retrospect, multiple factors contributed to the conflicting results of the candidate gene studies, including the high degree of polymorphism intrinsic to this region, the lack of resolution inherent in older genotyping technology, and small cohort size. Moreover, the risk imparted by the MHC locus is modest, and there are at least three independent loci within this region, as well as multiple risk alleles within each of these three loci. These factors complicate the identification of specific causal alleles. In fact, all three GWAS identify broad regions on Chr. 6p21, indicating that high-resolution mapping will be required to resolve the signals in the region. Conditional analysis has identified three distinct loci on Chr. 6p21 [12], and these will be discussed in more detail below.

6p21 (top SNPs: rs9275596, rs9275224, rs2856717) (Candidate Genes: *HLA-DRB1, -DQA1, -DQB1*)

The GWAS conducted by Gharavi et al. [12] identified three independent loci in the MHC region. The locus with the strongest association signal encompassed *HLA-DRB1, -DQA1,* and *-DQB1*, three genes that code for α and β subunits of MHC II molecules. This locus had a particularly strong association signal and contained three independent risk alleles (rs9275596, rs9275224, rs2856717) [14]. Further implicating this locus, Yu et al. found this region to be associated with variation in serum IgA levels in their GWAS [13]. Interestingly, the risk haplotype, *DQB1*0602-DQA1*0102-DRB1*1501m* has been associated with an increased risk of multiple sclerosis, systemic lupus erythematous, narcolepsy (DQB1), and liver injury from COX2 inhibitors, but is thought to be highly protective from type 1 diabetes mellitus [1, 15].

6p21 (top SNPs: rs1883414) (Candidate Genes: *HLA-DPB2, -DPB1, -DPA1, and COL11A2*)

A second locus in the MHC region was found to harbor *HLA-DPA1, -DPB1,* and *-DPB2,* which also code for α and β chains of MHC class II molecules. Of note, these genes are paralogs of *HLA-DRB1, -DQA1, and -DQB1.* Genetic

Table 1. Genetic loci associated with IgAN

Locus	Marker (minor allele)	Chr	Chr position (37.3)	Type of SNP	Risk allele frequency
					HapMap (HCB)
HLA-DRB1 HLA-DQA1 HLA-DQB1	rs9275596 (C) rs9275224 (A) rs2856717 (T)	6p21	32789609 32681631 32670308	noncoding noncoding noncoding	0.13 0.38 0.19
HLA-DPB2 HLA-DPB1 HLA-DPA1 COL11A2	rs1883414 (T)	6p21	33086448	noncoding	0.2
PSMB8 PSMB9 TAP1 TAP2	rs9357155 (A)	6p21	32809848	noncoding	0.15
CFH CFHR1-5	rs6677604 (A)	1q32	196686918	noncoding	0.09
HORMAD2 MTMR3 LIF OSM GATSL3 SF3A1	rs2412971 (A)	22q12	30494371	noncoding	0.39
DEFA gene cluster	rs2738048 (C)	8p23	6822785	noncoding	0.26
TNFSF13 MPDU1 EIF4A1 CD68 TP53 SOX15	rs3803800 (A)	17p13	7462969	coding	0.28

HCB = Han Chinese in Beijing; CEU = Utah residents with Northern and Eastern European ancestry from the CEPH collection; YRI = Yoruba in Ibadan, Nigeria; T1DM = type 1 diabetes; MS = multiple sclerosis; SLE = systemic lupus erythematous; AS = ankylosing spondylitis; GVHD = graft-versus-host disease; JRA = juvenile rheumatoid arthritis; SS = systemic sclerosis; AMD = age-related macular degeneration; HUS = hemolytic uremic syndrome; C3GN = C3 glomerulonephritis; IBD = inflammatory bowel disease; CVID = common variable immunodeficiency; RA = rheumatoid arthritis.

polymorphisms of *HLA-DPB1* are associated with chronic berylliosis, graft-versus-host disease, juvenile rheumatoid arthritis, insulin-dependent diabetes mellitus and sarcoidosis [16]. This region has also been associated with increased risk of systemic sclerosis, but the risk alleles associated with this phenotype are not in LD with any of the IgAN risk alleles [1]. This locus also contains *COL11A2*, coding of a component of type XI collagen.

HapMap (CEU)	HapMap (YRI)	Disease associations	Effect size odds ratio (95% CI)	Function
0.33 0.48 0.37	0.37 0.53 0.39	MS, narcolepsy, liver injury from COX2 inhibitors, SLE, T1DM	0.60 (0.52, 0.69) 0.73 (0.67, 0.80) 1.53 (1.31, 1.78)	antigen presentation
0.33	0.15	Chronic berylliosis, GVHD, JRA, T1DM, sarcoidosis	0.83 (0.78, 0.90)	antigen presentation
0.17	0.06	AS, T1DM, celiac disease	0.75 (0.69, 0.82)	antigen digestion and processing
0.23	0.47	HUS, C3GN, AMD	0.61 (0.53, 0.71)	innate immunity (alternative complement pathway)
0.39	0.79	Thymic atrophy, autoimmune glomerulonephritis, IBD	0.75 (0.70, 0.81)	involved in mucosal immunity and inflammation
0.29	0.19	Crohn's disease	0.79 (0.74, 0.84)	innate immune system (antimicrobial)
0.22	0.79	CVID, SLE, celiac disease, RA	1.21 (1.14, 1.28)	B cell homeostasis and immunoglobulin class switching

6p21 (top SNPs: rs9357155) (Candidate Genes: *PSMB8, PSMB9, TAP1, TAP2*)

A third independent locus found within the MHC region contains *TAP2*, *TAP1*, *PSMB8*, and *PSMB9*, which are interferon-regulated genes involved in the degradation and processing of antigens for presentation by MHC-I molecules. *PSMB8* codes for an immunoproteasome involved in processing class

I MHC peptides, and its expression is increased in peripheral blood mononuclear cells of some IgAN patients [17]. *TAP1* and *TAP2* code for proteins that are involved in the transport of peptides from the cytoplasm to the endoplasmic reticulum where they can be loaded onto MHC molecules; they have also been associated with ankylosing spondylitis [18], type 1 diabetes mellitus [19], and celiac disease [20]. Four of these five loci were consistently replicated across all cohorts [14]. *TAP2/PSMB9* locus, however, displayed heterogeneity across cohorts. The heterogeneity may be explained by the presence of a recombination hotspot in this region, which may result in differing linkage disequilibrium patterns across populations [1].

1q32 (top SNPs: rs6677604) (Candidate Genes: *CFH* and *CFHR* Gene Cluster)

The region of chromosome 1 associated with IgAN contains *CFH* (complement factor H) and five *CFH*-related genes (*CFHR3, CFHR1, CFHR4, CFHR2* and *CFHR5*) [1]. *CFH* codes for factor H, a key inhibitor of the alternative complement pathway. The complement system is involved in opsonizing and lysing pathogenic cells, as well as modulating the adaptive immune response. Inappropriate complement activation leads to self-injury. Factor H prevents uncontrolled complement activation by binding C3b and targeting it for factor I-mediated degradation; this prevents the formation and accelerates the decay of C3/C5 convertase [21].

Factor H is composed of 20 short consensus repeat domains and has two main functional regions, located at the opposite ends of the protein. The C-terminus mediates binding to cell surfaces, while the N-terminus, the regulatory site, mediates the decay of the C3 convertase [21]. The function of the *CFH*-related proteins is unknown but a high degree of homology exists between them and FH, especially at the C- and N-terminus, indicating that these proteins may bind the same ligands [21].

The top SNP associated with the *CFH* locus, rs6677604, is in strong linkage disequilibrium with a deletion of *CFHR3* and *CFHR1*, and imparts a decreased risk of developing IgAN (OR = 0.68) [12]. Of interest, the *CFHR3,1*-del is rare in the Chinese, explaining why this locus was not discerned in the GWAS exclusively conducted in this population [13]. It is theorized that the protective effect of the *CFHR3,1*-del may be the result of competition between FH and CFHR1, such that a loss of CFHR1 enhances the effect of FH, leading to decreased complement activation. Although the *CFHR3,1*-del has a protective effect in age-related macular degeneration and IgAN [1], it is associated with an increased risk of developing systemic lupus erythematous [22] and atypical acquired forms of hemolytic uremic syndrome, the latter further mediated by the formation of anti-FH antibodies [23]. Additionally, it has recently been shown that specific variants of *CFH* and *CFHR5* are associated with membranoproliferative glomerulonephritis type II (dense deposit disease) [1]. These data clearly implicate modulation of the alternative complement pathway in the pathogenesis of IgAN and other glomerular diseases.

22q12 (top SNPs: rs2412971) (Candidate Genes: *HORMAD2, MTMR3, LIF, OSM, GATSL3, SF3A1*)

The fifth locus, located on chromosome 22q12, contains *HORMAD2*, *MTMR3*, *LIF*, and *OSM*. The strongest signal from this locus, rs2412971, is located in an intron of *HORMAD2*. Among the genes in this locus, both *OSM* and *LIF* encode cytokines and are involved in mucosal immunity and inflammation. In particular, inactivation of *OSM* produces thymic atrophy and autoimmune glomerulonephritis in the mouse [24]. The functions of other genes within this interval have not been as well characterized. *HORMAD2* is involved in DNA double-stranded break repair [25], and *MTMR3* (encoding myotubularin-related protein-3) is a phosphatase implicated in autophagy [26]. Interestingly, the rs2412973 allele is protective against IgAN, and is associated with lower serum IgA levels among patients. Despite its protective effect on IgAN, this SNP is associated with an increased risk of inflammatory bowel disease [27]. Moreover, in the larger follow-up study, we observed a genetic interaction of this locus with *CFHR3,1-del*: the protective effect of rs2412971-A (*HORMAD2*) is reversed among *CFHR3,1-del* homozygotes [14]. The biological basis of this statistical finding remains unknown. The GWAS performed by Yu et al. [13] confirmed the association of this locus but in that study the strongest signal came from rs12537, which is located in the 3′ UTR of the *MTMR3* gene. These two SNPs are in strong LD ($r^2 = 0.64$, $D' = 0.96$) and most likely signal the same causal variant.

8p23 (top SNPs: rs2738048) (Candidate Genes: *DEFA* Gene Cluster)

The most recent GWAS, conducted by Yu et al. [13], identified a significant signal at rs2738048, centered on the α-defensin (*DEFA*) gene cluster on chromosome 8p23. Defensins are short, cysteine-rich, cationic peptides found in vertebrates, invertebrates and plants. They are natural antimicrobials, and as such, they play an important role in the innate immune system. Mammalian defensins are predominantly produced by the epithelium of mucosal surfaces (the skin, respiratory airways, gastrointestinal and genitourinary tracts) and by leukocytes (mostly neutrophils) [28]. Humans have six *DEFA* genes: DEFA1–4 are expressed in neutrophils, dendritic cells, and natural killer cells, while DEFA5–6 are expressed in Paneth cells of the gut. Interestingly, reduced expression of α-defensins in Paneth cells has been associated with an increased risk of Crohn's disease [1].

17p13 (top SNPs: rs3803800) (Candidate Genes: *TNFSF13, MPDU1, EIF4A1, CD68, TP53, SOX15*)

The seventh locus, containing *TNFSF13* and *TNFSF12,* was also detected in the GWAS conducted by Yu et al. [13]. The protein encoded by *TNFSF13,* also known as APRIL, is a tumor necrosis factor ligand, expressed primarily on monocytes/macrophages, and is important for B-cell homeostasis and immunoglobulin production. Specifically, APRIL is involved in T-cell-independent generation of IgA-positive B cells, as well as IgA1-to-IgA2 class switching [29]. Of note, the top SNP in this locus, rs3803800, is a missense variant that may affect posttranslational processing of APRIL. Recent studies have shown that serum levels of APRIL are elevated in patients with IgAN, and also in patients with systemic lupus erythematous and rheumatoid arthritis [1]. Additionally, the IgAN risk variant of *TNFSF13* has been associated with increased IgA levels in IgAN patients but decreased levels in controls [1]. Another recent study on BAFF (B-cell activation factor), a member of the TNF superfamily that is closely related to APRIL, showed that mice overexpressing BAFF produce high levels of aberrantly glycosylated IgA and subsequently develop fatal nephritis with mesangial deposits of IgA [30].

Risk Score and Geospatial Model

Each of the IgAN susceptibility alleles can additively affect an individual's risk of developing disease. Kiryluk et al. [12, 14] created a genetic risk score based on the seven loci identified in their GWAS, which accurately predicted the relative risk of disease between all replication cohorts. Subsequently, the geographic variation of the susceptibility loci, and thus the genetic risk score, was investigated using genomes of over 6,000 healthy individuals, as well as genomes from HapMap-III and the Human Genome Diversity Project [14]. The results accurately reflected current IgAN prevalence data, predicting the greatest burden of disease in Asians and the lowest in African-Americans. A geospatial model of IgAN genetic risk reflected the known East-West cline in disease prevalence. In general, the genetic risk score increased with distance from Africa, also predicting increased risk in Northern Europe [14]. The predicted North-South gradient was validated within Europe by published prevalence data of IgAN-attributable kidney failure. Thus, the genetic data uncovered a novel aspect of IgAN epidemiology.

A New Pathogenesis Model

The results of the three GWAS offered new insights into the pathogenesis of IgAN and have resulted in molecular candidates for the multi-hit hypothesis [1]. In this new pathogenesis model, the first hit is the inherited defect in IgA1 glycosylation in IgA1-producing cells. The production of Gd-IgA1 may be influenced by variants of *LIF, OSM,* or *TNFSF13,* which influence IgA1 production and class switching. Gd-IgA1 is not automatically pathogenic but may lead to IgAN in the presence of additional susceptibility factors. The second hit is defective allorecognition, namely the production of anti-Gd-IgA1 antibodies, which is likely mediated by MHC-II alleles in susceptible individuals. This is followed by the formation of immune-complexes that deposit in the kidney. In this setting, variants in the *TNFSF13* and *DEFA* loci may promote inflammation or increase the production of the anti-glycan antibodies. Finally, an overly active alternative complement pathway, mediated by variants in the *CFH* locus, may increase glomerular injury, initiated by immune-complex deposition.

Conclusion

During the past decade, the developments of new genetic tools have enabled novel insight into the pathogenesis of IgAN. The GWAS have identified common variants with modest effect on the risk of disease, presenting molecular candidates for disease. However, much remains to be discovered. Only ~5% of IgAN variance is explained by the genetic risk score based on seven IgAN susceptibility loci [14]. Larger GWAS cohorts can undoubtedly uncover additional loci, increasing the catalog of candidates. Moreover, susceptibility loci likely harbor multiple risk alleles, some of which probably impart a large effect on disease. Thus, high-resolution mapping or comprehensive sequencing of these intervals will likely identify variants underlying the GWAS signals and uncover additional independent risk alleles. Moreover, the availability of cost-effective Next-Generation Sequencing will enable direct interrogation of whole genome and provide further insight into pathogenesis.

Acknowledgments

This work was made possible by a fellowship award from the International Society of Nephrology (to J.X.), a student award from the American Society of Nephrology (to S.S.), and support from the Columbia Center for Glomerular disease, R01DK082753 and R01DK095510 (to A.G.G.) and National Basic Research Program of China 973, Program No. 2012CB517600 (No. 2012CB517604).

References

1 Kiryluk K, Novak J, Gharavi AG: Pathogenesis of immunoglobulin a nephropathy: recent insight from genetic studies. Ann Rev Med 2013;64:339–356.

2 Scolari F, Amoroso A, Savoldi S, Mazzola G, Prati E, Valzorio B, Viola BF, Nicola B, Movilli E, Sandrini M, Campanini M, Maiorca R: Familial clustering of IgA nephropathy: further evidence in an Italian population. Am J Kidney Dis 1999;33:857–865.

3 Suzuki H, Moldoveanu Z, Hall S, Brown R, Vu HL, Novak L, Julian BA, Tomana M, Wyatt RJ, Edberg JC, Alarcon GS, Kimberly RP, Tomino Y, Mestecky J, Novak J: IgA1-secreting cell lines from patients with IgA nephropathy produce aberrantly glycosylated IgA1. J Clin Invest 2008;118:629–639.

4 Gharavi AG, Moldoveanu Z, Wyatt RJ, Barker CV, Woodford SY, Lifton RP, Mestecky J, Novak J, Julian BA: Aberrant IgA1 glycosylation is inherited in familial and sporadic IgA nephropathy. J Am Soc Nephrol 2008;19: 1008–1014.

5 Kiryluk K, Moldoveanu Z, Sanders JT, Eison TM, Suzuki H, Julian BA, Novak J, Gharavi AG, Wyatt RJ: Aberrant glycosylation of IgA1 is inherited in both pediatric IgA nephropathy and Henoch-Schönlein purpura nephritis. Kidney Int 2011;80:79–87.

6 Bisceglia L, Cerullo G, Forabosco P, Torres DD, Scolari F, Di Perna M, Foramitti M, Amoroso A, Bertok S, Floege J, Mertens PR, Zerres K, Alexopoulos E, Kirmizis D, Ermelinda M, Zelante L, Schena FP: Genetic heterogeneity in Italian families with IgA nephropathy: suggestive linkage for two novel IgA nephropathy loci. Am J Hum Genet 2006;79:1130–1134.

7 Gharavi AG, Yan Y, Scolari F, Schena FP, Frasca GM, Ghiggeri GM, Cooper K, Amoroso A, Viola BF, Battini G, Caridi G, Canova C, Farhi A, Subramanian V, Nelson-Williams C, Woodford S, Julian BA, Wyatt RJ, Lifton RP: IgA nephropathy, the most common cause of glomerulonephritis, is linked to 6q22-23. Nat Genet 2000;26:354–357.

8 Paterson AD, Liu XQ, Wang K, Magistroni R, Song X, Kappel J, Klassen J, Cattran D, St George-Hyslop P, Pei Y: Genome-wide linkage scan of a large family with IgA nephropathy localizes a novel susceptibility locus to chromosome 2q36. J Am Soc Nephrol 2007; 18:2408–2415.

9 Kiryluk K, Julian BA, Wyatt RJ, Scolari F, Zhang H, Novak J, Gharavi AG: Genetic studies of IgA nephropathy: past, present, and future. Pediatr Nephrol 2010;25:2257–2268.

10 Klein RJ, Zeiss C, Chew EY, Tsai JY, Sackler RS, Haynes C, Henning AK, SanGiovanni JP, Mane SM, Mayne ST, Bracken MB, Ferris FL, Ott J, Barnstable C, Hoh J: Complement factor h polymorphism in age-related macular degeneration. Science 2005;308:385–389.

11 Feehally J, Farrall M, Boland A, Gale DP, Gut I, Heath S, Kumar A, Peden JF, Maxwell PH, Morris DL, Padmanabhan S, Vyse TJ, Zawadzka A, Rees AJ, Lathrop M, Ratcliffe PJ: HLA has strongest association with IgA nephropathy in genome-wide analysis. J Am Soc Nephrol 2010;21:1791–1797.
12 Gharavi AG, Kiryluk K, Choi M, Li Y, Hou P, Xie J, Sanna-Cherchi S, Men CJ, Julian BA, Wyatt RJ, Novak J, He JC, Wang H, Lv J, Zhu L, Wang W, Wang Z, Yasuno K, Gunel M, Mane S, Umlauf S, Tikhonova I, Beerman I, Savoldi S, Magistroni R, Ghiggeri GM, Bodria M, Lugani F, Ravani P, Ponticelli C, Allegri L, Boscutti G, Frasca G, Amore A, Peruzzi L, Coppo R, Izzi C, Viola BF, Prati E, Salvadori M, Mignani R, Gesualdo L, Bertinetto F, Mesiano P, Amoroso A, Scolari F, Chen N, Zhang H, Lifton RP: Genome-wide association study identifies susceptibility loci for IgA nephropathy. Nat Genet 2011;43:321–327.
13 Yu XQ, Li M, Zhang H, Low HQ, Wei X, Wang JQ, Sun LD, Sim KS, Li Y, Foo JN, Wang W, Li ZJ, Yin XY, Tang XQ, Fan L, Chen J, Li RS, Wan JX, Liu ZS, Lou TQ, Zhu L, Huang XJ, Zhang XJ, Liu ZH, Liu JJ: A genome-wide association study in Han Chinese identifies multiple susceptibility loci for IgA nephropathy. Nat Genet 2012;44:178–182.
14 Kiryluk K, Li Y, Sanna-Cherchi S, Rohanizadegan M, Suzuki H, Eitner F, Snyder HJ, Choi M, Hou P, Scolari F, Izzi C, Gigante M, Gesualdo L, Savoldi S, Amoroso A, Cusi D, Zamboli P, Julian BA, Novak J, Wyatt RJ, Mucha K, Perola M, Kristiansson K, Viktorin A, Magnusson PK, Thorleifsson G, Thorsteinsdottir U, Stefansson K, Boland A, Metzger M, Thibaudin L, Wanner C, Jager KJ, Goto S, Maixnerova D, Karnib HH, Nagy J, Panzer U, Xie J, Chen N, Tesar V, Narita I, Berthoux F, Floege J, Stengel B, Zhang H, Lifton RP, Gharavi AG: Geographic differences in genetic susceptibility to IgA nephropathy: GWAS replication study and geospatial risk analysis. PLoS Genet 2012;8:e1002765.
15 Erlich H, Valdes AM, Noble J, Carlson JA, Varney M, Concannon P, Mychaleckyj JC, Todd JA, Bonella P, Fear AL, Lavant E, Louey A, Moonsamy P: HLA DR-DQ haplotypes and genotypes and type 1 diabetes risk: analysis of the type 1 diabetes genetics consortium families. Diabetes 2008;57:1084–1092.
16 Gilchrist FC, Bunce M, Lympany PA, Welsh KI, du Bois RM: Comprehensive HLA-DP typing using polymerase chain reaction with sequence-specific primers and 95 sequence-specific primer mixes. Tissue Antigens 1998;51: 51–61.
17 Coppo R, Camilla R, Alfarano A, Balegno S, Mancuso D, Peruzzi L, Amore A, Dal Canton A, Sepe V, Tovo P: Upregulation of the immunoproteasome in peripheral blood mononuclear cells of patients with IgA nephropathy. Kidney Int 2009;75:536–541.
18 Feng M, Yin B, Shen T, Ma Q, Liu L, Zheng J, Zhao Y, Qian K, Liu D: TAP1 and TAP2 polymorphisms associated with ankylosing spondylitis in genetically homogenous Chinese Han population. Hum Immunol 2009;70:257–261.
19 Qu HQ, Lu Y, Marchand L, Bacot F, Frechette R, Tessier MC, Montpetit A, Polychronakos C: Genetic control of alternative splicing in the TAP2 gene: possible implication in the genetics of type 1 diabetes. Diabetes 2007;56:270–275.
20 Djilali-Saiah I, Caillat-Zucman S, Schmitz J, Chaves-Vieira ML, Bach JF: Polymorphism of antigen processing (TAP, LMP) and HLA class II genes in celiac disease. Hum Immunol 1994;40:8–16.
21 Rodriguez de Cordoba S, Esparza-Gordillo J, Goicoechea de Jorge E, Lopez-Trascasa M, Sanchez-Corral P: The human complement factor H: functional roles, genetic variations and disease associations. Mol Immunol 2004; 41:355–367.
22 Zhao J, Wu H, Khosravi M, Cui H, Qian X, Kelly JA, Kaufman KM, Langefeld CD, Williams AH, Comeau ME, Ziegler JT, Marion MC, Adler A, Glenn SB, Alarcon-Riquelme ME, Network B, Network G, Pons-Estel BA, Harley JB, Bae SC, Bang SY, Cho SK, Jacob CO, Vyse TJ, Niewold TB, Gaffney PM, Moser KL, Kimberly RP, Edberg JC, Brown EE, Alarcon GS, Petri MA, Ramsey-Goldman R, Vila LM, Reveille JD, James JA, Gilkeson GS, Kamen DL, Freedman BI, Anaya JM, Merrill JT, Criswell LA, Scofield RH, Stevens AM, Guthridge JM, Chang DM, Song YW, Park JA, Lee EY, Boackle SA, Grossman JM, Hahn BH, Goodship TH, Cantor RM, Yu CY, Shen N, Tsao BP: Association of genetic variants in complement factor H and factor H-related genes with systemic lupus erythematosus susceptibility. PLoS Genet 2011;7:e1002079.

23 Jozsi M, Licht C, Strobel S, Zipfel SL, Richter H, Heinen S, Zipfel PF, Skerka C: Factor H autoantibodies in atypical hemolytic uremic syndrome correlate with CFHR1/CFHR3 deficiency. Blood 2008;111:1512–1514.

24 Esashi E, Ito H, Minehata K, Saito S, Morikawa Y, Miyajima A: Oncostatin M deficiency leads to thymic hypoplasia, accumulation of apoptotic thymocytes and glomerulonephritis. Eur J Immunol 2009;39: 1664–1670.

25 Wojtasz L, Daniel K, Roig I, Bolcun-Filas E, Xu H, Boonsanay V, Eckmann CR, Cooke HJ, Jasin M, Keeney S, McKay MJ, Toth A: Mouse HORMAD1 and HORMAD2, two conserved meiotic chromosomal proteins, are depleted from synapsed chromosome axes with the help of TRIP13 AAA-ATPASE. PLoS Genet 2009; 5:e1000702.

26 Taguchi-Atarashi N, Hamasaki M, Matsunaga K, Omori H, Ktistakis NT, Yoshimori T, Noda T: Modulation of local PtdIns3P levels by the PI phosphatase MTMR3 regulates constitutive autophagy. Traffic 2010;11:468–478.

27 Imielinski M, Baldassano RN, Griffiths A, Russell RK, Annese V, Dubinsky M, Kugathasan S, Bradfield JP, Walters TD, Sleiman P, Kim CE, Muise A, Wang K, Glessner JT, Saeed S, Zhang H, Frackelton EC, Hou C, Flory JH, Otieno G, Chiavacci RM, Grundmeier R, Castro M, Latiano A, Dallapiccola B, Stempak J, Abrams DJ, Taylor K, McGovern D, Silber G, Wrobel I, Quiros A, Barrett JC, Hansoul S, Nicolae DL, Cho JH, Duerr RH, Rioux JD, Brant SR, Silverberg MS, Taylor KD, Barmuda MM, Bitton A, Dassopoulos T, Datta LW, Green T, Griffiths AM, Kistner EO, Murtha MT, Regueiro MD, Rotter JI, Schumm LP, Steinhart AH, Targan SR, Xavier RJ, Libioulle C, Sandor C, Lathrop M, Belaiche J, Dewit O, Gut I, Heath S, Laukens D, Mni M, Rutgeerts P, Van Gossum A, Zelenika D, Franchimont D, Hugot JP, de Vos M, Vermeire S, Louis E, Cardon LR, Anderson CA, Drummond H, Nimmo E, Ahmad T, Prescott NJ, Onnie CM, Fisher SA, Marchini J, Ghori J, Bumpstead S, Gwillam R, Tremelling M, Delukas P, Mansfield J, Jewell D, Satsangi J, Mathew CG, Parkes M, Georges M, Daly MJ, Heyman MB, Ferry GD, Kirschner B, Lee J, Essers J, Grand R, Stephens M, Levine A, Piccoli D, Van Limbergen J, Cucchiara S, Monos DS, Guthery SL, Denson L, Wilson DC, Grant SF, Daly M, Hakonarson H: Common variants at five new loci associated with early-onset inflammatory bowel disease. Nat Genet 2009;41: 1335–1340.

28 Lehrer RI, Lu W: Alpha-defensins in human innate immunity. Immunol Rev 2012;245: 84–112.

29 Stein JV, Lopez-Fraga M, Elustondo FA, Carvalho-Pinto CE, Rodriguez D, Gomez-Caro R, de Jong J, Martinez AC, Medema JP, Hahne M: April modulates B and T cell immunity. J Clin Invest 2002;109:1587–1598.

30 McCarthy DD, Kujawa J, Wilson C, Papandile A, Poreci U, Porfilio EA, Ward L, Lawson MA, Macpherson AJ, McCoy KD, Pei Y, Novak L, Lee JY, Julian BA, Novak J, Ranger A, Gommerman JL, Browning JL: Mice overexpressing BAFF develop a commensal flora-dependent, IgA-associated nephropathy. J Clin Invest 2011;121:3991–4002.

Ali Gharavi
Department of Medicine
College of Physicians and Surgeons, Columbia University
New York, NY 10032 (USA)
E-Mail ag2239@columbia.edu

Chen N (ed): New Insights into Glomerulonephritis.
Contrib Nephrol. Basel, Karger, 2013, vol 181, pp 65–74 (DOI: 10.1159/000348457)

Immunopathological Predictors of Prognosis in IgA Nephropathy

Yasuhiko Tomino

Division of Nephrology, Department of Internal Medicine, Juntendo University Faculty of Medicine, Tokyo, Japan

Abstract

IgA nephropathy (IgAN) is characterized by the expansion of the glomerular mesangial matrix with mesangial cell proliferation and/or mononuclear cell infiltration. Glomeruli typically contain generalized diffuse granular mesangial deposits of IgA (mainly galactose-deficient polymeric IgA1), IgG and C3. Electron-dense deposits are observed in the glomerular mesangial area and glomerular basement membrane. Therefore, this disease is considered to be an immune complex-mediated glomerulonephritis. The detailed observations of electron-dense deposits are of value for the evaluation of the disease activity. The evidence- and lumped-system-based histological classification can identify the magnitude of the risk of disease progression and is useful for predicting long-term renal outcome in this disease. A study of IgAN patients showed that the number of angiotensin-II-positive cells was correlated with mast cells containing both tryptase and chymase and containing only tryptase in the interstitial lesions with the most severe pathological changes. Hypercomplementemia occurs in the progression of IgAN and is controlled by an increase of complement regulatory proteins. The measurement of urinary levels of membrane attack complex and factor H and extraglomerular C3 deposition could be useful indicators of renal injury in patients with IgAN. Development of glomerulosclerosis in IgAN patients is associated with podocytopenia and the alteration of the podocyte components, i.e. podocalyxin and dendrin. It appears that the number of urinary podocytes and levels of urinary podocalyxin are useful for predicting histological changes in IgAN patients. A positive correlation was observed between acute extracapillary changes and the number of dendrin-positive nuclei per glomerulus in patients with IgAN. It is concluded that there are many immunopathological predictors of prognosis, including genetic background, in this disease. Thus, the early diagnostic screening of prognosis predictors and subsequent intervention are important for the good prognosis in this disease.

Copyright © 2013 S. Karger AG, Basel

Table 1. Clinical predictors of prognosis in IgAN [4]

1	Heavy proteinuria
2	Mild hematuria
3	Low serum albumin
4	Renal dysfunction at the time of renal biopsy
5	Diastolic hypertension
6	Male sex
7	Age <30 years
8	Urinary findings: many dysmorphic red blood cells, various cellular casts, IL-6, MCP-1
9	Serum IgA/C3 ratio: complement activation

IgA nephropathy (IgAN), first described in 1968 by Berger and Hinglais [1], is the most common primary chronic glomerulonephritis worldwide. Glomeruli typically contain generalized diffuse granular mesangial deposits of IgA (mainly galactose-deficient polymeric IgA1), IgG and C3. IgAN is generally considered to be an immune complex-mediated glomerulonephritis. Recently, Zhao et al. [2] have reported that the elevated serum levels of galactose-deficient IgA1 are associated with a poor prognosis of IgAN. Clinically, patients with IgAN show microscopic and macroscopic hematuria and/or proteinuria. The majority of patients show no symptoms although occasionally acute nephritic syndrome or nephrotic syndrome occurs. In Japan, about 30 and 5–10% of IgAN patients develop end-stage kidney disease (ESKD) within 15–20 and 5 years, respectively. Conversely, about 60% of patients can avoid ESKD. Since early medical intervention may lead to a better renal prognosis, it is indicated that early diagnostic screening and subsequent intervention are important for a good prognosis in IgAN patients [3]. Although the clinical course is generally gradual in patients with IgAN, progression to renal hypertension, renal anemia and ESKD is not as rare as originally thought. The clinical markers for poor prognosis in this disease are outlined in table 1 [4]. Many dysmorphic red blood cells, various cellular casts and activated platelets in urinary sediments are frequently observed in the advanced stage of this disease [5, 6]. Furthermore, levels of urinary IL-6, a T-cell-derived lymphokine, and MCP-1, a chemotactic protein for monocytes and T cells, are marked in the advanced stage of IgAN, characterized by severe proliferative glomerular changes [7, 8]. We have already reported [5, 6] the importance of four clinical markers in the diagnosis of patients with IgAN or in differential diagnosis from other types of primary chronic glomerulonephritis, as follows: (1) more than 5 red blood cells in urinary sediments, (2) persistent proteinuria (urinary protein >0.3 g/day), (3) serum IgA level >315 mg/dl, and (4) a serum IgA/complement 3 (C3) ratio >3.01. Since the serum IgA/C3 ratio decreases with improvement of pro-

Table 2. Histopathological predictors of prognosis in IgAN

1	Glomerular and tubulointerstitial injuries
2	Expressions of glomerular cytokines and/or ECM
3	Podocyte injury (podocytopenia)
4	Mast cell infiltration in the interstitium
5	Others

teinuria and/or hematuria, a high ratio might be one of the prognostic markers in this disease [9].

The objective of this review is to explain the predictors of prognosis in IgAN based on immunopathological findings.

Immunopathological Predictors of Prognosis

Histological Changes (table 2)

Glomerular Changes

Histopathologically, IgAN is characterized by the expansion of the glomerular mesangial matrix with mesangial cell proliferation and/or mononuclear cell infiltration. Paramesangial deposits and electron-dense deposits (EDD) are also observed in the glomeruli. The histological marker for poor prognosis of IgAN is generally considered to be glomerular and/or a tubulointerstitial injury (fig. 1). Several cytokines and/or growth factors, such as IL-1, IL-2, IL-6, TNF-α, PDF and TGF-β, have been shown to be involved in glomerular mesangial cell proliferation and the hyperproduction of extracellular matrices, i.e. fibronectin, type IV collagen and laminin. Intercellular adhesion molecule-1 (ICAM-1) is a glycoprotein of 90–110 kDa, which belongs to the immunoglobulin superfamily. Glomeruli that showed increases of ICAM-1 expression using immunofluorescence had marked infiltration of lymphocytes and monocytes. The average number of these cells in glomeruli was increased in advanced-stage patients who showed an increase of ICAM-1 expression in the glomeruli. It appears that the expression of ICAM-1 is closely linked to glomerular cell infiltration of lymphocytes and monocytes in patients with IgAN [10].

Lee et al. [11] indicated that subendothelial and subepithelial deposits were more likely in IgAN patients with severe glomerular lesions. Kusaba et al. [12] reported that the frequency of various EDD in IgAN and suggested that the presence of EDD in regions other than the glomerular paramesangial areas is associated with histological injury and renal status. It appears that the detailed observations of EDD are of value for the evaluation of the disease activity of this disease.

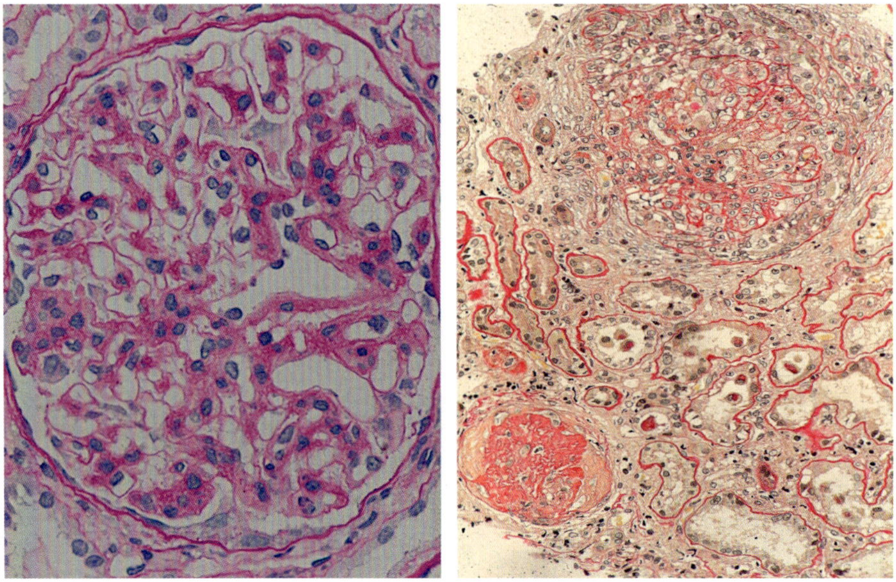

Fig. 1. Glomerular and tubulointerstitial changes in a patient with IgAN.

Among numerous histological grading systems for predicting the renal outcome of this disease, the recently published Oxford classification identifies prognostic pathologic features, providing substantial evidence that histological grading systems can be used to predict renal outcome of IgAN [13]. When developing a histological classification of IgAN, some issues need to be considered. In Japan, the Special IgAN Study Group of the Progressive Renal Diseases Study Committee organized by the Ministry of Health, Labor and Welfare conducted a multicenter retrospective case-control study on IgAN in 2004 to develop an evidence- and lumped-system-based clinicopathological classification of IgAN for predicting the long-term risk of progression to ESKD [14]. The study enrolled 287 patients with a median follow-up of 9.3 years (range 0.7–34.0) after renal biopsy. During follow-up, 49 patients (17%) progressed to ESKD. Multivariate logistic regression analysis showed that the independent pathological variables that predicted progression to ESKD were global sclerosis, segmental sclerosis and fibrous crescents for patients who progressed to ESKD within 5 years after biopsy, and global sclerosis and cellular/fibrocellular crescents for those who progressed to ESKD within 5–10 years after biopsy. Four histological grades, i.e. histological grades (HG) 1, HG 2, HG 3 and HG 4, were established corresponding to <25, 25–49, 50–74 and ≥75% of glomeruli exhibiting cellular or fibrocellular crescents, global sclerosis, segmental sclerosis or fibrous crescents. Eleven (7%) patients in HG 1, 12 (16%)

in HG 2, 13 (31%) in HG 3 and 13 (68%) in HG 4 progressed to ESKD. Multivariate logistic analysis revealed that the risk of progression to ESKD was significantly higher in HG 2, HG 3 and HG 4 than in HG 1 (odds ratios 2.4, 5.7 and 27.6 vs. 1.0). It appears that this evidence- and lumped-system-based histological classification can identify the magnitude of the risk of disease progression and is useful for predicting long-term renal outcome in patients with IgAN [14].

Tubulointerstitial Changes

Mast cells are derived from hematopoietic progenitors and migrate into inflammatory lesions. Human mast cells can be classified into two types according to their protease composition: those containing only tryptase (MCT) and those containing both tryptase and chymase (MCTC). MCT are involved in immunological responses, whereas MCTC seem instead to play a role in angiogenesis and tissue remodeling. Some reports have suggested that mast cells play an important role in the development of fibrosis and the degradation of extracellular matrices. Kurusu et al. [15] reported that the number of MCT in the fibrotic and non-fibrotic areas may be one of the predictive factors for the prognosis of patients with IgAN. Sakamoto-Ihara et al. [16] also reported that IgAN patients had more MCTC than MCT in interstitial lesions. Mast cells were observed around, but not in, the conglomerate of angiotensin II (Ang-II)-positive cells. The number of Ang-II-positive cells was correlated with MCTC and MCT in patients with IgAN with the most severe pathological changes. It appears that chymase-dependent Ang-II synthesis due to human mast cells may be involved in the inflammatory and fibrotic processes of IgAN [16].

Renal Complement Activation

Glomerular injury in IgAN patients is mediated by complement activation via the alternative and lectin pathways. Activation of C3 and C3 convertase production are the key causes of histological injury induced in the membrane attack complex (MAC; C5b-9) formation (fig. 2). MAC is produced via the activated common terminal pathway of all three complement pathways. Ross et al. [17] hypothesized that serum levels of complement components and regulatory proteins in IgAN patients are correlated with its pathogenesis. Onda et al. [18] have clarified the variations in serum complement components and regulatory proteins in patients with IgAN. It appears that hypercomplementemia occurs in the progression of IgAN and is controlled by an increase of complement regulatory proteins. In particular, higher levels of serum C4-binding protein may indicate the severity of histological injury in this disease [18]. They also reported that

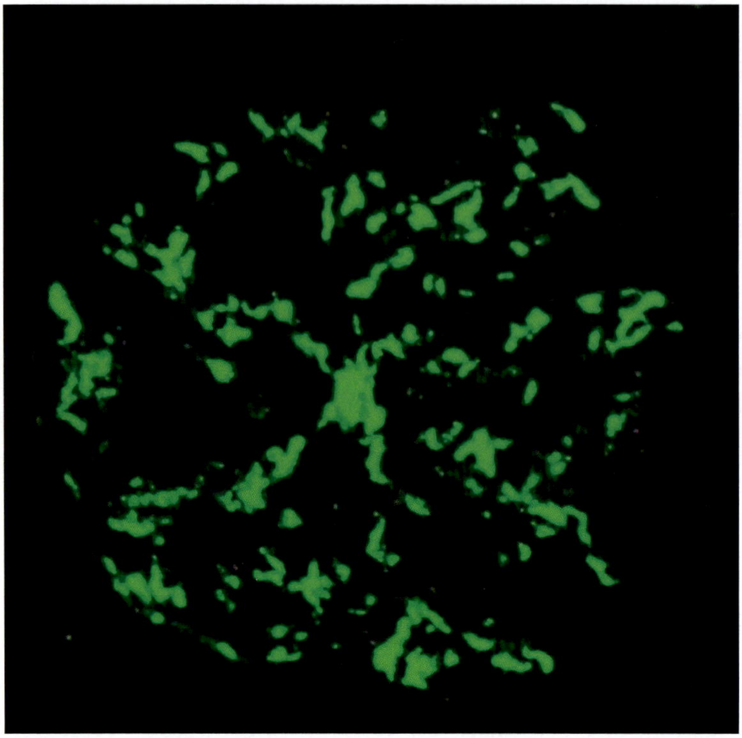

Fig. 2. Complement C3 deposition in a glomerulus in a patient with IgAN.

urinary levels of MAC, factor H and properdin in IgAN patients were significantly higher than those in healthy controls. It appears that complement activation occurs in the urinary space in IgAN and the measurement of levels of MAC and factor H in the urine could be a useful indicator of renal injury in this disease [19].

Ohsawa et al. [20] recently explored the significance of extraglomerular (Bowman's capsule and/or arteriole) C3 deposits in IgAN. The origin of C3 located in Bowman's capsule, where the deposit is away from the glomerular hilar portion, has yet to be defined. However, they speculated that an immune complex that consists of C3 is able to translocate from the paramesangial area to Bowman's capsule through the adhesive lesion between glomerular capillaries and Bowman's capsule. It appears that IgAN patients who have extraglomerular C3 deposits have worse clinical outcomes [20].

Podocyte Injury

The number of podocytes per glomerulus should serve as a podocyte injury parameter and provide prognostic information in patients with IgAN. Development of glomerulosclerosis in IgAN patients is associated with podocyte injury, which may be caused by apoptosis, necrosis, detachment from the glomerular basement membrane or autophagy of these cells. Podocytes in adults do not undergo mitosis, and normally the only way to respond to injury is by cell hypertrophy. Thus, increased podocyte surface area reflects the extent of podocyte hypertrophy. In severe patients, segments of the glomerular basement membrane become denuded. Direct contact of the naked portion of the glomerular basement membrane with the parietal epithelium of Bowman's capsule might lead to adhesion to Bowman's capsule and potentially to segmental sclerosis [21].

Lemley et al. [22] reported that podocyte loss, i.e. podocytopenia, is concurrent with increasing disease severity in patients with IgAN. They observed no corresponding correlations between the clinical indices of injury and the number of mesangial and endothelial cells in this disease. Endothelial and mesangial cell damage is followed by regeneration, but this is not the case with podocytes. Hishiki et al. [23] also reported that podocyte injuries, i.e. a reduction of the absolute number per glomerulus and an increase of glomerular surface area covered by one podocyte, were related to the disease progression in patients with IgAN. Thus, podocyte injury might provide additional prognostic information about such patients.

Podocalyxin (PCX) is present on the apical cell membrane of podocytes and is shed in the urine from injured podocytes. Urinary PCX (u-PCX) is associated with the severity of active glomerular injury in patients with various glomerular diseases. Hara et al. [24] reported that urinary podocytes are associated with the severity of active glomerular injury in children with glomerulonephritis including IgAN. Recently, Asao et al. [25] examined the relationship between the number of urinary podocytes, the levels of u-PCX and glomerular injury in adult IgAN patients. Histological analysis based on Shigematsu's classification [26] revealed a significant correlation between the levels of u-PCX and the severity of acute extracapillary changes. The incidence of proteinuria and the number of urinary podocytes in patients with segmental sclerosis were significantly higher than those in patients without segmental sclerosis. It appears that the levels of u-PCX and the number of urinary podocytes are useful for predicting histological changes in adult IgAN patients [25].

Dendrin is a novel component localized at the slit diaphragm in podocytes. Asanuma et al. [27] already showed that dendrin accumulates in the nucleus of injured podocytes in experimental nephritis and nuclear dendrin promotes podocyte apoptosis. Since it was unknown whether dendrin translocates from the slit membrane to podocyte nucleus in patients with IgAN, Kodama et al. [28] investigated the presence of nuclear dendrin and the relationship between the relocation of dendrin to the podocyte nucleus and disease progression. A positive correlation was observed between acute extracapillary changes and the number of dendrin-positive nuclei per glomerulus in IgAN. The location of dendrin in nuclei was observed in urinary podocytes of this disease. Urinary podocytes, which expressed apoptosis marker annexin V, were also detected. Translocation of dendrin to the podocyte nucleus as well as strong cathepsin L staining was detected in the glomeruli of IgAN patients. An increasing number of dendrin-positive nuclei in glomeruli suggests acute glomerular injury in this disease. It appears that the translocation of dendrin to podocyte nuclei enhances podocyte apoptosis in acute glomerular injury and leads to podocytopenia in patients with IgAN [28].

Conclusion

There are several immunopathological predictors for prognosis in patients with IgAN. Thus, it is suggested that early diagnostic screening of prognosis predictors and subsequent intervention are important for good prognosis in patients with IgAN.

Acknowledgments

I sincerely thank my colleagues in the Division of Nephrology, Department of Internal Medicine at Juntendo University Faculty of Medicine, Tokyo, Japan.

Disclosure Statement

The author has no conflicts of interest to disclose.

References

1 Berger J, Hinglais N: Les dépôts intercapillaries d'IgA-IgG. J Urol Nephrol 1968;74:694–695.
2 Zhao N, Hou P, Lv J, Moldoveanu Z, Kiryluk K, Gharavi AG, et al: The level of galactose-deficient IgA1 in the sera of patients with IgA nephropathy is associated with disease progression. Kidney Int 2012;82:790–796.
3 Okazaki K, Suzuki Y, Kobayashi T, Kodama F, Horikoshi S, Tomino Y: Influence of the period between onset of IgA nephropathy and medical intervention on renal prognosis. Health 2011;3:401–406.

4 Goto M, Wakai K, Kawamura T, Ando M, Endoh M, Tomino Y: A scoring system to predict renal outcome in IgA nephropathy: a nationwide 10-year prospective cohort study. Nephrol Dial Transplant 2009;24:3068–3074.
5 Maeda A, Gohda T, Funabiki K, Horikoshi S, Shirato I, Tomino Y: Significance of serum IgA levels and serum IgA/C3 ratio in diagnostic analysis of patients with IgA nephropathy. J Clin Lab Anal 2003;17:73–76.
6 Nakayama K, Ohsawa I, Maeda-Ohtani A, Murakoshi M, Satoshi Horikoshi S, Tomino Y: Prediction of diagnosis of immunoglobulin A nephropathy prior to renal biopsy and correlation with urinary sediment findings and prognostic grading. J Clin Lab Anal 2008;22:114–118.
7 Saitoh A, Suzuki Y, Takeda M, Kubota K, Tomino Y: Urinary levels of chemoattractant protein (MCP)-1 and disease activity in patients with IgA nephropathy. J Clin Lab Anal 1998;12:1–5.
8 Tomino Y, Funabiki K, Ohmuro H, Shimizu M, Yokoyama K, Shirato I, et al: Urinary levels of interleukin-6 and disease activity in patients with IgA nephropathy. Am J Nephrol 1991;11:459–464.
9 Tomino Y: Predictors of prognosis in IgA nephropathy. Kaohsiung J Med Sci 2012;28:517–520.
10 Tomino Y, Ohmuro H, Kuramoto T, Shirato I, Eguchi K, Sakai H, et al: Expression of intercellular adhesion molecule-1 and infiltration of lymphocytes in glomeruli of patients with IgA nephropathy. Nephron 1994;67:302–307.
11 Lee HS, Choi Y, Lee HS, Yu BH, Koh HI: Ultrastructural changes in IgA nephropathy in relation to histologic and clinical data. Kidney Int 1989;35:880–886.
12 Kusaba G, Ohsawa I, Ishii M, Inoshita H, Takagi M, Tanifuji C, et al: Significance of broad distribution of electron-dense deposits in patients with IgA nephropathy. Med Mol Nephrol 2012;45:29–34.
13 Cattran DC, Coppo R, Cook HT, et al: Working Group of the International IgA Nephropathy Network and the Renal Pathology Society the Oxford Classification of IgA Nephropathy: Rationale, clinicopathological correlations, and classification. Kidney Int 2009;76:534–545.
14 Kawamura T, Joh K, Okonogi H, Koike K, Utsunomiya Y, Miyazaki Y, et al: A histologic classification of IgA nephropathy for predicting long-term prognosis: emphasis on end-stage renal disease. J Nephrol 2012 (E-pub ahead of print).
15 Kurusu A, Suzuki Y, Horikoshi S, Shirato I, Tomino Y: Relationship between mast cells in the tubulointerstitium and prognosis of patients with IgA nephropathy. Nephron 2001;89:391–397.
16 Sakamoto-Ihara T, Suzuki Y, Kurusu M, Yamashita M, Horikoshi S, Tomino Y: Possible involvement of mast cells in renal fibrosis in patients with IgA nephropathy. Inflamm Res 2007;56:421–427.
17 Ross A, Rastaldi MP, Calvaresi N, Oortwijn BD, Schlagwein N, van Gijlwik-Janssen DJ, et al: Glomerular activation of the lectin pathway of complement in IgA nephropathy. J Am Soc Nephrol 2006;17:1724–1734.
18 Onda K, Ohi H, Tamano M, Ohsawa I, Wakabayashi M, Horikoshi S, et al: Hypercomplementemia in adult patients with IgA nephropathy. J Clin Lab Anal 2007;21:77–84.
19 Onda K, Ohsawa I, Ohi H, Tamano M, Mano S, Wakabayashi M, et al: Excretion of complement proteins and its activation marker C5b-9 in IgA nephropathy in relation to renal function. BMC Nephrol 2011;12:64–71.
20 Ohsawa I, Kusaba G, Ishii M, Sato N, Inoshita H, Onda K, et al: Extraglomerular C3 deposition and metabolic impacts in patients with IgA nephropathy. Nephrol Dial Transplant 2012 (E-pub ahead of print).
21 Kritz W, Kretzler M, Nagata M, Provoost AP, Shirato I, Uiker S, et al: A frequent pathway to glomerulosclerosis: detection of tuft architecture-podocyte damage-segmental sclerosis. Kidney Blood Press Res 1996;18:245–253.
22 Lemley KV, Lafayette RA, Safai M, Derby G, Blouch K, Squarer A, et al: Podocytopenia and disease severity in IgA nephropathy. Kidney Int 2002;61:1475–1485.
23 Hishiki T, Shirato I, Takahashi Y, Funabiki K, Horikoshi S, Tomino Y: Podocyte injury predicts prognosis in patients with IgA nephropathy using a small amount of renal biopsy tissue. Kidney Blood Press Res 2001;24:99–104.
24 Hara M, Yanagihara T, Kihara I: Urinary podocytes in primary focal segmental glomerulonephritis. Nephron 2001;89:342–347.

25 Asao R, Asanuma K, Kodama F, Akiba-Takagi M, Nagai Y, Seki T, et al: Relationship between levels of urinary podocalyxin, number of urinary podocytes, and histologic injury in adult patients with IgA nephropathy. Clin J Am Soc Nephrol 2012;7:1385–1393.
26 Shigematsu H: Histological grading and staging of IgA nephropathy. Pathol Int 1997;47: 194–202.
27 Asanuma K, Akiba-Tagagi M, Kodama F, Asao R, Nagai Y, Lydia A, et al: Dendrin location in podocytes is associated with the disease progression in animal and human glomerulopathy. Am J Nephrol 2011;33: 537–549.
28 Kodama F, Asanuma K, Takagi M, Hidaka T, Asanuma E, Fukuda H, et al: Translocation to the podocyte nucleus in acute glomerular injury in patients with IgA nephropathy. Nephrol Dial Transplant 2012 (E-pub ahead of print).

Yasuhiko Tomino, MD, PhD
Division of Nephrology, Department of Internal Medicine
Juntendo University Faculty of Medicine
2-1-1 Hongo, Bunkyo-ku, Tokyo 813-8421 (Japan)
E-Mail yasu@juntendo.ac.jp

Chen N (ed): New Insights into Glomerulonephritis.
Contrib Nephrol. Basel, Karger, 2013, vol 181, pp 75–83 (DOI: 10.1159/000348460)

Treatment of Progressive IgA Nephropathy: An Update

Weiming Wang · Nan Chen

Department of Nephrology, Ruijin Hospital, School of Medicine, Shanghai Jiao Tong University, Shanghai, PR China

Abstract

IgA nephropathy (IgAN) is the most common primary glomerular disease worldwide. About 25–30% of IgAN patients will progress to end-stage kidney disease in 20–25 years. Early-onset symptoms that are highly suggestive of progressive IgAN include massive proteinuria, hypertension, renal damage, glomerular sclerosis, crescent formation, and tubulointerstitial fibrosis. Progressive IgAN may progress to renal failure in a short time. Optimized supportive therapy is the fundamental treatment for progressive IgAN patients, and includes renin-angiotensin system blockers, blood pressure control, antiplatelet and anticoagulant drugs, statins, and allopurinol. In progressive IgAN patients whose clinical and pathological manifestations are more severe, active therapy may be considered including glucocorticoid therapy, cyclophosphamide, azathioprine, mycophenolate mofetil, tacrolimus, and other immunosuppressants. However, there are currently controversies on the definition and treatment of progressive IgAN.

Copyright © 2013 S. Karger AG, Basel

IgA nephropathy (IgAN) is the most common primary glomerulonephritis in the world. In Mediterranean countries, Asia and Australia, the IgAN prevalence among primary glomerulonephritis is 20–45%, whereas in Great Britain, Canada, and the USA, it is 2–10% [1, 2]. In our study, IgAN was found to account for 39.6–44.6% of primary glomerulonephritis cases. In addition, IgAN is one of the most important causes of end-stage renal disease (ESRD). A long-term follow-up study showed that about 25–30% of IgAN patients will progress to ESRD in 20–25 years [3, 4].

The pathogenesis of IgAN is complicated and remains to be fully elucidated [5, 6]. The common pathology of primary IgAN is characterized by a diffuse deposition of IgA in the glomerular mesangial area, but the clinical manifesta-

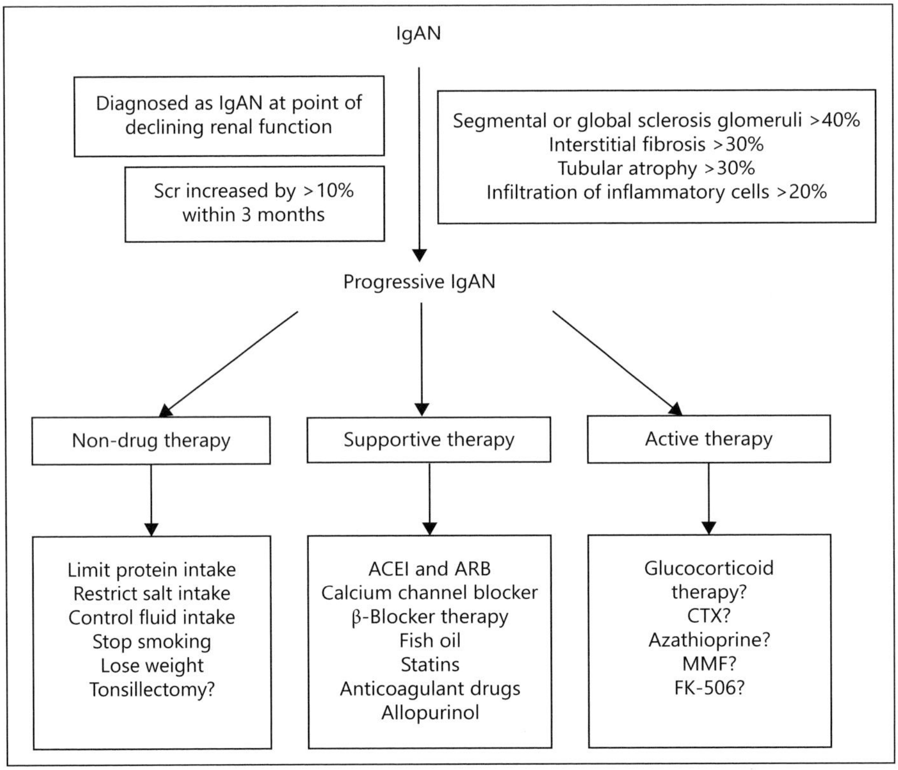

Fig. 1. Schematic diagram for the treatment of progressive IgAN.

tions and histology are different. Subtypes of patients are clinically distinct in their immune pathology, clinical presentation, response to treatment, and prognosis. The majority of IgAN patients have a benign prognosis, however some IgAN patients exhibit progressive or advanced renal failure. These patients should be monitored more closely by nephrologists and treated accordingly. In recent years, epidemiological, clinical, serological, and histological evidence have demonstrated that risk factors for progressive IgAN are mainly associated with proteinuria, hypertension, renal function impairment, the severity of glomerulosclerosis and tubulointerstitial fibrosis in renal pathology, and gene polymorphism. Early-onset markers of progressive IgAN include massive proteinuria (>1 g/day), hypertension, advanced renal function (an increase in serum creatinine (Scr) by >10% within 3 months), glomerular sclerosis, crescent formation, fibrosis, and vascular sclerosis [1, 7]. If not actively treated, progressive IgAN may progress to renal failure in a short interval. Here, we review the research results published in recent years on the treatment of progressive IgAN (fig. 1).

Proteinuria, Hypertension, Glomerular Filtration Rate, and Renal Pathological Lesions Are the Main Factors in Determining the Treatment Scheme for Progressive IgAN

The degree of proteinuria is the strongest predictor of IgAN [6–8]. Most studies indicate albuminuria levels >1 g/day are indicative of increased risk of renal disease progression to ESRD. Uncontrolled blood pressure is another factor leading to the progression of IgAN [8–10]. The glomerular filtration rate (GFR) at a patient's first visit is one of the most important prognostic factors [9]. From the histological index, the IgAN Oxford classification has provided important prognostic evidence, such as chronic fibrotic changes, especially the glomerulosclerosis, tubulointerstitial fibrosis, mesangial cell proliferation, capillary endothelium cells proliferation, among other symptoms [11, 12].

There are several definitions for progressive IgAN. The primary purpose of the ongoing STOP-IgAN trial [13] was to determine the efficacy of an immunosuppressive therapy in combination with a comprehensive supportive therapy to induce clinical remission in patients at risk for progressive IgAN. Patients were diagnosed as primary IgAN with typical mesangioproliferative features, proteinuria of ≥0.75 g/day, and the presence of at least one more risk factor for the development of ESRD. These risk factors included: (a) arterial hypertension, defined as ambulatory blood pressure >140/90 mm Hg or the use of antihypertensive medication, or (b) impaired renal function, defined by creatinine clearance or estimated GFR <90 ml/min [12]. In other clinical trials, patients with progressive IgAN were defined by the following characteristics: (a) an increase in Scr by >10% within 3 months [14], (b) histologically diagnosed as IgAN at the point of declining renal function [15], (c) GFR estimated by the Cockroft-Gault formula within 30–60 ml/min/1.73 m^2 and proteinuria >1 g/day [16], or (d) impaired (Scr >130 μmol/l) and declining renal function (rising by at least 15% in the year before entry) [17]. However, a mass interstitial infiltration of inflammatory cells should be considered a progressive indicator and active therapy should be given immediately.

Supportive Therapy

Without a doubt, designing and implementing the best support treatment regimen is fundamental for promoting the prognosis and survival of progressive IgAN patients. Supportive care therapies include, but are not limited to, blood pressure control (recommend application of angiotensin-converting enzyme inhibitor (ACEI) and angiotensin receptor blocker (ARB), gradually increasing the dosage), non-dihydropyridine calcium channel blocker therapy, β-blocker therapy, aldosterone antagonist therapy (unless the control of blood pressure is required, dihydropyridine drugs should not be used), limiting protein intake, restricting salt intake, controlling fluid intake, stopping smoking, and hyperuricemia therapy [18].

Non-Pharmacological Therapy

A treatment that remains controversial is tonsillectomies for IgAN patients with recurrent tonsillitis and hematuria. In Japan, this treatment is recommended in combination with immunosuppressive treatment. This treatment is mostly based on data and conclusions from retrospective studies [19, 20]. However, these studies have several weaknesses, including small sample sizes, non-randomized nature, and other confounding factors that influence proteinuria and hematuria, and not on renin-angiotensin system inhibitors (RAS). Therefore, a larger, randomized, controlled and well-designed clinical trial is required to provide better evidence for tonsillectomy in the treatment of progressive IgAN patients.

Pharmacological Therapy

RAS Blockers

An evidence-based medical review studied the effect of non-immunosuppressive IgAN treatment [21]. This review included 56 randomized controlled studies with 2,838 cases of adult and pediatric IgAN cases and evaluated the effect of tonsillectomy, fish oil, antihypertensive drugs, and anticoagulation drugs as compared to placebo or other immunosuppressive drugs. After a comprehensive analysis, the authors suggested that antihypertensive drugs are the most effective non-immunosuppressive treatments, especially ACEI and ARB. When used in IgAN, ACEI and ARB can not only control blood pressure but also protect kidney function. ACEI and ARB treatments have also been shown to significantly reduce microalbuminuria and delay the deterioration of renal function in IgAN patients. A meta-analysis [22] of 11 randomized controlled studies, representing 585 IgAN patients, showed that ACEI/ARB treatments played a significant role in the decrease of microalbuminuria and protection of renal function. Reich et al. [8] followed 542 IgAN patients and found that the decline of renal function was associated with the increase of proteinuria excretion. The rate of renal function impairment was 25 times faster in patients with persistent proteinuria >3 g/day, as opposed to those with levels <1 g/day. When proteinuria was reduced from >3 to <1 g/day, the patients experienced a similar disease course as those patients with proteinuria levels <1 g/day and the prognosis was much better than patients with proteinuria levels >3 g/day. However, the renal outcomes were not significantly different between patients with proteinuria >0.3 g/day and those with levels between 0.3 and 1 g/day.

The Kidney Disease: Improving Global Outcomes (KDIGO) guidelines suggest the following for IgAN treatment: when proteinuria is >1 g/day, use long-term treatment with ACEI/ARB; when proteinuria is between 0.5 and 1 g/day, use ACEI/ARB treatment with the dosage gradually increasing to a tolerated level; when proteinuria is <1 g/day, the blood pressure of IgAN patients should be controlled at <130/80 mm Hg; when proteinuria is >1 g/day, blood pressure should be controlled at <125/75 mm Hg. Therefore, when the proteinuria is >0.5 g/day and/or blood pressure is >140/90 mm Hg in IgAN patients, the ACEI/ARB is the first line of treatment. The current evidence does not support the combination therapy with ACEI/ARB in patients with IgAN.

Antiplatelet and Anticoagulant Drugs

The roles of antiplatelet and anticoagulant drugs in the treatment of IgAN have been studied in different patients and in different combinations, resulting in differing results. Ishii et al. [15] conducted a prospective trial treating patients with progressive IgAN with a combined therapy regimen of heparin/warfarin and RAS inhibitors. The patients had a marked decline in loss of GFR, indicating that the combined therapy can slow the progressive decline in renal function through its marked antiproteinuric and anti-inflammatory effects.

Currently, antiplatelet and anticoagulation therapy remains controversial. Because of the toxicity of cyclophosphamide, triple therapies (cyclophosphamide, warfarin, and aspirin) are not recommended [18, 23].

Glucocorticoid Therapy

Lv et al. [24] searched the Medline, Embase, and Cochrane Library databases for randomized controlled trials of corticosteroid therapy for IgAN that had been published between 1966 and March 2011. Analysis of all the relevant studies showed that steroid therapy was associated with a lower risk of kidney failure and a reduction in proteinuria. The authors concluded that steroids appear to provide renal protection in patients with IgAN but increase the risk for adverse events. Several randomized controlled clinical studies have shown that if patients' GFR is >50 ml/min and they were given 6 months of glucocorticoid therapy, proteinuria and the subsequent risk of ESRD were reduced [25–27]. However, a clinical study conducted in the USA found no benefit of steroid therapy in IgAN patients [28]. A Japanese randomized controlled clinical study using a low-dose glucocorticoid treatment of 20 mg/day including a gradual tapering regimen and maintenance for over 2 years showed beneficial effect [29]. In summary, the first and foremost treatment of progressive IgAN is optimal support therapy. If these support treatments do not reduce proteinuria levels to <1 g/day and GFR remains >50 ml/min, patients could be given 6 months of glucocorticoid treatment even though the findings are still controversial in this field. The ongoing STOP-IgAN study will hopefully provide us more firm evidence on the effect of glucocorticoid therapy in patients with progressive IgAN.

Immunosuppressant Therapy

Glucocorticoid Combined with Cyclophosphamide (CTX)

Many retrospective studies or case reports from Asia have shown that treatment with glucocorticoid and CTX or azathioprine in progressive IgAN patients is significantly better than other treatment regimens [30–34]. However, issues of case enrollment, observation bias, and lack of best supportive care measures in these patients make these studies difficult to interpret. When compared to single therapy, combination therapy is generally more prone to side effects such as glaucoma, cataracts, childhood aseptic necrosis of the femoral head, bone marrow suppression, secondary diabetes, and tuberculosis [18]. The KDIGO organization guidelines do not recommend immunosuppressive combination therapy, unless the patient has rapidly progressive glomerulonephritis and/or vasculitis [18]. Ballardie and Roberts [17] studied 38 patients with progressive IgAN randomized to treatment with prednisolone and cytotoxic agents, to study therapy with low-dose CTX then azathioprine, and to control groups. The follow-up period lasted 2–6 years. Proteinuria, present in all patients at the time of entry into the trial, was reduced by treatment at 12 months, compared with pretreatment levels or controls. There was a significant preservation of function at 3 years in the treatment group.

Glucocorticoid Combined with Mycophenolate Mofetil (MMF)

The effect of MMF treatment in patients with progressive IgAN has been evaluated by researchers from China, Belgium, and the USA. One Chinese study reported that compared with high-dose oral glucocorticoid, taking glucocorticoid with 1–1.5 g/day MMF over a period of 12 months can reduce proteinuria to 0.5–0.75 g/day [35]. However, it was difficult to interpret the results due to unclear supportive measures and blood pressure control. In another study, 20 IgAN patients were given 2 g/day MMF combined with high-dose ACEI. These patients showed reduced proteinuria levels and maintained stable renal function, as compared to those treated with ACEI [36, 37]. In contrast, two separate studies in Belgium and the USA found that proteinuria was not affected by MMF treatment in Caucasian patients [36–39], although the exact role of ethnicity remains unclear. One possible side effect of MMF treatment in the Chinese population, however, is *Pneumocystis carinii* infection [40].

Glucocorticoid Combined with Tacrolimus (FK506)
To date, there are few clinical studies investigating tacrolimus treatment of progressive IgAN. Zhang et al. [41] enrolled 14 refractory IgAN patients with each receiving tacrolimus (0.05–0.1 mg/kg/day) and prednisone (0.5 mg/kg/day) for at least 6 months. Of the 14 patients enrolled, 3 were withdrawn because their Scr had increased 30% over baseline, 9 showed complete or partial remission, and 7 achieved remissions within 1 month. Tacrolimus treatment led to a rapid proteinuria remission in refractory IgAN patients. However, Scr should be monitored. Hyperglycemia may be developed in some patients during the FK506 treatment.

Other Immunosuppressants
Several clinical studies have been conducted to determine the effects of several other immunosuppressive medications in IgAN patients, including cyclosporin A and mizoribine. However, evidence is lacking to support their use in progressive IgAN patients [42, 43].

Conclusion

IgAN was previously considered as a benign condition and did not require special treatment. However, recent evidence suggests that IgAN remains a major cause of ESRD and treatment of progressive IgAN could reduce proteinuria and slow disease progression. Currently, glucocorticoid with or without immunosuppressive agents has been used to treat progressive IgAN patients, but more carefully designed clinical trials are required to further validate the efficacy of these regimens.

Acknowledgments

This work was supported by grants from the National Basic Research Program of China, 973 Program (Nos. 2012CB517700 and 2012CB517600, 2012CB517604), the National Key Technology R&D Program (12-5) (Nos. 2011BAI10B00 and 2011BAI10B06), the National Nature Science Foundation of China (No. 81000295), and the Key Program of Shanghai Science and Technology Commission (No. 08dz1900502).

References

1 Donadio JV, Grande JP: IgA nephropathy. N Engl J Med 2002;347:738–748.

2 Li LS, Liu ZH: Epidemiologic data of renal diseases from a single unit in China: analysis based on 13,519 renal biopsies. Kidney Int 2004;66:920–923.

3 Manno C, Strippoli GF, D'Altri C, et al: A novel simpler histological classification for renal survival in IgA nephropathy: a retrospective study. Am J Kidney Dis 2007;49: 763–775.

4 Barratt J, Feehally J: IgA nephropathy. J Am Soc Nephrol 2005;16:2088–2097.

5 Chen N, Nusbaum P, Halbwachs-Mecarelli L, et al: Light-chain composition of serum IgA1 and in vitro IgA1 production in IgA nephropathy. Nephrol Dial Transplant 1991;6: 846–850.

6 Gharavi AG, Kiryluk K, Choi M, et al: Genomewide association study identifies susceptibility loci for IgA nephropathy. Nat Genet 2011;43:321–327.

7 Xie J, Kiryluk K, Wang W, et al: Predicting progression of IgA nephropathy: new clinical progression risk score. PLoS One 2012; 7:e38904.

8 Reich HN, Troyanov S, Scholey JW, et al: Remission of proteinuria improves prognosis in IgA nephropathy. J Am Soc Nephrol 2007;18: 3177–3183.

9 D'Amico G: Natural history of idiopathic IgA nephropathy and factors predictive of disease outcome. Semin Nephrol 2004;24:179–196.

10 Berthoux F, Mohey H, Laurent B, et al: Predicting the risk for dialysis or death in IgA nephropathy. J Am Soc Nephrol 2011;22: 752–761.

11 Cattran DC, Coppo R, Cook HT, et al: The Oxford classification of IgA nephropathy: rationale, clinicopathological correlations, and classification. Kidney Int 2009;76:534–545.

12 Roberts IS, Cook HT, Troyanov S, et al: The Oxford classification of IgA nephropathy: pathology definitions, correlations, and reproducibility. Kidney Int 2009;76:546–556.

13 Eitner F, Ackermann D, Hilgers RD, et al: Supportive versus immunosuppressive therapy of progressive iga nephropathy (STOP) IgAN trial: rationale and study protocol. J Nephrol 2008;21:284–289.

14 Rasche FM, Keller F, von Müller L, et al: Sequential immunosuppressive therapy in progressive IgA nephropathy. Contrib Nephrol. Basel, Karger, 2007, 157, pp 109–113.

15 Ishii T, Kawamura T, Tsuboi N, et al: Prospective trial of combined therapy with heparin/warfarin and renin-angiotensin system inhibitors in progressive IgA nephropathy. Contrib Nephrol. Basel, Karger, 2007, vol 157, pp 114–119.

16 Cruzado JM, Poveda R, Ibernón M, et al: Low-dose sirolimus combined with angiotensin-converting enzyme inhibitor and statin stabilizes renal function and reduces glomerular proliferation in poor prognosis IgA nephropathy. Nephrol Dial Transplant 2011;26: 3596–3602.

17 Ballardie FW, Roberts IS: Controlled prospective trial of prednisolone and cytotoxics in progressive IgA nephropathy. J Am Soc Nephrol 2002;13:142–148.

18 Rosselli JL, Thacker SM, Karpinski JP, et al: Treatment of IgA nephropathy: an update. Ann Pharmacother 2011;45:1284–1296.

19 Xie Y, Nishi S, Ueno M, et al: The efficacy of tonsillectomy on long-term renal survival in patients with IgA nephropathy. Kidney Int 2003;63:1861–1867.

20 Hotta O, Miyazaki M, FurutaT, et al: Tonsillectomy and steroid pulse therapy significantly impact on clinical remission in patients with IgA nephropathy. Am J Kidney Dis 2001;38:736–743.

21 Reid S, Cawthon PM, Craig JC, et al: Non-immunosuppressive treatment for IgA nephropathy. Cochrane Database Syst Rev 2011;16:CD003962.

22 Cheng J, Zhang W, Zhang XH, et al: ACEI/ARB therapy for IgA nephropathy: a meta-analysis of randomised controlled trials. Int J Clin Pract 2009;63:880–888.

23 Woo KT, Lee GS, Pall AA: Dipyridamole and low-dose warfarin without cyclophosphamide in the management of IgA nephropathy. Kidney Int 2000;57:348–349.

24 Lv J, Xu D, Perkovic V, et al; TESTING Study Group: Corticosteroid therapy in IgA nephropathy. J Am Soc Nephrol 2012;23:1108–1116.

25 Lv J, Zhang H, Chen Y, et al: Combination therapy of prednisone and ACE inhibitor versus ACE-inhibitor therapy alone in patients with IgA nephropathy: a randomized controlled trial. Am J Kidney Dis 2009;53: 26–32.

26 Manno C, Torres DD, Rossini M, et al: Randomized controlled clinical trial of corticosteroids plus ACE inhibitors with long-term follow-up in proteinuric IgA nephropathy. Nephrol Dial Transplant 2009;24:3694–3701.

27 Pozzi C, Andrulli S, Del Vecchio L, et al: Corticosteroid effectiveness in IgA nephropathy: long-term results of a randomized, controlled trial. J Am Soc Nephrol 2004;15:157–163.

28 Hogg RJ, Lee J, Nardelli N, et al: Clinical trial to evaluate omega-3 fatty acids and alternate-day prednisone in patients with IgA nephropathy: report from the Southwest Pediatric Nephrology Study Group. Clin J Am Soc Nephrol 2006;1:467–474.

29 Katafuchi R, Ikeda K, Mizumasa T, et al: Controlled, prospective trial of steroid treatment in IgA nephropathy: a limitation of low-dose prednisolone therapy. Am J Kidney Dis 2003;41:972–983.

30 Sato M, Hotta O, Tomioka S, et al: Cohort study of advanced IgA nephropathy: efficacy and limitations of corticosteroids with tonsillectomy. Nephron Clin Pract 2003;93:c137–c145.

31 Goumenos DS, Davlouros P, El Nahas AM, et al: Prednisolone and azathioprine in IgA nephropathy – a ten-year follow-up study. Nephron Clin Pract 2003;93:C58–C68.

32 Chen KJ, Cheng CH, Wu MJ, et al: Effect of corticosteroid and cyclophosphamide in IgA nephropathy patients with heavy proteinuria and/or moderate-severe pathological changes. J Chin Med Assoc 2003;66:263–270.

33 Hotta O, Taguma Y, Yoshizawa N, et al: Long-term effects of intensive therapy combined with tonsillectomy in patients with IgA nephropathy. Acta Otolaryngol Suppl.1996; 523:165–168.

34 Tsuruya K, Harada A, Hirakata H, et al: Combination therapy using prednisolone and cyclophosphamide slows the progression of moderately advanced IgA nephropathy. Clin Nephrol 2000;53:1–9.

35 Chen X, Chen P, Cai G, et al: A randomized control trial of mycophenolate mofetil treatment in severe IgA nephropathy. Zhonghua Yi Xue Za Zhi 2002;82:796–801.

36 Tang S, Leung JC, Chan LY, et al: Mycophenolate mofetil alleviates persistent proteinuria in IgA nephropathy. Kidney Int 2005;68: 802–812.

37 Tang SC, Tang AW, Wong SS, et al: Long-term study of mycophenolate mofetil treatment in IgA nephropathy. Kidney Int 2010; 77:543–549.

38 Maes BD, Oyen R, Claes K, et al: Mycophenolate mofetil in IgA nephropathy: results of a 3-year prospective placebo-controlled randomized study. Kidney Int 2004;65:1842–1849.

39 Frisch G, Lin J, Rosenstock J, et al: Mycophenolate mofetil vs. placebo in patients with moderately advanced IgA nephropathy: a double-blind randomized controlled trial. Nephrol Dial Transplant 2005;20:2139–2145.

40 Lv J, Zhang H, Cui Z, et al: Delayed severe pneumonia in mycophenolate mofetil-treated patients with IgA nephropathy. Nephrol Dial Transplant 2008;23:2868–2872.

41 Zhang Q, Shi SF, Zhu L, et al: Tacrolimus improves the proteinuria remission in patients with refractory IgA nephropathy. Am J Nephrol 2012;35:312–320.

42 Chabova V, Tesar V, Zabka J, et al: Long-term treatment of IgA nephropathy with cyclosporin A – a preliminary report. Nephrol Dial Transplant 1997;12:2206–2207.

43 Yoshikawa N, Nakanishi K, Ishikura K, et al: Combination therapy with mizoribine for severe childhood IgA nephropathy: a pilot study. Pediatr Nephrol 2008;23:757–763.

Prof. Nan Chen, MD
Department of Nephrology, Ruijin Hospital
School of Medicine, Shanghai Jiao Tong University
No. 197 Ruijin Er Road, Shanghai 200025 (PR China)
E-Mail chen-nan@medmail.com.cn

Chen N (ed): New Insights into Glomerulonephritis.
Contrib Nephrol. Basel, Karger, 2013, vol 181, pp 84–90 (DOI: 10.1159/000348637)

Different Targets for Treating Focal Segmental Glomerular Sclerosis

Rosanna Coppo

Nephrology Dialysis and Transplantation, City of Health and Science of Turin,
Regina Margherita Children's Hospital, Turin, Italy

Abstract

Various drugs have been used for the treatment of focal segmental glomerular sclerosis (FSGS) or minimal change disease (idiopathic nephrotic syndrome, INS) including methylprednisolone pulses, alkylating agents and calcineurin inhibitors, often without a strong rationale. For some drugs the rationale has been recently provided by the identification of mechanisms regulating proteinuria. The characterization of molecules acting as permeability factors, including hemopexin, soluble urokinase receptor and cardiotrophin-like cytokine-1, supports plasma exchange in severe cases of INS, particularly in patients at high risk of recurrence of FSGS after transplantation. The collaboration of B and T cells for the production of permeability factors has provided a rationale for targeting B cells using rituximab. Several studies in children and adults, mostly in steroid-dependent cases, or after recurrence in grafted kidneys, have reported improved outcomes after rituximab. Podocyte actin cytoskeleton is a new target for therapy. Calcineurin inhibitors block the dephosphorylation of the cytoskeleton component synaptopodin, and steroids increase actin polymerization and stability. High doses of methylprednisolone and cyclosporine proved to be beneficial in patients with a high risk of FSGS recurrence. An interesting new target for blunting the INS pathogenetical mechanisms is the transcription factor nuclear factor-κB (NF-κB) which was reported to be activated in circulating mononuclear cells of these patients. NF-κB is regulated by the proteasome and saquinavir, a HIV protease inhibitor, is provided with antiproteasome activity. Using saquinavir associated with small doses of calcineurin inhibitors, we treated a small series of very difficult cases of INS with insufficient response to steroid therapy and multiple immunosuppressive drugs. Saquinavir allowed a significant reduction of steroid cumulative doses and disappearance of features of steroid toxicity. In conclusion, recent reports have allowed a new insight into the pathogenetical mechanism regulating proteinuria in INS, offering new targets for treating severe cases.

Copyright © 2013 S. Karger AG, Basel

Focal segmental glomerulosclerosis (FSGS) is a protean disease. Genetic or acquired changes in slit diaphragm structure and function and podocyte cytoskeleton reorganization can be responsible for development of severe proteinuria [1, 2]. The clinical course is highly variable, but in general worse in patients with FSGS than in those with minimal change disease (MCD). However, no clear cut exists in clinical features between the two histologic entities, and children with FSGS and MCD are considered to belong to a single clinical syndrome characterized by severe proteinuria and hypoalbuminemia in the absence of systemic diseases, defined idiopathic nephrotic syndrome (INS) [3]. Most children with INS respond to steroid therapy and have a limited number of relapses, however others experience frequent relapses and develop steroid dependency (SDNS) or are primary or secondary steroid-resistant (SRNS) [4]. The most difficult cases of INS have features of FSGS or present with mild mesangial proliferation and IgM deposits (IgMN), but some have true MCD and renal biopsy in early cases is not predictive of outcome [3]. Treatment of FSGS and SRNS or SDNS MCD is challenging both in children and in adults.

Various drugs have been used for the treatment of difficult cases of INS, including high doses of steroids and alkylating agents, but often without a strong theoretical rationale. For some of them, studies subsequent to early clinical applications showed that a rationale in support of a specific targeted effect could be envisaged. Investigations in this area have produced very interesting data in recent years and promise further developments in the near future.

The search for new treatments is driven by the recent identification of underlying mechanisms regulating proteinuria. The first strategy was to inhibit T-lymphocyte activation [5], considered the source of toxic lymphokines for the glomerular filtration barrier [6, 7]. As early as in 1977, Shaloub et al. focused for the first time on the search for a pathogenetical factor of massive proteinuria in INS on lymphocytes: a clone was supposed to produce a circulating permeability factor acting on podocytes and increasing the protein passage into the urinary space, and this factor was detected in patients with MCD [6]. Several cytokines were supposed to play the role of permeability factor (including IL-2, IL-18 and the vascular permeability factor) but with weak evidence in humans [2].

Interest has been recently focused on hemopexin, a protease which activates protein kinase B and induces nephrin-dependent reorganization of actin cytoskeleton of podocytes and proteinuria when injected in rats [8, 9]. Another interesting permeability factor has been recently detected in sera of patients with FSGS, and identified as soluble urokinase receptor (suPAR) [10]. Induction of urokinase receptor (uPAR) in podocytes leads to integrin activation and foot process effacement [11]. Savin et al. [12] have recently reported that cardiotrophin-like cytokine (CLC-1), a member of the IL-6 family, is present in plasma of recurrent FSGS after kidney transplantation, and acts as a permeability factor on isolated glomeruli increasing their permeability to albumin.

These recent publications indicating the presence of permeability factors in sera of patients with severe FSGS have provided a significant advancement in understanding the benefits reported by several authors in using plasmapheresis to remove theoretical previously unknown circulating factors. This treatment and absorption of staphylococcal protein A columns was proposed by Dantal et al. [13] in patients with FSGS in native kidneys or in grafted organs. However, the results were often limited in time, with relapse at withdrawal and progression to ESRD also in patients with antiproteinuric response. Plasmapheresis or immunoadsorbance on protein A in association with cyclophosphamide was successfully adopted as prevention or early treatment of FSGS recurrence in children with transplanted kidney [14]. Intensive plasma exchange in preparation to living related renal transplantation in high-risk patients, followed by posttransplant plasmapheresis, was effective in preventing recurrence in case series [15]. Perioperative and prophylactic plasmapheresis seem to be promising in high-risk subjects with FSGS in native kidneys.

The synthesis of a permeabilizing factor has addressed the INS treatment target towards T lymphocytes for decades. Therefore, high steroid doses, cytotoxic/antiproliferative drugs and calcineurin inhibitors have been adopted to depress T lymphocytes [16–18]. More recently the additional role of B cells in collaborating with T cells [19] which may favor the production of permeability factors has provided a rationale for therapies targeting B cells, and rituximab has been the most widely used drug [20, 21]. Rituximab is a chimeric monoclonal antibody that selectively targets CD20-positive B cells while sparing plasma cells. Several recent reports indicate a positive effect of rituximab in about 60% of SDNS, 20–30% of SRNS, with a good outcome – up to 40% of the cases – also in posttransplant recurrence of FSGS in children [19] Rituximab can induce independence from calcineurin inhibitors in patients with INS needing continuous administration of these drugs. In the published reports, rituximab was well tolerated, however its use may be complicated by antigenicity, infusion reactions and hypogammaglobulinemia. Rare cases of posterior leukoencephalopathy, probably caused by reactivation of J virus, have been reported [22].

Recent studies have indicated a new target for therapy in FSGS and in general for INS, the actin cytoskeleton of kidney podocytes [23]. The interaction between synaptopodin and serine/treonine binding protein maintains F-actin stability, normal cytoskeleton structure and absence of proteinuria. Under the effect of serine/treonine proteases, protein kinase A or calcineurin, the binding protein is dephosphorylated, its protective effect on synaptopodin is lost, leaving synaptopodin exposed to degradation under the effect of cathepsin. The degradation of synactopodin affects cytoskeleton stability and induces foot process effacement [23]. Calcineurin inhibitors can inhibit the degradation of podocytic synaptopodin and actin. Therefore, podocytes are a direct target of cyclosporine independent of nuclear factor AT inhibition in T cells, since this drug blocks the calcineurin-mediated dephosphorylation of the cytoskeleton component synaptopodin [24].

Similar conclusions were drawn for the effect of steroids, since glucocorticoid receptors are present on podocytes and translocate to the podocyte nucleus upon dexamethasone treatment increasing actin polymerization and the stability of actin filaments [25]. These findings suggested that the antiproteinuric effect of glucocorticoids is, at least in part, mediated by a direct effect on the podocyte actin. An example of a positive effect of immunosuppression and stabilization of cytoskeleton may be the favorable effects of association of high doses of methyprednisolone and cyclosporin A to prevent and treat recurrence of FSGS after renal transplantation [26].

An interesting new target for blunting the pathogenetical mechanisms leading to INS is the transcription factor nuclear factor-κB (NF-κB). This ubiquitous transcription factor governs the expression of genes encoding for cytokines, chemokines, growth factors, adhesion molecules, and other factors involved in the immune and inflammatory response. An upregulation of NF-κB has been detected also in cultured podocytes in correlation with nephrin reduction [27]. NF-κB was recently reported to be activated in circulating mononuclear cells of patients with INS [28, 29] and transcriptome analysis of T lymphocytes confirmed this finding [30]. NF-κB is regulated by the proteasome, since its active subunits p50 and p65 are sequestered in cytoplasm as inactive forms by the inhibiting protein IκBα and translocate to the nucleus to initiate transcriptional activity only after IκBα degradation by the ubiquitin-proteasome pathway [31]. Bortezomib is the most known proteasome inhibitor, which has been successfully used in diseases with B-cell hyperactivity and NF-κB activation, including multiple myeloma and antibody-mediated renal rejection, however it has severe neurotoxic side effects [32].

We previously showed that saquinavir, a HIV protease inhibitor, is provided with antiproteasome activity [33, 34]. Since this drug is commonly used for long-lasting treatments of seropositive subjects of any age, we tested its effects in a small group of few very difficult cases of INS with insufficient response to steroid therapy and multiple immunosuppressive drugs [35]. Saquinavir allowed a significant reduction of steroid cumulative doses over 1 year of treatment and disappearance of features of steroid toxicity. Calcineurin inhibitors were used in association at lower doses, as expected since saquinavir is an inhibitor of cytochrome P450-3A4 protein, known to metabolize calcineurin inhibitors [36], hence CyA and TAC dosages were adjusted to maintain the same blood levels as before introducing saquinavir. A favorable outcome was observed in children with difficult long-lasting SDNS, some with secondary steroid resistance and toxic side effects due to steroids or other immunosuppressant drugs.

Saquinavir interferes with the catalytic proteases of the proteasome [34] preventing the degradation of IκB and the nuclear translocation of NF-κB. The NF-κB activation is supposed to control synthesis of circulating T-cell factors that causes podocyte cytoskeleton disorganization leading to increased glomerular permeability and proteinuria. We observed a downregulation of nuclear binding of NF-κB p50 and p65 subunits in circulating peripheral blood mononuclear cells of the patients treated with saquinavir. NF-κB signaling in podocytes promotes the development of proteinuria in mice after injection of toll-like receptor 4 ligand, LPS [37]. Activation of toll-like receptors on cultured human podocytes leads to actin reorganization and foot processes rearrangement via a NF-κB-dependent mechanism which can be inhibited by proteasome inhibitors [38]. We observed that saquinavir significantly blunted the NF-κB upregulation induced in cultured human podocytes with TNF-α and LPS, suggesting a possible direct impact of this drug on podocyte protein synthesis [35].

Glucocorticoids and calcineurin inhibitors also induce repression of NF-κB function [39, 40]. These observations suggest that the combination of saquinavir to small doses of steroids and calcineurin inhibitors may result in an adequate cumulative immunosuppressive effect, strengthening the final effect and limiting adverse effects of individual drugs.

In conclusion, exciting recent reports have allowed a new insight into the pathogenetical mechanism regulating proteinuria in INS, offering new targets for treating severe cases with insufficient response to generally prescribed drugs, with hopefully less severe side effects. The research in this field will contribute to further improve the outcome of difficult cases of disease in native kidney of a recurrence after kidney transplantation.

References

1 Hildebrandt F: Genetic kidney diseases. Lancet 2010;375:1287–1295.

2 McCarthy ET, Sharma M, Savin VJ: Circulating permeability factors in idiopathic nephrotic syndrome and focal segmental glomerulosclerosis. Clin J Am Soc Nephrol 2010; 5:2115–2121.

3 Habib R, Churg J: Minimal change disease, mesangial proliferative glomerulonephritis and focal sclerosis: individual entities of a spectrum of disease; in Robison RR (ed): Nephrology Today. Proceedings of the Ninth Congress of Nephrology. New York, Springer, 1984, pp 634–644.

4 Niaudet P: Long-term outcome of children with steroid-sensitive idiopathic nephrotic syndrome. Clin J Am Soc Nephrol 2009;4: 1547–1548.

5 Cunard R, Kelly CJ: T cells and minimal change disease. J Am Soc Nephrol 2002;13: 1409–1411.
6 Lagrue G, Xheneumont S, Branellec A, Hirbec G, Weil B: A vascular permeability factor elaborated from lymphocytes. Demonstration in patients with nephrotic syndrome. Biomedicine 1975;23:37–40.
7 Matsumoto K, Kanmatsuse K: Transforming growth factor-β_1 inhibits vascular permeability factor release by T cells in normal subjects and in patients with minimal change nephrotic syndrome. Nephron 2001;87:111–117.
8 Lennon R, Singh A, Welsh GI, Coward RJ, Satchell S, Ni L, Mathieson PW, Bakker WW, Saleem MA: Hemopexin induces nephrin-dependent reorganization of the actin cytoskeleton in podocytes. J Am Soc Nephrol 2008;19:2140–2149.
9 Bakker WW, Borghuis T, Harmsen MC, van den Berg A, Kema IP, Niezen KE, Kapojos JJ: Protease activity of plasma hemopexin. Kidney Int 2005;68:603–610.
10 Wei C, Saleem M, Goes N, Reiser J: Soluble urokinase receptor is a circulating glomerular disease recurrence factor. J Am Soc Nephrol 2008;19:103A.
11 Wei C, Moller CC, Altintas MM, Li J, Schwarz K, Zacchigna S, Xie L, Henger A, Schmid H, Rastaldi MP, Cowan P, Kretzler M, Parrilla R, Bendayan M, Gupta V, Nikolic B, Kalluri R, Carmeliet P, Mundel P, Reiser J: Modification of kidney barrier function by the urokinase receptor. Nat Med 2008;14: 55–63.
12 Savin VJ, Sharma M: Plasma "factors" in recurrent nephrotic syndrome after kidney transplantation: causes or consequences of glomerular injury? Am J Kidney Dis 2009;54:406–409.
13 Dantal J, Bigot E, Bogers W, Testa A, Kriaa F, Jacques Y, Hurault de Ligny B, Niaudet P, Charpentier B, Soulillou JP: Effect of plasma protein adsorption on protein excretion in kidney-transplant recipients with recurrent nephrotic syndrome. N Engl J Med 1994;33: 7–14.
14 Gohh RY, Yango AF, Morrissey PE, Monaco AP, Gautam A, Sharma M, McCarthy ET, Savin VJ: Preemptive plasmapheresis and recurrence of FSGS in high-risk renal transplant recipients. Am J Transplant 2005;12: 2907–2912.
15 Gonzalez E, Ettenger R, Rianthavorn P, Tsai E, Malekzadeh M: Preemptive plasmapheresis and recurrence of focal segmental glomerulosclerosis in pediatric renal transplantation. Pediatr Transplant 2011;15:495–501.
16 International Study of Kidney Disease in Children: Cyclophosphamide therapy in focal segmental glomerular sclerosis: a controlled clinical trial. Eur J Pediatr 1983;140:149.
17 Mendoza SA, Reznik VM, Griswold WR, Krensky AM, Yorgin PD, Tune BM: Treatment of steroid-resistant focal segmental glomerulosclerosis with pulse methylprednisolone and alkylating agents. Pediatr Nephrol 1990;4:303–307.
18 Niaudet P, Habib R: Cyclosporine in the treatment of idiopathic nephrosis. J Am Soc Nephrol 1994;5:1049–1056.
19 Mättö M, Raunio AR, Postila V, Huttunen K, Hirvonen MR, Pelkonen J: Human B cells and macrophages cooperate in T-cell-independent type 2 response. Scand J Immunol 2008;67:209–217.
20 Prytuła A, Iijima K, Kamei K, Geary D, Gottlich E, Majeed A, Taylor M, Marks SD, Tuchman S, Camilla R, Ognjanovic M, Filler G, Smith G, Tullus K: Rituximab in refractory nephrotic syndrome. Pediatr Nephrol 2010; 25:461–468.
21 Munyentwali H, Bouachi K, Audard V, Remy P, Lang P, Mojaat R, Deschênes G, Ronco PM, Plaisier EM, Dahan KY: Rituximab is an efficient and safe treatment in adults with steroid-dependent minimal change disease. Kidney Int 2013 (E-pub ahead of print).
22 Carson KR, Evens AM, Richey EA, Habermann TM, Focosi D, Seymour JF, Laubach J, Bawn SD, Gordon LI, Winter JN, Furman RR, Vose JM, Zelenetz AD, Mamtani R, Raisch DW, Dorshimer GW, Rosen ST, Muro K, Gottardi-Littell NR, Talley RL, Sartor O, Green D, Major EO, Bennett CL: Progressive multifocal leukoencephalopathy after rituximab therapy in HIV-negative patients: a report of 57 cases from the Research on Adverse Drug Events and Reports project. Blood 2009;113:4834–4840.
23 Faul C, Donnelly M, Merscher-Gomez S, Chang YH, Franz S, Delfgaauw J, Chang JM, Choi HY, Campbell KN, Kim K, Reiser J, Mundel P: The actin cytoskeleton of kidney podocytes is a direct target of the antiproteinuric effect of cyclosporine A. Nat Med 2008; 14:931–938.

24 Sharma R, Sharma M, Ge X, McCarthy ET, Savin VJ: Cyclosporine protects glomeruli from FSGS factor via an increase in glomerular cAMP. Transplantation 1996;62:1916–1920.
25 Ransom RF, Lam NG, Hallett MA, et al: Glucocorticoids protect and enhance recovery of cultured murine podocytes via actin filament stabilization. Kidney Int 2005;68:2473–2483.
26 Canaud G, Zuber J, Sberro R, Royale V, Anglicheau D, Snanoudj R, Gaha K, Thervet E, Lefrère F, Cavazzana-Calvo M, Noël LH, Méjean A, Legendre CH, Martinez F: Intensive and prolonged treatment of focal and segmental glomerulosclerosis recurrence in adult kidney transplant recipients: a pilot study. Am J Transplant 2009;9:1081–1086.
27 Ristola M, Arpiainen S, Saleem MA, et al: Regulation of Neph3 gene in podocytes – key roles of transcription factors NF-κB and Sp1. BMC Mol Biol 2009;10:83.
28 Sahali D, Pawlak A, Le Gouvello S, Lang P, Valanciuté A, Remy P, Loirat C, Niaudet P, Bensman A, Guellaen G: Transcriptional and post-transcriptional alterations of IκBα in active minimal-change nephrotic syndrome. J Am Soc Nephrol 2001;12:1648–1658.
29 Audard V, Pawlak A, Candelier M, Lang P, Sahali D: Upregulation of nuclear factor-related κB suggests a disorder of transcriptional regulation in minimal change nephrotic syndrome. PLoS One 2012;7:e30523.
30 Mansour H, Cheval L, Elalouf JM, Aude JC, Alyanakian MA, Mougenot B, Doucet A, Deschênes G: T-cell transcriptome analysis points up a thymic disorder in idiopathic nephrotic syndrome. Kidney Int 2005;67:2168–2177.
31 Ruland J: Return to homeostasis: downregulation of NF-κB responses. Nat Immunol 2011;12:709–714.
32 Fuchs O: Transcription factor NF-κB inhibitors as single therapeutic agents or in combination with classical chemotherapeutic agents for the treatment of hematologic malignancies. Curr Mol Pharmacol 2010;3:98–122.
33 La Porte CJ: Saquinavir, the pioneer antiretroviral protease inhibitor. Expert Opin Drug Metab Toxicol 2009;5:1313–1322.
34 Piccinini M, Rinaudo MT, Chiapello N, Ricotti E, Baldovino S, Mostert M, Tovo PA: The human 26S proteasome is a target of antiretroviral agents. AIDS 2002;16:693–700.
35 Coppo R, Camilla R, Porcellini MG, Peruzzi L, Gianoglio B, Amore A, Daprà V, Loiacono E, Fonsato V, Dal Canton A, Esposito C, Esposito P, Tovo PA: Saquinavir in steroid-dependent and -resistant nephrotic syndrome: a pilot study. Nephrol Dial Transplant 2012;27:1902–1910.
36 Brinkman K, Huysmans F, Burger DM: Pharmacokinetic interaction between saquinavir and cyclosporine. Ann Intern Med 1998;129:914–915.
37 Reiser J, von Gersdorff G, Loos M, Oh J, Asanuma K, Giardino L, Rastaldi MP, Calvaresi N, Watanabe H, Schwarz K, Faul C, Kretzler M, Davidson A, Sugimoto H, Kalluri R, Sharpe AH, Kreidberg JA, Mundel P: Induction of B7-1 in podocytes is associated with nephrotic syndrome. J Clin Invest 2004;113:1390–1397.
38 Bruggeman LA, Drawz PE, Kahoud N, Lin K, Barisoni L, Nelson PJ: TNFR2 interposes the proliferative and NF-κB-mediated inflammatory response by podocytes to TNF-α. Lab Invest 2011;91:413–425.
39 Garside H, Stevens A, Farrow S, Normand C, Houle B, Berry A, Maschera B, Ray D: Glucocorticoid ligands specify different interactions with NF-κB by allosteric effects on the glucocorticoid receptor DNA binding domain. J Biol Chem 2004;279:50050–50059.
40 Meyer S, Kohler NG, Joly A: Cyclosporine A is an uncompetitive inhibitor of proteasome activity and prevents NF-κB activation. FEBS Lett 1997;413:354–358.

Rosanna Coppo
Nephrology Dialysis and Transplantation, City of Health and Science of Turin
Regina Margherita Children's Hospital
Piazza Polonia 94, I–10127 Turin (Italy)
E-Mail rosanna.coppo@unito.it

Chen N (ed): New Insights into Glomerulonephritis.
Contrib Nephrol. Basel, Karger, 2013, vol 181, pp 91–100 (DOI: 10.1159/000348471)

Screening of ACTN4 and TRPC6 Mutations in a Chinese Cohort of Patients with Adult-Onset Familial Focal Segmental Glomerulosclerosis

Qianying Zhang · Jun Ma · Jingyuan Xie · Zhaohui Wang · Bin Zhu · Xu Hao · Li Yang · Hong Ren · Nan Chen

Department of Nephrology, Ruijin Hospital, School of Medicine, Shanghai Jiao Tong University, Shanghai, PR China

Abstract

Objectives: Familial focal segmental glomerulosclerosis (FSGS) has been widely reported as having an underlining genetic component to its pathogenesis. Recently, mutations in TRPC6 and ACTN4 were identified to be associated with familial FSGS, however few studies have reported the mutation rates of these two genes in Chinese familial FSGS patients. The aim of this study was to determine the prevalence of TRPC6 and ACTN4 mutations in Chinese adult-onset familial FSGS. ***Methods:*** 80 FSGS pedigrees with Chinese ancestry who were admitted to our hospital from September 1997 to January 2012 were screened for TRPC6 and ACTN4 mutations. There was at least one biopsy-proven FSGS in each family. An additional family member with renal function insufficiency, obvious proteinuria, or end-stage renal disease was required if there was only 1 biopsy-proven FSGS case in the family. FSGS secondary to systemic diseases, such as obesity, hypertension, diabetes, or virus infection, were excluded. Other hereditary renal diseases, such as Alport's disease and Fabry disease, were also excluded by related tests. Genomic DNA was extracted from peripheral blood cells, and Sanger sequencing was performed for all exons and exon-intron boundaries of TRPC6 and ACTN4 in the probands of all FSGS pedigrees enrolled in this study. ***Results:*** Of the total index patients, 33 were female and 47 were male. The median age at disease onset was 39 years (range 15–67). The median proteinuria level was 1,324 mg/day and the median serum creatinine level was 114.5 μmol/l (range 41–1,040) at diagnosis. A missense mutation (Q889K) of TRPC6 was found in two independent families. No mutation was found in ACTN4. Ac-

Q.Z. and J.M. contributed equally to this work.

cordingly, the mutation rate of TRPC6 was 2.5% and of ACTN4 it was 0%. ***Conclusion:*** Mutations of TRPC6 and ACTN4 occur in only a minor portion of Chinese familial FSGS patients. Further genetic studies conducted in these patients will be helpful to increase our understanding of the pathogenesis of FSGS. Copyright © 2013 S. Karger AG, Basel

Focal segmental glomerulosclerosis (FSGS) is considered one of the most important causes of glomerulonephritis and nephrotic syndrome in adults. FSGS has been reported in 5.6–37.8% of the biopsy-proven primary glomerulonephritis [1–4], and it represents ~14% of all primary glomerulonephritis cases encountered in our hospital. In addition, the prevalence of FSGS is known to vary among different areas of China.

The typical clinical manifestations of FSGS include proteinuria (often as nephrotic syndrome), hypertension, and progression to end-stage renal disease (ESRD). Patients with massive proteinuria and who are unresponsive to steroid therapy often have a poor prognosis; moreover, up to 50% of these patients progress to ESRD within 3–8 years [5]. Alarmingly, the incidence of FSGS continues to steadily increase [3, 6, 7].

FSGS is categorized as primary, secondary, and familial. Familial FSGS is often associated with poor renal survival and a low response rate to steroid treatment [8, 9]. Both autosomal recessive and autosomal dominant inheritance patterns have been reported for familial FSGS [9, 10]. Currently, genetic variations in the podocyte genes are generally believed to be the major cause of familial FSGS [11–13]. Among the podocyte genes associated with this disease, ACTN4, which encodes α-actinin-4 [10], and TRPC6, which encodes the transient receptor potential channel 6 [14], have been found in patients with adult-onset familial FSGS.

Although recognition of familial FSGS has expanded in recent years, few studies have reported the mutation rates of ACTN4 and TRPC6 in Chinese familial FSGS patients. The aim of this study was to determine the prevalence of ACTN4 and TRPC6 mutations within a cohort of Chinese patients with adult-onset familial FSGS.

Materials and Methods

Patients

Eighty pedigrees of adult-onset familial FSGS were recruited from patients who were admitted to the Department of Nephrology at Shanghai Ruijin Hospital between September 1997 and January 2012. The clinical and laboratory data were collected. The ages of presentation were >12 years for all patients included. There was at least 1 biopsy-proven FSGS case in each family. An additional family member with renal function insufficiency, obvious proteinuria, or ESRD was required if there was only 1 biopsy-proven FSGS in the family [15, 16]. FSGS secondary to systemic diseases, such as obesity, hypertension, diabetes, or virus infection, were excluded from the study. Patients with other hereditary renal diseases, such as Alport's disease, Fabry disease, or thin basement membrane nephropathy (as identified by skin and renal tissue type IV collagen staining, electronic microscopy findings, and various hearing and visual tests) were also excluded from the study. Family members were invited to participate in the study. All participants provided informed consent. The study was carried out with approval from the Human Research Ethics Committee of Shanghai Ruijin Hospital (Shanghai Jiao Tong University, School of Medicine, Shanghai, PR China).

Methods

The clinical information recorded included sex, age of disease onset, brief medical histories, and familial histories. Laboratory results, such as 24-hour urine protein and serum creatinine at diagnosis, were also gathered.

Genomic DNA was isolated from peripheral white blood cells by standard laboratory protocols. Sanger sequencing of all 21 exons of the ACTN4 gene, 13 exons of the TRPC6 gene, and the exon-intron boundaries of both was performed for all probands after amplification by polymerase chain reaction. Chromatograms were analyzed by the Sequencher software (Gene Codes Corp., Ann Arbor, Mich., USA). Genetic reference data were obtained from the PubMed database and analyzed by the NCBI single nucleotide polymorphisms (SNPs) database.

Statistical Analysis

All statistical analyses were carried out by SPSS version 16.0 statistical software (SPSS, Chicago, Ill., USA) and Microsoft Excel 2007 spreadsheet software. Clinical data were described using mean ± SD or median (range) for continuous variables and using percentages for categorical data.

Results

Clinical Characteristics

Eighty independent FSGS families of Chinese descent were enrolled in the study. Among all the probands, 33 were female and 47 were male. The median age at onset was 39 years (range 15–67). The median proteinuria was 1,324 mg/day and the median serum creatinine was 114.5 μmol/l (range 41–1,040) at diagnosis.

Mutational Analysis for ACTN4 and TRPC6

ACTN4 and TRPC6 were first screened in the probands of each family. Two index patients from two unrelated families were found to have a missense heterozygous change, p.Q889K (m.2664C>A) in TRPC6, which had previously been confirmed as a gain-of-function missense mutation by Zhu et al. [17]. The posi-

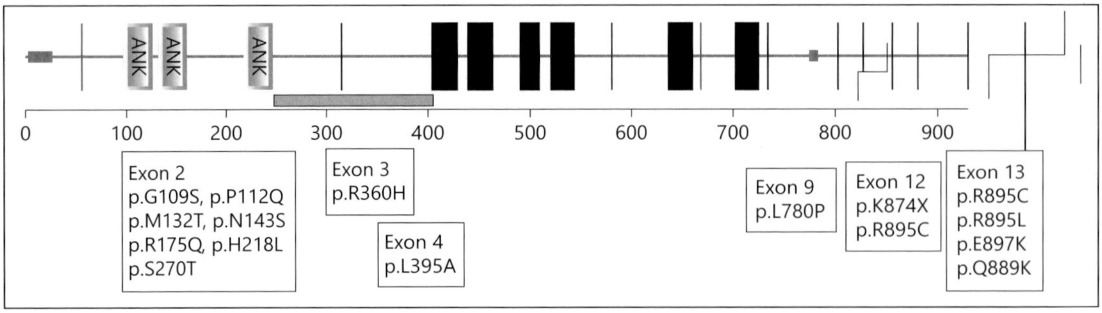

Fig. 1. Location of the mutations identified in the TRPC6 gene.

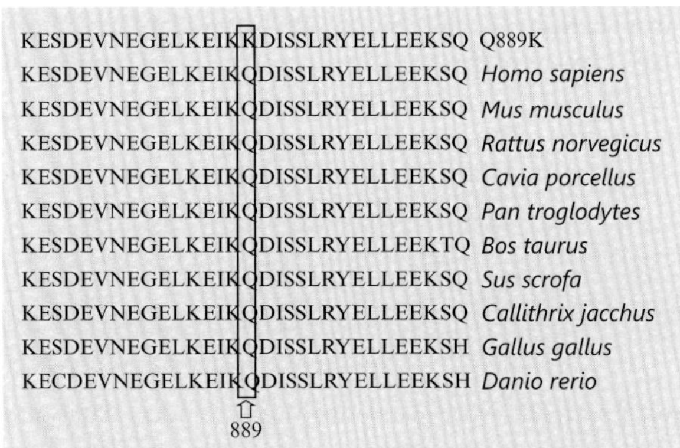

Fig. 2. Mutation m.2664C>A in TRPC6 leads to a glutamine to lysine substitution at amino acid position 889, where the amino acid sequence is highly conserved.

tion of the Q889K mutation is marked in figure 1, along with other mutations reported in the literature for TRPC6 [11, 14, 17–23]. Multispecies alignment showed high conservation at position 889 (fig. 2). Both patients appeared to be late-onset. The first patient carried a heterozygous Q889K mutation of TRPC6 and was a 35-year-old man with proteinuria of 2,700 mg/24 h and serum creatinine of 70 μmol/l at diagnosis. He had a younger female sibling with biopsy-proven FSGS, who had shown no response to steroid therapy. The patient was treated with angiotensin-receptor blockers after the diagnosis but the proteinuria remained. The second patient who carried the heterozygous Q889K mutation was a 48-year-old man. He had been diagnosed with proteinuria 5 years previously and had experienced hypertension for the previous year. His proteinuria was 1,031 mg/24 h and serum creatinine was 90 μmol/l at renal biopsy.

Table 1. Prevalence of ACTN4 and TRPC6 mutations in familial FSGS

Study	Number of families	Patients	Mutation rate of ACTN4	Mutation rate of TRPC6	Remarks
Weins et al. 2005 [24]	141	Caucasian/African/ Middle Eastern/ Asian (2%)	3.5%	NA	
Reiser et al. 2005 [18]	71	Caucasian/African	NA	7.0%	
Heeringa et al. 2009 [20]	21	Predominantly Caucasian	NA	4.8%	Early-onset FSGS; median age at onset 17.5 years (range 2–70); autosomal dominant mode of inheritance
Santin et al. 2009 [19]	12	Caucasian	NA	8.3%	
Hofstra et al. 2013 [23]	5	Caucasian	NA	20.0%	Autosomal dominant mode of inheritance; mutations in ACTN4 and INF2 had been excluded
Barua et al. 2012 [28]	213	Caucasian/African/ Asian	3.3%	2.3%	Testing ACTN4 (exons 1–10)
Santin et al. 2011 [25]	7	Caucasian	0%	14.3%	Familial adult-onset SRNS, testing ACTN4 (exons 1–10)
Present study 2013	80	Asian	0%	2.5%	

NA = Data not available.

No mutations were found in the ACTN4 gene. Thus, the observed disease-causing mutation rates of TRPC6 and ACTN4 were 2.5 and 0% in this cohort of Chinese patients with adult-onset familial FSGS (table 1). 22 SNPs of ACTN4 and 14 SNPs of TRPC6 were also found in this group of patients (table 2).

Discussion

Genetic variations in podocytes and slit membranes are closely related to FSGS. Buscher et al. [11] reported a mutation rate of 8% in podocyte genes from their single-center cohort of adult patients with FSGS and renal failure. The ACTN4 and TRPC6 podocyte genes are well recognized as important causes of familial FSGS. The ACTN4 gene-encoded α-actinin-4 is a cytoskeletal protein that plays an essential role in maintaining the integrity of the podocyte cytoskeleton. Mu-

Table 2. SNPs of ACTN4 and TRPC6 in familial FSGS

ACTN4			TRPC6		
db SNP#	position	frequencies of ancestral alleles in Asians	db SNP#	position	frequencies of ancestral alleles in Asians
rs17523223	5′-UTR	0.89	rs3824934	5′-UTR	0.56
rs2303040	intron1	0.69	rs3802829	exon 1	0.94
rs2112649	intron 2	0.63	rs17096918	intron 1	0.59
rs3786844	intron 3	0.86	rs4262694	intron 1	0.13
rs3786845	intron 3	0.59	rs112727405	intron 1	NA
rs35436155	intron 3	0.96	rs10501986	intron 1	0.86
rs2287730	intron 3	1.00	rs10501981	intron 1	0.87
rs3217235	intron 3	NA	rs4542378	intron 2	0.89
rs2287729	intron 3	NA	rs7931830	intron 3	0.01
rs2287728	intron 3	0.93	rs12366144	exon 6	0.98
rs78078603	intron 4	0.94	rs12362848	intron 7	0.01
rs3745859	exon 5	0.59	rs7105083	intron 10	0.94
rs35729716	intron 5	0.06	rs10570417	intron 12	NA
rs12984794	intron 7	0.96	rs7931399	intron 13	0.00
rs7248577	intron 11	0.94			
rs2306197	intron 12	0.93			
rs62120068	intron 13	0.94			
rs12986337	exon 16	0.07			
rs34134906	intron 16	0.94			
rs12979947	intron 16	0.05			
rs12974733	intron 19	0.94			
rs1136956	exon 20	0.94			

NA = Data not available.

tations in the ACTN4 gene may affect regulation of the actin cytoskeleton in glomerular podocytes, and often cause an autosomal dominant form of FSGS in adults [10]. The relationship of ACTN4 mutation and familial FSGS was first described in 2000 by Kaplan et al. [10]. In 2005, Weins et al. [24] performed a genetic analysis of probands from 141 pedigrees with familial FSGS and found five disease-causing ACTN4 mutations (W59R, I149del, K255E, T259I, and S262I). In their study, the ACTN4 mutations accounted for ~3.5% of familial FSGS. Santin et al. [25] performed genetic testing in patients from 110 families with congenital nephrotic syndrome or primary FSGS/diffuse mesangial sclerosis or/and steroid-resistant nephrotic syndrome. Among them, 24 were familial cases, including 7 adult-onset cases, but no ACTN4 mutation was found, similar to our study.

The TRPC6 gene-encoded transient receptor potential cation channel, subfamily C, member 6, interacts with several proteins in podocytes, including essential slit diaphragm proteins, such as podocin and nephrin [26]. The gain-of-function mutations in TRPC6 are associated with increased calcium amplitudes, and thus disturbed podocyte function [14]. Since most patients with TRPC6 mutations have an advanced onset age, it is presumed that these mutations may cause only minor changes in podocyte function and that irreversible changes occur after a long time or possibly after the accumulation of other renal insults [26]. The relationship of a TRPC6 mutation and familial FSGS was first described in 2005 by Winn et al. [14], and since then 17 mutations of TRPC6 have been identified. All of these mutations are located in the non-transmembrane domains, with many being in exon 2, exon 12, and exon 13 (fig. 1). The TRPC6 mutation Q889K and its function were first described in 2009 by Zhu et al. [17], whereby in vitro functional research demonstrated it to be a gain-of-function mutation. Reports of the prevalence of TRPC6 in familial FSGS have varied from 2.3 to 20%. When Reiser et al. [18] screened the probands of 71 pedigrees with familial FSGS for TRPC6 gene mutations, the overall mutation rate was ~7.0%. In our study, one TRPC6 mutation (Q889K) was detected in two pedigrees, and the overall mutation rate was 2.5%. Meanwhile, we have not found any mutations in ACTN4.

The mutation rates of both ACTN4 and of TRPC6 were obviously lower than reported in previous studies (table 1). We theorized that this discrepancy might be related to the following reasons: the mutation rates of these two genes may differ among patients of different ethnicities. Our study represents the first published effort to screen these two genes simultaneously in a large cohort of Chinese FSGS patients. Hofstra et al. [23] reported a mutation rate of TRPC6 as high as 20% in familial FSGS, but mutations in ACTN4 and INF2 had been excluded prior to enrollment. Considering the high prevalence of INF2 mutation in familial FSGS, the prevalence of TRPC6 mutation might have been overestimated after the exclusion of patients carrying INF2 mutations. In addition, the sample sizes of some of the previous studies have been relatively small, which might have led to bias in the estimated mutation rates.

A consensus has not been reached on whether screening for gene mutations is necessary at the initial diagnosis of FSGS. The guidelines published by the Kidney Disease Improving Global Outcomes (KDIGO) Foundation discourages performing genetic testing routinely at the initial evaluation of FSGS [27]. We agree with this recommendation for sporadic FSGS because genetic testing remains an expensive practice and the mutation rate is expected to be very low in these cases. However, for FSGS patients with family histories, especially those who are resistant to steroid therapies, genetic testing should be considered for the following reasons: First, confirming the familial FSGS would play an important role in predicting prognosis of a patient and in choosing the initial treatment. Familial FSGS may have clinical characteristics that are similar to sporadic FSGS, but the prognosis is different. Patients with familial FSGS tend to be unresponsive to steroid therapy. Thus, when abundant proteinuria is present, the prognosis of familial FSGS is poor. Conlon et al. [8] reported a series of familial FSGS patients, among which 25 of the 31 patients progressed to ESRD within 10 years of diagnosis. In that study, no patient responded to the prednisone treatment, and the immunosuppressive therapy was generally of little help. Genetic testing may benefit patients with familial FSGS because once the diagnosis is confirmed, physicians can avoid exposing patients to long-term use of corticosteroids which would provide little to no benefit but pose a significant risk of detrimental side effects. In addition, calcineurin inhibitors, such as CsA and FK506, might be considered for use in patients found to carry TRPC6 mutations [19]. Second, for familial FSGS with ESRD, genetic testing may contribute to the decision of renal transplantation. Since familial FSGS seldom recurs after renal transplantation [8], confirming of the disease by genetic screening may provide the impetus for a hospital to provide and for a patient to accept the transplantation procedure. Third, determining the causal variant for FSGS families may be helpful when members are considering having a baby.

In conclusion, the mutation rates of TRPC6 and ACTN4 were very low in Chinese adult-onset FSGS families, which indicated that other, yet unrecognized genes may contribute to this disease onset and pathogenesis. Further genetic studies should be conducted in this population.

Acknowledgements

We are grateful to all of the patients for their participation in the present study. This study was supported by grants from the National Basic Research Program of China 973 Program (Nos. 2012CB517600 and 2012CB517604), and the National Natural Science Foundation of China (Nos. 81200488 and 81000295). J. Xie is sponsored by the Shanghai Pujiang Program (No. 11PJ1406700).

References

1 Chang JH, Kim DK, Kim HW, Park SY, Yoo TH, Kim BS, et al: Changing prevalence of glomerular diseases in Korean adults: a review of 20 years of experience. Nephrol Dial Transplant 2009;24:2406–2410.

2 Braun N, Schweisfurth A, Lohofener C, Lange C, Grundemann C, Kundt G, et al: Epidemiology of glomerulonephritis in Northern Germany. Int Urol Nephrol 2011;43: 1117–1126.

3 Dragovic D, Rosenstock JL, Wahl SJ, Panagopoulos G, deVita MV, Michelis MF: Increasing incidence of focal segmental glomerulosclerosis and an examination of demographic patterns. Clin Nephrol 2005;63:1–7.

4 Polito MG, de Moura LA, Kirsztajn GM: An overview on frequency of renal biopsy diagnosis in Brazil: clinical and pathological patterns based on 9,617 native kidney biopsies. Nephrol Dial Transplant 2010;25:490–496.

5 Korbet SM: Clinical picture and outcome of primary focal segmental glomerulosclerosis. Nephrol Dial Transplant 1999;14(suppl 3): 68–73.

6 Kitiyakara C, Eggers P, Kopp JB: Twenty-one-year trend in ESRD due to focal segmental glomerulosclerosis in the United States. Am J Kidney Dis 2004;44:815–825.

7 Bahiense-Oliveira M, Saldanha LB, Mota EL, Penna DO, Barros RT, Romao-Junior JE: Primary glomerular diseases in Brazil (1979–1999): is the frequency of focal and segmental glomerulosclerosis increasing? Clin Nephrol 2004;61:90–97.

8 Conlon PJ, Butterly D, Albers F, Rodby R, Gunnells JC, Howell DN: Clinical and pathologic features of familial focal segmental glomerulosclerosis. Am J Kidney Dis 1995;26: 34–40.

9 Boute N, Gribouval O, Roselli S, Benessy F, Lee H, Fuchshuber A, et al: NPHS2, encoding the glomerular protein podocin, is mutated in autosomal recessive steroid-resistant nephrotic syndrome. Nat Genet 2000;24:349–354.

10 Kaplan JM, Kim SH, North KN, Rennke H, Correia LA, Tong HQ, et al: Mutations in ACTN4, encoding α-actinin-4, cause familial focal segmental glomerulosclerosis. Nat Genet 2000;24:251–256.

11 Buscher AK, Konrad M, Nagel M, Witzke O, Kribben A, Hoyer PF, et al: Mutations in podocyte genes are a rare cause of primary FSGS associated with ESRD in adult patients. Clin Nephrol 2012;78:47–53.

12 Woroniecki RP, Kopp JB: Genetics of focal segmental glomerulosclerosis. Pediatr Nephrol 2007;22:638–644.

13 Dai S, Wang Z, Pan X, Chen X, Wang W, Ren H, et al: ACTN4 gene mutations and single nucleotide polymorphisms in idiopathic focal segmental glomerulosclerosis. Nephron Clin Pract 2009;111:c87–c94.

14 Winn MP, Conlon PJ, Lynn KL, Farrington MK, Creazzo T, Hawkins AF, et al: A mutation in the TRPC6 cation channel causes familial focal segmental glomerulosclerosis. Science 2005;308:1801–1804.

15 Winn MP, Conlon PJ, Lynn KL, Howell DN, Slotterbeck BD, Smith AH, et al: Linkage of a gene causing familial focal segmental glomerulosclerosis to chromosome 11 and further evidence of genetic heterogeneity. Genomics 1999;58:113–120.

16 Winn MP: Approach to the evaluation of heritable diseases and update on familial focal segmental glomerulosclerosis. Nephrol Dial Transplant 2003;18(suppl 6): vi14–vi20.

17 Zhu B, Chen N, Wang ZH, Pan XX, Ren H, Zhang W, et al: Identification and functional analysis of a novel TRPC6 mutation associated with late onset familial focal segmental glomerulosclerosis in Chinese patients. Mutat Res 2009;664:84–90.

18 Reiser J, Polu KR, Moller CC, Kenlan P, Altintas MM, Wei C, et al: TRPC6 is a glomerular slit diaphragm-associated channel required for normal renal function. Nat Genet 2005;37:739–744.

19 Santin S, Ars E, Rossetti S, Salido E, Silva I, Garcia-Maset R, et al: TRPC6 mutational analysis in a large cohort of patients with focal segmental glomerulosclerosis. Nephrol Dial Transplant 2009;24:3089–3096.

20 Heeringa SF, Moller CC, Du J, Yue L, Hinkes B, Chernin G, et al: A novel TRPC6 mutation that causes childhood FSGS. PloS One 2009; 4:e7771.

21 Mir S, Yavascan O, Berdeli A, Sozeri B: TRPC6 gene variants in Turkish children with steroid-resistant nephrotic syndrome. Nephrol Dial Transplant 2012;27:205–209.
22 Gigante M, Caridi G, Montemurno E, Soccio M, d'Apolito M, Cerullo G, et al: TRPC6 mutations in children with steroid-resistant nephrotic syndrome and atypical phenotype. Clin J Am Soc Nephrol 2011;6:1626–1634.
23 Hofstra JM, Lainez S, van Kuijk WH, Schoots J, Baltissen MP, Hoefsloot LH, et al: New TRPC6 gain-of-function mutation in a non-consanguineous Dutch family with late-onset focal segmental glomerulosclerosis. Nephrol Dial Transplant 2013 (E-pub ahead of print).
24 Weins A, Kenlan P, Herbert S, Le TC, Villegas I, Kaplan BS, et al: Mutational and biological analysis of α-actinin-4 in focal segmental glomerulosclerosis. J Am Soc Nephrol 2005; 16:3694–3701.
25 Santin S, Bullich G, Tazon-Vega B, Garcia-Maset R, Gimenez I, Silva I, et al: Clinical utility of genetic testing in children and adults with steroid-resistant nephrotic syndrome. Clin J Am Soc Nephrol 2011;6:1139–1148.
26 Dryer SE, Reiser J: TRPC6 channels and their binding partners in podocytes: role in glomerular filtration and pathophysiology. Am J Physiol Renal Physiol 2010;299:F689–F701.
27 KDIGO Clinical Practice Guideline for Glomerulonephritis. Kidney Int Suppl 2012;2: 139–274.
28 Barua M, Brown EJ, Charoonratana VT, Genovese G, Sun H, Pollak MR: Mutations in the INF2 gene account for a significant proportion of familial but not sporadic focal and segmental glomerulosclerosis. Kidney Int 2013;83:316–322.

Prof. Nan Chen, MD
Department of Nephrology, Ruijin Hospital
School of Medicine, Shanghai Jiao Tong University
No. 197 Ruijin Er Road, Shanghai 200025 (PR China)
E-Mail chen-nan@medmail.com.cn

Chen N (ed): New Insights into Glomerulonephritis.
Contrib Nephrol. Basel, Karger, 2013, vol 181, pp 101–108 (DOI: 10.1159/000348462)

Increased Risk of Treatment Failure and End-Stage Renal Disease in Familial Focal Segmental Glomerular Sclerosis

Xu Hao · Jingyuan Xie · Jun Ma · Zhaohui Wang · Qiongxiu Zhou · Li Yang · Xiaoxia Pan · Hong Ren · Nan Chen

Department of Nephrology, Ruijin Hospital, School of Medicine, Shanghai Jiao Tong University, Shanghai, PR China

Abstract

Objective: Focal segmental glomerular sclerosis (FSGS) is one of the most important causes of end-stage renal disease (ESRD). The pathogenesis, clinical manifestation, pathological changes and treatment of FSGS differ in patients with and without a family history of the disease. Few studies have compared familial FSGS (FFSGS) and sporadic FSGS (SFSGS). The aim of this study was to assess the clinical and pathological features and the prognosis of FSGS in patients with and without a family history of the disease. ***Methods:*** We enrolled 124 FFSGS patients and 124 age- and sex-matched SFSGS patients in the study. All patients underwent a renal biopsy to determine FSGS. The mean follow-up time was 28.3 ± 12.5 months for the FFSGS group and 26.5 ± 19.5 months for the SFSGS group ($p > 0.05$). Baseline clinical characteristics were recorded for all participants. The primary outcomes of the study were ESRD and remission of proteinuria (defined as a 50% reduction of the baseline urine protein level). The pathological changes were assessed by focal/global glomerulosclerosis and the tubulointerstitial lesion score. ***Results:*** There were no age or gender differences between the two groups. A total of 43.75% of the FFSGS patients and 35.16% of the SFSGS patients had high blood pressure, but the difference was not statistically significant ($p = 0.079$). In addition, patients in the FFSGS group had a lower urine protein excretion rate (1.4 ± 1.4 vs. 2.0 ± 1.8 g/24 h) and a higher serum albumin value (3.6 ± 6.2 vs. 3.0 ± 1.1 g/dl) than patients in the SFSGS group ($p < 0.01$). A total of 62.9% of the FFSGS patients and 22.9% of the SFSGS patients had hematuria, and the difference was statistically significant ($p = 0.0001$). Nephrotic syndrome occurred less frequently in the FFSGS group than in the SFSGS

X.H. and J.X. contributed equally to this work.

group (13.3 vs. 22.6%, p = 0.003). The baseline serum creatinine, uric acid and eGFR values were similar in the two groups. When pathological changes were examined, the FFSGS patients showed more severe global glomerulosclerosis and tubular interstitial injury than the SFSGS patients. During the follow-up period, the FFSGS group had a lower proteinuria remission rate (23.08 vs. 48.39%, p = 0.006) and a lower median renal survival time (96 vs. 72 months, p = 0.04) than the SFSGS group. ***Conclusions:*** Compared to SFSGS patients, FFSGS patients displayed distinct clinicopathological features that were associated with less response to treatment and worse renal outcomes.

Copyright © 2013 S. Karger AG, Basel

Introduction

Focal segmental glomerulosclerosis (FSGS) is a pathological diagnosis that is characterized by mesangial sclerosis, the obliteration of capillaries, hyalinosis, foam cells and adhesion between the glomerular tuft and Bowman's capsule [1]. FSGS is characterized by idiopathic steroid-resistant nephrotic syndrome and progression to end-stage renal disease (ESRD). The incidence of FSGS was reported as cases per million individuals [2]. In the United States, FSGS is responsible for approximately 5–20% of all cases of ESRD [3]. There are several subtypes of FSGS, including primary, secondary, familial (FFSGS) and FSGS that is associated with congenital syndromes [4]. Primary FSGS is the most common form of FSGS; secondary FSGS is associated with another disease process. The inheritance pattern of FFSGS includes autosomal dominant and autosomal recessive forms [4–6]. In addition, FSGS has been associated with various congenital syndromes, such as Charcot-Marie-Tooth disease, Laurence-Moon-Biedl syndrome and craniomandibular dermatodysostosis [7–9].

The diagnosis of FSGS is based on the focal and segmental areas of glomerular sclerosis and tuft collapse. The common clinical findings include proteinuria, nephrotic syndrome and progression to ESRD. The pathogenesis indicates that FSGS risk factors include genetic, toxic, infectious and inflammatory factors. FFSGS is primarily caused by mutations in podocyte genes. Mutations in *NPHS1*, *NPHS2*, *CD2AP*, *PLCε1* and *MYO1E* can cause FFSGS through an autosomal recessive pattern of inheritance, and mutations in *ACTN4*, *TRPC6* and *INF2* are associated with autosomal dominant FSGS. Because of the different molecular basis and pathogenesis of familial and primary FSGS, there may be some differences in their clinical manifestation and prognosis. Few studies have compared familial and primary FSGS. In the present study, we investigated the clinical and pathological features of familial versus sporadic FSGS (SFSGS) and compared the prognosis of FSGS patients with and without a family history.

Patients and Methods

The present study was approved by the Institutional Review Board of Shanghai Ruijin Hospital, and written informed consent was obtained from all participants. Participants were recruited from the Ruijin Hospital School of Medicine at Shanghai Jiao Tong University. Inclusion criteria for the FFSGS patients were the following: each family had at least one individual with biopsy-proven FSGS, and each family had a second member with biopsy-proven FSGS, obvious proteinuria, progressive renal dysfunction or ESRD with an unknown cause [4, 10]. Patients with Alport syndrome, Fabry disease or other hereditary kidney diseases were excluded from the study; in addition to relying on electronic microscopy, eye examinations and hearing tests, these patients were identified by screening for type IV collagen in renal and skin biopsy tissue and by detecting α-galactosidase, an enzyme in the blood.

The mean follow-up time was 28.3 ± 12.5 months for the FFSGS group and 26.5 ± 19.5 months for the primary FSGS group ($p > 0.05$). Baseline and clinical characteristics were recorded for all patients. The primary outcomes of this study were ESRD and partial remission of proteinuria (defined as a 50% reduction of the baseline urine protein). One senior pathologist was responsible for reviewing all of the renal biopsy slides. A semi-quantitative scoring method was adopted to evaluate the pathological lesions. A percentage of the focal and global glomerulosclerosis was calculated, and a score of 0 (0%), 1 (<25%), 2 (25–50%) or 3 (>50%) was assigned to each slide to indicate the severity of tubular atrophy, inflammatory cell infiltration and interstitial fibrosis [11].

For normally distributed quantitative variables, the distributions were summarized as means and standard deviations; for non-normally distributed quantitative variables, the distributions were summarized as medians and ranges. For continuous variables, Student's t test or a Mann-Whitney U test was used to compare the two groups. Categorical variables were expressed as frequencies and percentages, and the group proportions were compared with a Pearson χ^2 test.

Results

The FFSGS group included 71 men and 53 women from a total of 83 families. The SFSGS group included 76 men and 48 women ($p > 0.05$). The difference in the median age of onset between the groups was statistically significant ($p = 0.035$; FFSGS group: 34 years, range 10–62 years vs. SFSGS group: 40 years, range 12–80 years). Compared to the SFSGS group, the FFSGS group had a lower urine protein excretion rate (1.1 vs. 1.5 g) (fig. 1a) and a higher serum albumin value (3.6 vs. 3.0 g/dl, $p < 0.01$). Furthermore, fewer patients had NS in the FFSGS group compared to the primary FSGS group (13.3 vs. 22.6%, $p = 0.029$). We also found more FFSGS patients than SFSGS patients with hematuria (62.9 vs. 22.9%, $p = 0.0001$), but the incidence of high blood pressure was similar between the two groups (43.75% in the FFSGS group vs. 35.16% in the SFSGS group, $p = 0.079$). Moreover, we found no differences between the groups in the baseline

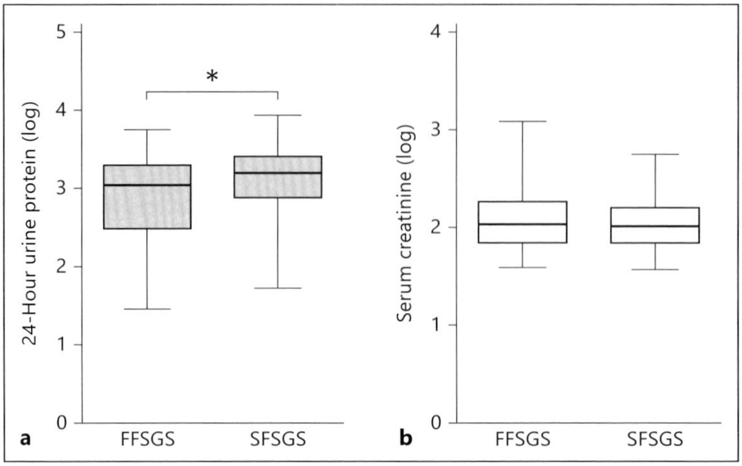

Fig. 1. 24-Hour urine protein (**a**) and serum creatinine (**b**) levels in the FFSGS and SFSGS groups. * $p < 0.01$.

serum creatinine (fig. 1b), uric acid or eGFR values. The clinical and pathological data are shown in table 1.

We compared several pathological features between the familial and sporadic FSGS patients. There was no significant difference in focal glomerulosclerosis between the FFSGS and SFSGS groups ($p = 0.141$), but more global glomerulosclerosis was observed in familial patients ($p = 0.0007$). Additionally, we found more serious tubular atrophy and interstitial fibrosis in the familial patients ($p = 0.0004$) (table 1).

During the follow-up period, partial remission was observed in 23.08% of the familial patients and 48.39% of the primary patients ($p = 0.01$). In the SFSGS group, the proteinuria level significantly decreased after treatment ($\log_{proteinuria}$: 3.12–2.66), but no proteinuria response was detected in the FFSGS group ($\log_{proteinuria}$: 2.87–2.88) (fig. 2a). The median ESRD-free survival times were 72 months for the familial group and 96 months for the primary group (log-rank, $p = 0.04$) (fig. 2b). Considering at least a 50% eGFR decline, more outcome events were found in the FFSGS group than in the SFSGS group (27.42 vs. 0%).

Discussion

The global incidence of FSGS has been increasing. FSGS accounts for up to 64% of glomerular lesions in Black individuals who undergo renal biopsy and between 15 and 20% of glomerular lesions in other ethnic groups [12]. Because

Table 1. Clinicopathological characteristics of FFSGS and SFSGS

	FFSGS	SFSGS	p value
Number	124	124	
Male/female	71/53	76/48	0.096
Follow-up, months	28.3±12.5	26.5±19.5	0.081
Age of onset	34 (10–62)	40 (12–80)	0.035
Hypertension, %	43.75	35.16	0.079
Serum creatinine, μmol/l	174.3±186.6	153.0±144.2	0.086
Uric acid	415.6±118.1	397.9±109.4	0.275
eGFR, ml/min per 1.73 m^2	69.49±38.27	81.76±50.47	0.069
Hematuria, %	62.9	22.9	0.0001
Proteinuria, g/day	1.1 (0.03–5.6)	1.5 (0.05–8.7)	0.003
Nephrotic syndrome, %	13.3	22.6	0.029
Focal glomerulosclerosis, %	13.23±11.83	15.71±13.49	0.141
Global glomerulosclerosis, %	25.56±21.4	16.52±20.96	0.0007
Tubulointerstitial lesion score	4.07±0.28	3.00±0.1	0.0004
Remission of proteinuria, %	23.08	48.39	0.042
End-stage renal disease, %	27.42	2.42	0.003

Remission of proteinuria was defined as a proteinuria decline of at least 50% of the baseline level. Hematuria was defined as more than 3 red blood cells in each high-power field.

FSGS has become a primary cause of ESRD, additional research should focus on its pathogenesis. In 1957, Vernier et al. [13] described 4 siblings with proteinuria whose parents had no evidence of kidney disease, which suggests that the trait is transmitted through the family in an autosomal recessive pattern. The cases in this study manifested as steroid-resistant nephrotic syndrome and progressive renal failure, and the patients died 7 years later. In some cases, the pathology showed minimal changes; in other cases, the changes were consistent with FSGS. Autosomal dominant and recessive forms of FFSGS have been reported in various ethnic groups and in all ages [14]. Recent studies have found mutations in genes (e.g. *ACTN4, TRPC6, NPHS2* and *INF2*) that code for podocyte-related molecules that are involved in the pathogenesis of FFSGS, but these genetic variants can only explain a small portion of FSGS cases [15]. In the present study, the FFSGS patients generally did not have mutations in known podocyte genes. The renal pathology and clinical characteristics of FFSGS are virtually identical to those of the more common sporadic forms, and few studies have compared the clinical and prognostic features of FFSGS and SFSGS.

Based on recent studies, we suggest that different pathogeneses of familial and primary FSGS may lead to various clinical manifestations. In a previous study, Conlon et al. [5] investigated the clinical and pathologic features of FFSGS in 31 patients from 8 families; the authors concluded that FFSGS carried a poor

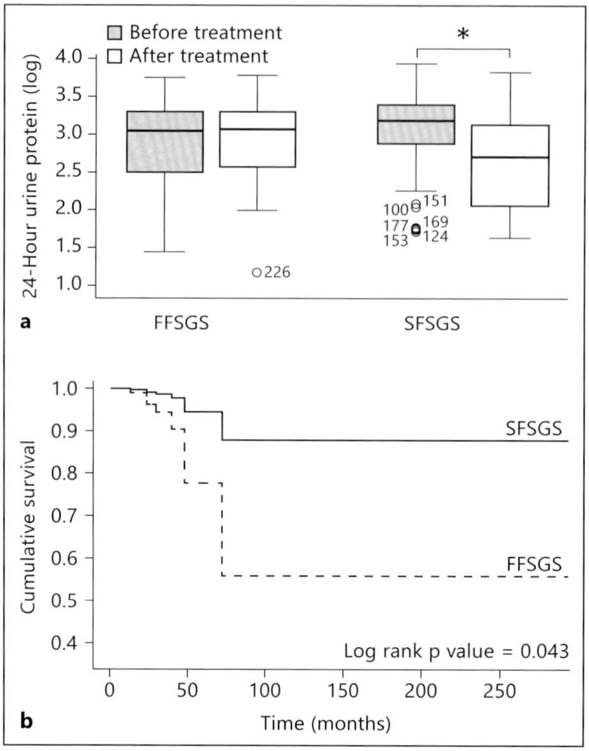

Fig. 2. Response to treatment and prognosis for FFSGS and SFSGS. **a** 24-Hour urine protein before and after treatment in the FFSGS and SFSGS groups. **b** Kaplan-Meier survival curves for FSGS patients in the FFSGS group (green) and the SFSGS group (blue). * $p < 0.01$.

prognosis, and histologic examination of the renal tissue revealed a variety of changes that have been previously described as occurring in FSGS. Tomero et al. [16] reported a Spanish family with FFSGS that spanned three generations. The authors reported the occurrence of ESRD in the fourth and fifth decades of life.

The present study enrolled 124 FFSGS patients and 124 SFSGS patients. Patients with FFSGS presented at a younger age than patients with primary FSGS, which indicates that familial patients had a longer course of the disease before the renal biopsy. More than half of the familial patients and 22.9% of the sporadic patients had hematuria, though the underlying reason was unknown. Hematuria may reflect a difference in the pathogenesis of familial and sporadic FSGS. Additional clinical features were found in familial patients, such as a lower urine protein excretion rate, a higher serum albumin value and fewer cases of nephrotic syndrome. These findings may be explained by the more severe pathological injuries, such as focal glomerulosclerosis, tubular atrophy, inflammatory cell infiltration and interstitial fibrosis that were found in FFSGS patients. During the follow-up period, the SFSGS group had more cases of partial remission of proteinuria than the FFSGS group, and the median ESRD-free survival time for the FFSGS group was significantly longer than the SFSGS group. At the end of the follow-up period, 34 familial patients (27.42%) and 3 primary patients (2.42%) had progressed to ESRD. These results suggest that, when compared to SFSGS patients, FFSGS patients experience more severe pathological changes (focal glomerulosclerosis and tubulointerstitial lesions) that are associated with less response to treatment and a worse prognosis.

Conclusion

Compared to SFSGS patients, FFSGS patients experience more severe pathological changes, less response to treatment, and a worse renal prognosis. While further validation by an independent cohort is required, our results indicate a potentially distinct pathogenesis and different response to treatment in FFSGS patients compared to SFSGS patients. Additional studies should focus on the pathogenesis of and treatments for FFSGS.

Acknowledgements

We are grateful to the patients who participated in the present study. This study was supported by grants from the National Basic Research Program of China 973 (Program Nos. 2012CB517600 and 2012CB517604) and the National Natural Science Foundation of China (Nos. 81000295 and 81070568). J. Xie is sponsored by the Shanghai Pujiang Program (11PJ1406700).

References

1 Rood IM, Deegens JK, Wetzels JF: Genetic causes of focal segmental glomerulosclerosis: implications for clinical practice. Nephrol Dial Transplant 2012;27:882–890.

2 Kitiyakara C, Kopp JB, Eggers P: Trends in the epidemiology of focal segmental glomerulosclerosis. Semin Nephrol 2003;23:172–182.

3 Kitiyakara C, Eggers P, Kopp JB: Twenty-one-year trend in ESRD due to focal segmental glomerulosclerosis in the United States. Am J Kidney Dis 2004;44:815–825.

4 Winn MP: Approach to the evaluation of heritable diseases and update on familial focal segmental glomerulosclerosis. Nephrol Dial Transplant 2003;18(suppl 6):vi14–vi20.

5 Conlon PJ, Butterly D, Albers F, Rodby R, Gunnells JC, Howell DN: Clinical and pathologic features of familial focal segmental glomerulosclerosis. Am J Kidney Dis 1995;26: 34–40.

6 Walker R, Bailey RR, Lynn KL, Burry AF: Focal glomerulosclerosis – another familial renal disease? NZ Med J 1982;95:686–688.

7 Barakat AJ, Arianas P, Glick AD, Butler MG: Focal sclerosing glomerulonephritis in a child with Laurence-Moon-Biedl syndrome. Child Nephrol Urol 1990;10:109–111.
8 Pedagogos E, Flanagan G, Francis DM, Becker GJ, Danks DM, Walker RG: A case of craniomandibular dermatodysostosis associated with focal glomerulosclerosis. Pediatr Nephrol 1995;9:354–356.
9 Lemieux G, Neemeh JA: Charcot-Marie-Tooth disease and nephritis. Can Med Assoc J 1967;97:1193–1198.
10 Winn MP, Conlon PJ, Lynn KL, et al: Linkage of a gene causing familial focal segmental glomerulosclerosis to chromosome 11 and further evidence of genetic heterogeneity. Genomics 1999;58:113–120.
11 Cattran DC, Coppo R, Cook HT, et al: The Oxford classification of IgA nephropathy: rationale, clinicopathological correlations, and classification. Kidney Int 2009;76:534–545.
12 Korbet SM, Genchi RM, Borok RZ, Schwartz MM: The racial prevalence of glomerular lesions in nephrotic adults. Am J Kidney Dis 1996;27:647–651.
13 Vernier RL, Brunson J, Good RA: Studies on familial nephrosis. I. Clinical and pathologic study of four cases in a single family. AMA J Dis Child 1957;93:469–485.
14 Ichikawa I, Fogo A: Focal segmental glomerulosclerosis. Pediatr Nephrol 1996;10:374–391.
15 Buscher AK, Konrad M, Nagel M, et al: Mutations in podocyte genes are a rare cause of primary FSGS associated with ESRD in adult patients. Clin Nephrol 2012;78:47–53.
16 Sanchez Tomero JA, Arrieta J, Alguacil A, Alonso A, Macias JF, Tabernero JM: Focal segmental glomerular sclerosis in three generations of a single family. Int J Pediatr Nephrol 1985;6:199–204.

Prof. Nan Chen, MD
Department of Nephrology, Ruijin Hospital
School of Medicine, Shanghai Jiao Tong University
No. 197 Ruijin Er Road, Shanghai 200025 (PR China)
E-Mail chen-nan@medmail.com.cn

Chen N (ed): New Insights into Glomerulonephritis.
Contrib Nephrol. Basel, Karger, 2013, vol 181, pp 109–118 (DOI: 10.1159/000348468)

Treatment and Prognosis of Primary Focal Segmental Glomerulosclerosis

Hong Ren · Pingyan Shen · Xiao Li · Xiaoxia Pan · Qianying Zhang · Xiaobei Feng · Wen Zhang · Nan Chen

Department of Nephrology, Ruijin Hospital, School of Medicine, Shanghai Jiaotong University, Shanghai, PR China

Abstract

This study aimed to analyze the treatment, clinical outcomes, and risk factors that affect the prognosis of patients with primary focal segmental glomerulosclerosis (FSGS) and to provide theoretical evidence for various treatment options in these patients. The study reviewed the clinical, laboratory, and pathological data of 168 patients with primary FSGS treated at Ruijin Hospital between January 2002 and October 2011. Of these patients, 108 were male (64.3%) and 60 were female (35.7%). The median age of disease onset was 38 years (range 12–78 years). The median case history was 10 months (range 4 days to 30 years). The mean proteinuria level was 2.3 ± 0.6 g/day. 75 (44.6%) patients had nephrotic syndrome. The mean serum creatinine was 108.1 ± 8.9 μmol/l. Over a follow-up period of 25.3 ± 11.4 months, end-stage renal failure occurred in 4 patients, and all 4 survived. In the group treated with angiotensin-converting enzyme inhibitor/angiotensin receptor blocker, the following factors were identified as risk factors for experiencing a 50% increase in serum creatinine over the baseline: a baseline eGFR <60 ml/min, proteinuria >1 g/day during the follow-up period, glomerular sclerosis >grade 1, and tubulointerstitial lesions >stage 1. In the group treated with steroids, patients who achieved a stable remission had better preserved renal function and milder glomerular sclerosis than steroid-dependent patients ($p < 0.01$). Steroid-resistant FSGS patients had a worse histological severity of glomerular sclerosis than steroid-dependent patients ($p < 0.01$). The prognosis of FSGS was correlated with the amount of proteinuria, the level of serum creatinine, and the severity of glomerular sclerosis and tubulointerstitial lesions. Steroids may be more effective in those who have better preserved renal function and milder glomerular sclerosis.

Copyright © 2013 S. Karger AG, Basel

According to the USRDS, primary focal segmental glomerulosclerosis (FSGS) is responsible for 3.3% of all cases of end-stage renal disease (ESRD) resulting from primary kidney disease [1]. Studies of survival with FSGS have helped identify patients at a high risk of progression to ESRD and have been important in determining the choice of treatment and predicting the prognosis of FSGS. Korbet et al. [2] reported that the amount of proteinuria was an important factor that influenced the prognosis of FSGS. The 10-year renal survival rate was >80% among FSGS patients with non-nephrotic-range proteinuria. However, 50% of patients with nephrotic-range proteinuria progressed to ESRD within 6–8 years. Gipson et al. [3] reported that FSGS patients with severe tubulointerstitial lesions had a poorer prognosis. Some researchers have also suggested that the response to therapy was the best prognostic indicator of outcomes for FSGS patients. Half of patients who do not respond to treatment have doubled serum creatinine levels within 4 years [4–6].

In this study, we retrospectively analyzed the data of 168 primary FSGS patients, summarized the treatment, clinical outcomes, and risk factors that affect a patient's prognosis, and provided theoretical evidence for better treatment options in primary FSGS patients.

Materials and Methods

Patients

The study included 168 patients who were admitted to the Nephrology Department of Shanghai Ruijin Hospital between January 2002 and October 2011 with biopsy-proven cases of primary FSGS and whose records contained complete follow-up data.

Methods

Clinical Studies, Laboratory Examination, and Pathologies

The patients' demographic characteristics and clinical features are summarized in table 1, including gender, age, height, weight, blood pressure, and brief medical history. The laboratory results for hemoglobin, 24-hour urine protein, serum albumin, serum creatinine, serum nitrogen, and serum uric acid at renal biopsy and at follow-up visits were also obtained. The treatment options, the patients' response to angiotensin-converting enzyme inhibitor/angiotensin receptor blocker (ACEI/ARBs), glucocorticoids, or immunosuppressants, the side effects, and the adjustments or withdrawal of the therapeutic drugs were recorded and reviewed.

Pathologies and indices of renal damage: The severity of glomerulosclerosis was expressed on an arbitrary scale from 0 to 3. Grade 0 indicated no glomerular sclerosis, grade 1 indicated that the percentage of glomerulosclerosis was <25%, grade 2 indicated that the percentage of glomerulosclerosis was 25–50%, and grade 3 indicated that the percentage of glomerulosclerosis was >50% [7–9]. Tubulointerstitial scores (TIS) were also used to evaluate the lesions of the tubulointerstitium according to the damage caused by tubular atrophy, interstitial inflammatory cell infiltration, and interstitial fibrosis. The

Table 1. Baseline clinical characteristics of FSGS patients

	ACEI/ARBs alone	Glucocorticoids alone	Glucocorticoids combined with immunosuppressants	p value
Number	79	75	62	
Age, years, median (range)	41 (28–80)	35 (12–78)	37 (16–77)	0.667
Females/males	30/49	33/42	24/38	0.325
Duration of disease, median (range)	10 (0.26–360)	7 (0.13–60)	10 (2–144)	0.203
Systolic blood pressure, mm Hg	127.7±7.6	124.3±4.4	124.7±3.3	0.106
Diastolic blood pressure, mm Hg	78.1±3.8	77.1±3.8	78.9±2.9	0.421
Hemoglobin, g/l	130.1±3.2	130.2±1.9	137.6±3.1	0.254
Urine protein, g/24 h	0.88±0.1	2.6±1.8	3.1±0.5	<0.001*,#
Serum albumin, g/l	37.1±0.7	23.5±10.2	25.3±2.1	<0.001*,#
Serum creatinine, μmol/l	138.1±11.8	82.3±9.3	98.1±6.1	0.628
eGFR, ml/min	67.3±10.9	84.7±8.9	86.3±7.4	0.540

Measurement data with skewed distribution are presented as the median (range).
MDRD formula eGFR = $175 \times SCr^{-1.154}$ $age^{-0.203} \times 1{,}212$ (if patient is Black) $\times$ 0.742 (if female).
* ACEI/ARB alone vs. glucocorticoids alone, $p < 0.001$.
# ACEI/ARB alone vs. glucocorticoids combined with immunosuppressants, $p < 0.001$.

scores were accessed by the severity and extent of the lesions: no lesion, mild lesion (<25%), moderate lesion (25–50%), and severe lesion (>50%) were assigned scores of 0, 1, 2, and 3, respectively. According to the total TIS (the summation of the 3 TIS), the TIS grading system was defined as follows: grade 0: score 0; grade 1: scores from 1 to 3; grade 2: scores from 4 to 6, and grade 3: scores from 7 to 9 [7–9].

Statistical Analyses

Statistical analyses were performed with SPSS version 11.0 (SPSS, Chicago, Ill., USA). Continuous variables are presented as the mean ± SD or median (range). t tests, non-parametric tests, and rank sum tests were used for comparisons between groups, depending on the distribution of the data. A multivariate logistic regression model (forward LR method) was used to detect the risk factors. Categorical data were analyzed using the χ^2 test. $p < 0.05$ was considered statistically significant.

Results

Baseline Clinical Characteristics of FSGS Patients

Among the 168 patients with primary FSGS, 108 (64.3%) were males and 60 (35.7%) were females. The male:female ratio was 1.78:1. The median age at onset was 38 years (range 12–78 years). The median duration of the disease was 10 months (range 4 days to 30 years). The mean urine protein was 2.3 ± 0.6 g/24 h. Of the 168 patients, 75 (44.6%) were diagnosed with nephrotic syndrome. The average serum creatinine was 108.1 ± 8.9 μmol/l, and the average eGFR (MDRD formula) was 78.8 ± 6.9 ml/min. The average follow-up period was 25.3 ± 11.4 months.

79 patients were treated with ACEI/ARBs alone, 75 patients were treated with glucocorticoids alone, and 62 patients received glucocorticoids combined with immunosuppressants. Among the latter two groups, 48 (64%) and 46 patients (74.2%), respectively, were simultaneously treated with ACEI/ARBs. At baseline, only urine protein excretion and serum albumin in the ACEI/ARB-alone group were significantly different from the glucosteroid-alone or glucosteroid and immunosuppressant group ($p < 0.001$) (table 1).

Prognosis of Patients Treated with ACEI/ARBs Alone

The median follow-up time for the 79 patients treated with ACEI/ARBs alone was 26 (3–104) months. Among these patients, 70 (88.6%) continued with the initial therapy during the follow-up period. 6 (7.6%) were given glucocorticoids together with ACEI/ARBs due to an increase in urine protein at a median follow-up time of 12 (7–32) months. 17 (21.5%) had a 50% increase in serum creatinine over the baseline level at a median follow-up time of 36 (13–90) months. 3 (3.8%) progressed to ESRD. The duration between the diagnosis of FSGS and the diagnosis of ESRD was 15, 21, and 44 months for these 3 patients, respectively. None of the patients in this group died during the follow-up period.

We used a univariate analysis and a multivariate logistic regression analysis (forward LR method) to detect the probable risk factors for a 50% increase in serum creatinine. We speculated that certain factors (i.e., gender, age, major comorbidities, certain laboratory indices, pathological features, and the total length of follow-up time) might have some correlation with the 50% increase in serum creatinine. The results of the univariate analysis are shown in table 2.

According to the results of the univariate analysis, we performed a binary multivariate logistic regression analysis. The results showed that the risk factors for a 50% increase of serum creatinine among patients treated with ACEI/ARBs alone were the following: the percentage of glomerulosclerosis at renal biopsy >grade 1 (regression coefficient: 3.29, $p < 0.001$, OR = 34.7); the severity of tubulointerstitial lesions >grade 1 (regression coefficient: 2.11, $p = 0.009$, OR = 9.2); eGFR (by MDRD formula) <60 ml/min (regression coefficient: 1.97, $p = 0.012$, OR = 7.8), and urine protein >1 g/24 h during the follow-up period (regression coefficient: 1.51, $p = 0.03$, OR = 5.6).

Prognosis of Patients Treated with Glucocorticoids Alone

The median follow-up time of the 75 patients treated with glucocorticoids alone was 8 (2–26) months. The median initial prednisone (prednisolone) dose was 55 (45–65) mg/day. The median total course of steroid treatment was 7

Table 2. Risk factors for a 50% increase in serum creatinine among patients treated with ACEI/ARBs alone: univariate analysis

Parameter	Serum creatinine increased by ≤50% % (n = 62)	Serum creatinine increased by >50% % (n = 17)	OR value	p value
Age >50 years	16 (26.5)	8 (47.1)	2.48	0.017
Males	39 (63.2)	9 (52.9)	0.81	0.788
Duration of the disease >12 months	27 (44.1)	8 (47.1)	1.15	0.067
Hypertension	23 (36.8)	11 (64.7)	3.41	0.003
Severity of glomerulosclerosis > grade 1	20 (32.3)	15 (88.2)	23.4	<0.001
Severity of tubulointerstitial lesions > grade 1	22 (35.3)	14 (82.3)	13.2	<0.001
Thickening of the renal arteriolar wall	7 (11.8)	7 (41.2)	3.17	0.021
Hyalinosis of the renal arterioles	23 (36.8)	7 (41.2)	1.46	0.322
Before treatment				
Urine protein >1 g/24 h	15 (23.5)	8 (47.1)	2.72	0.011
eGFR <60 ml/min	13 (20.6)	13 (76.5)	12.8	<0.001*
Uric acid >450 mmol/l	25 (41.2)	12 (70.6)	2.89	0.028
Hematuria (microscopic or gross)	37 (60.3)	8 (47.1)	0.69	0.781
During the follow-up period				
Urine protein >1 g/24 h	10 (16.1)	9 (52.9)	4.92	<0.001*
Uric acid >450 mmol/l	17 (27.9)	9 (52.9)	2.17	0.034
Hematuria (microscopic or gross)	36 (58.8)	7 (41.2)	1.19	0.693

MDRD formula eGFR = 175 × $SCr^{-1.154}$ $age^{-0.203}$ × 1,212 (if patient is Black) × 0.742 (if female). All data are presented as n (%).
* Subgroup in which serum creatinine increased by ≤50% vs. subgroup in which serum creatinine increased by >50%, p < 0.001.

(2–18) months. The remission rates at the end of the first and second years after steroid treatment were 34.6 and 20%, respectively, and the relapse rates were 25.3 and 34.6%, respectively. The median relapse time was 7 (3–19) months after remission. 42 patients (56%) were given immunosuppressants together with glucocorticoids due to steroid dependence or steroid resistance. None of the patients in this group progressed to ESRD or death during the follow-up period (table 3).

We compared the baseline clinical characteristics, renal pathologies, and laboratory indices of patients with different responses to therapies (continuing remission, steroid dependence, and steroid resistance) after 2 years of follow-up. Those who presented with continuous remission after steroid treatment had a lower baseline serum creatinine level (70.9 ± 12.1 vs. 110.7 ± 8.9 μmol/l, p = 0.002), a lower percentage of glomerulosclerosis >grade 1 (14.3 vs. 46.1%, p < 0.001), and a higher eGFR (97.3 ± 13.7 vs. 76.5 ± 8.9 ml/min, p = 0.009) when compared with those who were steroid-resistant. Patients who were steroid-dependent had a lower percentage of glomerulosclerosis >grade 1 than those

Table 3. Efficacy of glucocorticoids

Course months	CR n	PR n	NR n	Combined with immunosuppressant, n	Accumulative drop-off, n	Accumulative relapse, n	Efficacy %	Relapse %
2	22	18	35	8	0	0	53.3	0.0
3	24	17	26	21	0	1	54.6	1.3
6	20	15	13	26	6	6	46.6	8.0
12	19	7	0	36	16	19	34.6	25.3
24	12	3	0	42	24	26	20.0	34.6

Table 4. Baseline characteristics of patients with different responses after 2 years of glucocorticoid therapy

Clinical and pathological manifestations	Response		
	persistent remission	steroid-dependent	steroid-resistant
Cases	7	26	13
Age, median (range)	31 (12–44)	34 (17–68)	35 (14–78)
Male/female	4/3	16/10	6/7
Median course, months (range)	7 (0.5–48)	8 (0.13–36)	10 (3–60)
Hypertension, cases (%)	3 (42.8)	8 (30.8)	7 (53.8)
Microhematuria, cases (%)	4 (57.1)	13 (50)	6 (46.1)
Proteinuria, g/24 h	2.6±0.2	2.7±0.2	2.9±0.1
Serum albumin, g/l	27.1±0.7	26.5±1	23.5±0.6
Serum creatinine, μmol/l	70.9±12.1*	77.8±6.7	110.7±8.9
eGFR, ml/min	97.3±13.7*	89.9±7.4	76.5±8.9
Glomerulosclerosis > grade 1, cases (%)	1 (14.3)*	6 (23.1)#	6 (46.1)
Tubulointerstitial lesions > grade 1, cases (%)	1 (14.3)	4 (15.4)	3 (23.1)
Thickening of the renal arteriolar wall, cases (%)	0 (0)	1 (3.8)	1 (7.8)
Hyalinosis of the renal arterioles, cases (%)	1 (14.3)	3 (11.5)	2 (15.4)
Immunofluorescence positive, cases (%)	2 (28.6)	12 (46.1)	6 (46.1)

* Persistent remission vs. steroid-resistant, $p < 0.01$.
# Persistent remission vs. steroid-dependent, $p < 0.01$.

who were steroid-resistant (23.1 vs. 46.1%, $p < 0.001$). No differences were observed with respect to other clinical manifestations or pathological or laboratory indices between these two groups ($p > 0.05$) (table 4).

Prognosis of Patients Treated with Glucocorticoids Combined with Immunosuppressants

Sixty-two patients were treated with glucocorticoids combined with immunosuppressants due to steroid dependence or steroid resistance; 42 of these patients had been previously treated with glucocorticoids alone. In addition to steroid therapy, 38 patients also received cyclophosphamide, 19 also received tacrolimus, and 5 also received cyclosporine. After 6 months of follow-up, the remission rates of these 3 subgroups were 54, 70.6, and 80%, respectively. After 12 months of follow-up, the remission rates were 61.2, 76.9, and 66.6%, respectively. The median time to first remission in the cyclophosphamide, tacrolimus, and cyclosporine subgroups was 3 (1–11), 2 (1–6), and 2 (1–4) months, respectively. One of the patients treated with cyclophosphamide + steroids progressed to ESRD. None of the patients died during the follow-up period.

Discussion

The 5-year survival rate of primary FSGS is 45–75% in Western Europe [6, 10, 11]. Rydel et al. [6] retrospectively analyzed the clinicopathological data of 81 adult FSGS patients, among whom 53% were Black. In their study, 74% of the patients presented with nephrotic syndrome, and 25.9% of the patients had proteinuria >10 g/24 h at biopsy. The average total follow-up time was 62.4 ± 74.4 months, and the 5-year renal survival rate was 75.6%. Deegens et al. [10] performed a retrospective analysis of 93 Dutch adult patients with biopsy-proven primary FSGS. The average urine protein excretion was 8 ± 5.3 g/24 h, and the average serum creatinine level was 147 ± 99 μmol/l, with a 5-year renal survival rate of 66%. In our study, the mean urine protein excretion was 2.3 ± 0.6 g/24 h in the 168 primary FSGS patients. Of these patients, 44.6% presented with nephrotic syndrome. None of the patients' urine protein excretion exceeded 10 g/24 h. The mean serum creatinine level was 108.1 ± 8.9 μmol/l. The mean follow-up time was 25.3 ± 11.4 months. The 2-year renal survival rate was 97.6%, which was better than previously reported rates. This improved rate might have been due to the lower percentage of nephrotic syndrome, early diagnosis and treatment, ethnicity, and the relatively short follow-up time in this study. Because this was a retrospective study, we chose patients of the ACEI/ARB-alone group with less proteinuria and worse renal function when we considered their therapy options. This resulted in significant differences in the baseline proteinuria and serum albumin between the ACEI/ARB-alone group and the glucosteroid-alone and glucosteroid + immunosuppressant groups.

Patients with urine protein excretion >10 g/24 h typically progress to ESRD within 3–5 years. Nevertheless, the 10-year renal survival rate of patients with non-nephrotic-range proteinuria is >80% [2]. The control of proteinuria is the key to prevent the deterioration of renal function. The outcomes of FSGS are related to serum creatinine at biopsy, the percentage of glomerulosclerosis, and the severity of renal tubulointerstitial lesion [7, 12]. In this study, we used univariate and multivariate logistic regression analyses to evaluate the risk factors that are related to a poor prognosis in those treated with ACEI/ARBs alone. The results show that patients with urine proteinuria >1 g/24 h during the follow-up period, eGFR <60 ml/min before renal biopsy, severe glomerulosclerosis, or severe tubulointerstitial lesions tended to have worse renal function. These findings are similar to those presented in previous reports [13]. For this reason, early diagnosis, early treatment, and the control of proteinuria were essential for FSGS patients to delay the progression to ESRD and improve the long-term prognosis.

Today, cytotoxic agents (cyclophosphamide, chlorambucil, or mycophenolate mofetil) and calcineurin inhibitors (cyclosporin A or tacrolimus) are used for relapsing FSGS [14] and steroid therapy is still the first-line treatment for primary FSGS. A meta-analysis demonstrated that after a median treatment and follow-up time of 8 (2–24) months, the total remission rate was 54% (10–70%), including a complete remission rate of 35% (0–70%) and a partial remission rate of 19% (0–50%) [15]. The differences between the remission rates observed in various studies are mainly attributed to the courses of treatment. Cases in which the complete remission rate is >30% have an average course of treatment of between 5 and 8 months [6], and cases in which the remission rate is <20% have an average course of treatment of less than 2 months [6]. In our study, 75 primary FSGS patients were treated with glucocorticoids. The median total course of treatment was 7 (2–18) months. After 6 months of treatment, the total remission rate was 46.6%, including a complete remission rate of 26.7%, which was similar to that reported previously [13]. Further observation showed that the 1- and 2-year total response rates were 34.6 and 20%, respectively, and the complete remission rates were 25.3 and 16%. The relapse rates of primary FSGS are reportedly 19–46.4% [12]. Three-quarters of the patients who relapse can achieve remission again after repeated treatment. Fortunately, the patients who relapse have a similar prognosis to those who respond initially to treatment. In this study, with a follow-up time of 2 years, the relapse rate was 34.6%, and the median time to relapse was 7 (3–19) months after the first remission, which are similar to the data reported in the literature.

Further analysis revealed that those who had relatively lower serum creatinine and higher eGFR at renal biopsy tended to be sensitive to steroid therapy. Patients with a low percentage of glomerulosclerosis were likely to achieve persistent clinical remission. Meyrier [16] observed that the response to steroid therapy could be attributed to the severity and progression of primary FSGS. Those with a serum creatinine level >145 μmol/l (1.6 mg/dl) at renal biopsy tended to have poor therapeutic effects because >75% of the nephron had been damaged, and the response to the steroid therapy was therefore poor. Additionally, urine protein >14 g/24 h, collapsing FSGS, and Black ethnicity predict poor responses to steroid therapy in primary FSGS patients [16, 17].

In summary, our study revealed that the 2-year renal survival rate was 97.6% in 168 primary FSGS patients. At renal biopsy, glomerulosclerosis >grade 1, tubulointerstitial lesions >grade 1, eGFR <60 ml/min, and urine protein >1 g/24 h during the follow-up period were independent risk factors for a 50% increase in serum creatinine in non-nephrotic-syndrome patients. The total response rate to the steroid therapy was 20%, and the relapse rate was 34.6% after 2 years of follow-up. Lower serum creatinine and higher eGFR at renal biopsy led to better responses to steroid therapy. Patients with a low percentage of glomerulosclerosis were likely to achieve persistent clinical remission. Our retrospective study described several subordinate findings that are consistent with previous findings. Larger, multicenter, prospective controlled studies are needed to obtain more evidence about the effective therapeutic options for primary FSGS.

Acknowledgements

The authors gratefully acknowledge all study participants for their contributions. This work was supported by a grant from the National Basic Research Program of China 973 Program No. 2012CB517600 (No. 2012CB517604) and the National Natural Science Foundation of China (Nos. 81070560, 81070568, and 81200517).

References

1 Kitiyakara C, Eggers P, Kopp JB: Twenty-one-year trend in ESRD due to focal segmental glomerulosclerosis in the United States. Am J Kidney Dis 2004;44:815–825.

2 Korbet SM: Primary focal segmental glomerulosclerosis; in Brady HR, Wilcox CS (eds): Therapy in Nephrology and Hypertension. A Companion to Brenner and Rector's The kidney. Philadelphia, Saunders, 1999, pp 178–188.

3 Gipson DS, Chin H, Presler TP, et al: Differential risk of remission and ESRD in childhood FSGS. Pediatr Nephrol 2006;21:344–349.

4 Singh A, Tejani C, Tejani A: One-center experience with cyclosporine in refractory nephrotic syndrome in children. Pediatr Nephrol 1999;13:26–32.

5 Troyanov S, Wall CA, Miller JA, et al: Focal and segmental glomerulosclerosis: definition and relevance of a partial remission. J Am Soc Nephrol 2005;16:1061–1068.

6 Rydel JJ, Korbet SM, Borok RZ, et al: Focal segmental glomerulosclerosis in adults: presentation, course and response to treatment. Am J Kidney Dis 1995;25:534–542.

7 Pan B, Jang Z, Chang J, et al: Clinicopathological analysis of adult primary focal segmental glomerulosclerosis in 263 cases. Chin J Nephrol 2007;23:13–17.

8 Prayaga AK, Anuradha SV, Manjusha Y, et al: Morphologic evaluation of renal function using semiquantitative method in primary nonproliferative glomerular diseases. Indian J Pathol Microbiol 2011;54:42–46.

9 Barisoni L, Valeri A, D'Agati V, et al: FSGS: A 20-year epidemiologic study. J Am Soc Nephrol 1994;5:347.

10 Deegens JK, Steenbergen JE, Borm GF, et al: Pathologic variants of focal segmental glomerulosclerosis in an adult Dutch population – epidemiology and outcome. Nephrol Dial Transplant 2008;23:186–192.

11 Chun MJ, Korbet SM, Schwartz MM, et al: Focal segmental glomerulosclerosis in nephrotic adults: presentation, prognosis, and response to therapy of the histologic variants. J Am Soc Nephrol 2004;15:2169–2177.
12 Tang Xi, Liu Zhi-Hong, Xu Feng: Long-term outcome of adult patients with idiopathic focal segmental glomerulosclerosis Chin J Nephrol Dial Transplant 2010;19:23–29.
13 Korbet SM: Treatment of primary FSGS in adults. J Am Soc Nephrol 2012;23:1769–1776.
14 Ren H, Shen P, Li X, et al: Tacrolimus versus cyclophosphamide in steroid-dependent or steroid-resistant focal segmental glomerulosclerosis: a randomized controlled trial. Am J Nephrol 2013;37:84–90.
15 Korbet SM: Angiotensin antagonists and steroids in treatment of focal segmental glomerulosclerosis. Semin Nephrol 2003;23:219–228.
16 Meyrier A: An update on the treatment options for focal segmental glomerulosclerosis. Expert Opin Pharmacother 2009;10:615–628.
17 Crook ED, Habeeb D, Gowdy O, et al: Effects of steroids on focal segmental glomerulosclerosis in a predominantly African-American population. Am J Med Sci 2005;330:19–24.

Prof. Nan Chen, MD
Department of Nephrology, Ruijin Hospital
School of Medicine, Shanghai Jiao Tong University
No. 197 Ruijin Er Road, Shanghai 200025 (PR China)
E-Mail chen-nan@medmail.com.cn

Chen N (ed): New Insights into Glomerulonephritis.
Contrib Nephrol. Basel, Karger, 2013, vol 181, pp 119–130 (DOI: 10.1159/000348470)

New Insights into Treatment of Nephrotic Syndrome in Children

Xiaowen Wang · Hong Xu

Department of Nephrology, Children's Hospital of Fudan University, Shanghai, PR China

Abstract

Primary nephrotic syndrome (PNS) is the most common glomerular disease seen in children. The pathogenic basis of PNS, in context of witnessed immunological dysfunction and its relationship with slit diaphragm and the podocyte cytoskeleton, has been explored for years. The outcome and management of PNS in children is determined by the response to corticosteroids and the frequency of relapses. The first episode of nephrotic syndrome should be treated adequately, both in terms of dose and duration of steroids, aiming at reducing the risk for subsequent relapses. While patients with steroid-sensitive nephrotic syndrome have a favorable long-term outcome, almost half of them relapse frequently and are at risk of adverse effects of steroids. Immunosuppressive agents are suggested for patients with frequently relapsing nephrotic syndrome or steroid-dependent nephrotic syndrome. Calcineurin inhibitors are strongly recommended for patients who show resistance to steroids. Rituximab is a new therapy that could be a potential area but still requires further trials. The aim of therapy is to induce and maintain remission of proteinuria while avoiding medication-related adverse effects.

Copyright © 2013 S. Karger AG, Basel

Nephrotic syndrome is the most common glomerular disease seen in children. It is characterized by massive proteinuria, hypoalbuminemia, and edema. Nephrotic syndrome in children can be categorized into three groups: primary, secondary, and congenital. Secondary nephrotic syndrome is described as nephrotic syndrome associated with well-defined diseases such as lupus nephritis, IgA nephropathy, Henoch-Schönlein purpura, Alport's syndrome, etc. Congenital nephrotic syndrome occurring in the first year of life, particularly in the first 3 months, is more likely to be due to a gene mutation or associated with infections such as syphilis or toxoplasmosis.

The most usual form in childhood is primary nephrotic syndrome (PNS), which typically develops in the absence of features of nephritis or associated primary extrarenal disease. The annual incidence of PNS has been estimated to range from 1 to 3 new cases per 100,000 children, and the prevalence is about 16 cases per 100,000 children [1, 2].

Oral corticosteroids are still the most effective agents for inducing remission in patients with steroid-sensitive nephrotic syndrome. Approximately 80% of children achieve complete remission within 4 weeks of steroid therapy after their first presentation, but a similar proportion relapse and need repeated courses of steroid therapy. Repeated courses of steroid therapy have potential serious adverse effects such as obesity, poor growth, hypertension, osteoporosis, diabetes mellitus, and ophthalmological disorders. Steroid-sparing agents including calcineurin inhibitors (CNIs), cyclophosphamide (CTX), and mycophenolate mofetil (MMF) are often used to treat these children. From 10 to 20% of the patients are resistant to steroids (SRNS) and half of them will advance to end-stage renal failure. For children with refractory nephrotic syndrome it is a challenge to create optimal treatment in terms of more efficiency and fewer adverse events.

Definitions Used for Nephrotic Syndrome

The diagnosis of nephrotic syndrome requires severe proteinuria (>50 mg/kg/day, or a protein:creatinine ratio >2,000 mg/g), hypoalbuminemia (<30 mg/l), the presence of edema, and hyperlipidemia. The nephrotic range of proteinuria and hypoalbuminemia are imperative for making the diagnosis.

Classifications upon Response to Steroids [3, 4]
Corticosteroids have been used for the most vital treatment of PNS since the 1950s. The definitions of response to steroids have varied little over the years.

Steroid-Sensitive Nephrotic Syndrome (SSNS)
Patients who enter complete remission within initial 4–8 weeks of steroid therapy (dose: 1.5–2.0 mg/kg·day or 60 mg/m^2 to a maximum of 60 mg/day) are referred to as having *steroid-responsive* or *steroid-sensitive* nephrotic syndrome.

Steroid Resistance (SRNS)
Initial non-responder is defined as failure to achieve complete remission after 4–8 weeks of steroid therapy from the first presentation. Late non-responder is defined as persistent proteinuria during 4 or more weeks of steroids following one or more remissions.

Relapse/Infrequent Relapse/Frequent Relapse/Steroid Dependence [3, 4]
The majority of children with nephrotic syndrome (50–70%) develop one or more relapses of the disease. These relapses may be transient, with spontaneous remission occurring within 4–14 days in some. Despite prolonged initial steroid use, about 40–50% of those with complete remission of nephrotic syndrome subsequently develop either frequent relapses or steroid dependence.

Relapse of nephrotic syndrome is defined as recurrence of severe proteinuria (urine protein:creatinine ratio (uPCR) ≥2,000 mg/g (200 mg/mmol) or ≥3+ of protein on urine dipstick for 3 consecutive days), often with a recurrence of edema.

Infrequent relapse is stated one relapse within 6 months of initial response, or one to three relapses in any 12-month period.

Frequent relapse (FRNS) is defined as two or more relapses within 6 months of initial response, or four or more relapses in any 12-month period.

Steroid dependence (SDNS) represents two consecutive relapses during steroid therapy, or within 14 days of ceasing therapy.

Judgment of Response to Therapy
Complete remission is defined as uPCR <200 mg/g (20 mg/mmol) or <1+ of protein on urine dipstick for 3 consecutive days. *Partial remission* represents a proteinuria reduction of 50% or greater from the presenting value and absolute uPCR between 200 and 2,000 mg/g (20–200 mg/mmol). *No remission* is failure to reduce urine protein excretion by 50% from baseline or persistent excretion uPCR >2,000 mg/g (200 mg/mmol)

Pathology

In over 80% patients with PNS, the histology shows insignificant glomerular abnormalities on light microscopy, termed minimal change disease (MCD) [5]. A 31-year review of renal biopsy histology in our center showed that MCD accounted for 61.8% of patients with FRNS and SDNS [6]. The underlying pathology in patients with SRNS is focal segmental glomerulosclerosis (FSGS), although MCD can also be seen in SRNS. Data from Hong Kong indicated 37% of patients with SRNS had FSGS and its relative incidence was increased when compared to previous studies [7]. A small number of patients with SRNS showed mesangial proliferation, membranoproliferative glomerulonephritis, membranous nephropathy, or, rarely, diffuse mesangial sclerosis.

Indications for Renal Biopsy

Most children with PNS do not require a renal biopsy. Renal biopsy is required at the onset of nephrotic syndrome in situations where a cause other than minimal change nephrotic syndrome is suspected, such as: age at onset ≤1 or ≥10 years; gross hematuria, persistent microscopic hematuria or low serum C3; acute renal failure, not attributable to hypovolemia, and suspected secondary causes [1, 6]. A renal biopsy should be considered if diagnosis of steroid resistance is made, or CNIs are chosen for patients with FRNS or SDNS, or therapy of CNIs exceeds 2 years [6, 8].

Treatment of the Initial Episode of SSNS

The first widely accepted recommendations for the initial treatment of the childhood nephrotic syndrome came from the International Study of Kidney Disease in Children, and suggested that treatment begins with prednisone at a dose of 60 mg/m^2/day. Nowadays, steroids are still the first choice for the treatment of PNS.

The initial episode of nephrotic syndrome should be treated adequately, both in terms of dose and duration of corticosteroids, since this is an important determinant of the long-term course. Steroid therapy (prednisone or prednisolone) should be given for at least 12 weeks. Oral prednisone is administered as a single daily dose starting at 60 mg/m^2/day or 2 mg/kg/day to a maximum of 60 mg/day [3, 4]. Daily oral prednisone should be given for 4–6 weeks followed by alternate-day medication as a single daily dose starting at 40 mg/m^2 or 1.5 mg/kg (maximum 40 mg on alternate days) and continued for 2–5 months with tapering of the dose [4, 9].

Treatment for Relapse SSNS, SDNS, and SRNS

Corticosteroid Therapy

Infrequent Relapse

A single daily dose of prednisone 60 mg/m^2 or 2 mg/kg (maximum of 60 mg/day) is recommended until the child has been in complete remission for at least 3 days. After achieving complete remission, the child should be given prednisone as a single dose on alternate days (40 mg/m^2 per dose or 1.5 mg/kg per dose: maximum 40 mg on alternate days) for at least 4 weeks [3, 4, 9].

FRNS and SDNS

Relapses in children with FRNS or SSNS should be treated with daily prednisone until the child has been in remission for at least 3 days, followed by alternate-day prednisone for at least 3 months [3, 4, 9]. After achieving complete remission, the dose of prednisone can be taped by 0.25 mg/kg every 4 weeks. Prednisone is given on alternate days in the lowest dose to maintain remission without major adverse effects in children with FRNS and SDNS [3, 9]. Daily prednisone is suggested at the lowest dose to maintain remission without major adverse effects in children with SDNS where alternate-day prednisone therapy is not effective [10]. Steroids during infection: daily prednisone is proposed during episodes of upper respiratory tract infection and other infections to reduce the risk for relapse in children with FRNS and SDNS already on alternate-day prednisone [11, 12].

Intravenous Pulse Methylprednisolone for SRNS

Intravenous pulse methylprednisolone (MP) can be considered when the patient shows no response after 4 weeks of treatment with prednisone at a dose of 60 mg/m^2/day or 2 mg/kg/day. The dose of intravenous pulse MP is suggested 15–30 mg/kg/day to a maximum of 1 g/day, daily for 3 days [3]. If complete remission is achieved, oral prednisone should be given and taped in the same way as SSNS. If the patient remains non-responsive, immunosuppressants should be used in addition to alternate-day prednisone. The dose of alternate-day prednisone still needs to be taped by 5–10 mg every 2–4 weeks, followed by the lowest dose to maintain remission. For some patients, steroids can be completely stopped [9]. For patients who show resistance to CNIs, pulse MP can be an option. Intravenous high-dose MP is given for 3 days every 2 weeks and the course can be repeated six times. This treatment can be accompanied by oral prednisone and other immunosuppressive agents if needed [8].

Immunosuppressive Medications

Immunosuppressants should be considered in patients who failed to achieve complete remission with 8 weeks of adequate steroid therapy [13]. Patients who remain severely proteinuric after 4 weeks of steroid therapy can also be given immunosuppressants. When children with FRNS or SDNS develop a steroid-related adverse effect, immunosuppressants should be initiated to reduce the steroid dose. CNIs, tacrolimus or cyclosporine are considered to be the 'first line' for patients with SRNS. MMF, pulse MP and a combination of alternative immunosuppressive agents could be considered if patients show no response to the CNIs [11]. The advantages and disadvantages of corticosteroid-sparing agents in common use in nephrotic syndrome are shown in table 1.

Table 1. Advantages and disadvantages of corticosteroid-sparing agents in common use [modified from 31]

Agent	Advantages	Disadvantages
Cyclophosphamide	Prolonged remission off therapy Inexpensive	Less effective in SDNS Monitoring of blood count during therapy Potential serious short- and long-term adverse effects
Mycophenolate mofetil	Prolonged remissions in some children with FRNS and SDNS Few adverse effects	Continued treatment often required to maintain remission Probably less effective than CNIs Low level of evidence for SRNS Expensive
Cyclosporine	Prolonged remissions in some children with SDNS More effective in SRNS	Continued treatment often required to maintain remission Expensive Nephrotoxic Drug level monitoring is needed Cosmetic side effects
Tacrolimus	Prolonged remissions in some children with SDNS More effective in SRNS	Continued treatment often required to maintain remission Expensive Drug level monitoring is needed Nephrotoxic Risk of diabetes mellitus

Calcineurin Inhibitors

1. *FRNS/SDNS*: CNIs can be given as steroid-sparing agents in patients with FRNS/SDNS [9]. Cyclosporine is administered at a dose of 4–5 mg/kg/day (starting dose) in two divided doses [4, 9]. The principal side effects of cyclosporine are kidney dysfunction, hypertension, hyperkalemia, gum hypertrophy, and hypertrichosis. Tacrolimus 0.1 mg/kg/day (starting dose) given in two divided doses can be used instead of cyclosporine when the cosmetic side effects of cyclosporine are unacceptable [14]. CNIs levels should be monitored during therapy to limit toxicity. Most studies have used cyclosporine targeting 12-hour trough levels of 80–120 ng/ml with maintenance of lower levels after a child has been in stable remission for 3–6 months, aiming to minimize cyclosporine nephrotoxicity [8, 14]. The efficacy of tacrolimus would appear to be similar to that of cyclosporine based on an observational study in SDNS, and its dose should be adjusted to maintain the 12-hour trough levels in the range of 5–10

ng/ml [13, 15]. CNIs are suggested to be given for at least 12 months, as most children will relapse when CNIs are stopped [4, 15].

2. *SRNS:* CNIs are recommended as initial therapy for children with SRNS, and it should be continued for a minimum of 6 months and then stopped if a partial or complete remission of proteinuria is not achieved [16]. A study at our center indicated that children with SRNS receiving cyclosporine had a better response than those treated with CTX after 12 months of follow-up [17]. CNI is continued for a minimum of 12 months when at least a partial remission is achieved by 6 months. Low-dose corticosteroid therapy can be combined with CNI therapy [3]. Based on case studies [18, 19], complete and partial remissions are less common in the presence of nephrotic syndrome associated with gene mutations (WT1, NPHS2, PLCE1, etc.) that disrupt the podocyte cytoskeleton. Though infrequent, remissions have been reported in children with nephrotic syndrome with gene mutations, suggesting that a trial of CNI therapy may induce at least a partial remission in these patients. The underlying mechanism may be explained by its impact on the podocyte cytoskeleton instead of the immunosuppressive effect [20].

Alkylating Agents

1. *FRNS/SDNS*: Alkylating agents (CTX, chlorambucil) may result in prolonged remission off all therapy, though they may have significant adverse effects. CTX is considered beneficial to reduce the dose of steroids, aiming at fewer steroid side effects. Children older than 7–8 years and those with FRNS and a lower steroid threshold do better than those who are younger and have SDNS or a high steroid threshold. The dose of oral CTX is 2–2.5 mg/kg/day for 8–12 weeks. Its cumulative dose should not exceed 168 mg/kg, and repeat courses are avoided [4]. CTX can also be given by pulse intravenous infusion at the dosage of 10 mg/kg daily for 2 days every 2 weeks until the cumulative dose is 200 mg/kg or 500 mg/m^2 once a month for 6 months [3]. The chief adverse effects of CTX treatment include leukopenia, alopecia, nausea, vomiting and hemorrhagic cystitis; long-term toxicity includes the risk of gonadal toxicity and malignancies. Leukocyte counts should be monitored every 2 weeks and the medication discontinued if <3,000–4,000/mm^3. Fluid intake should be increased and the child is encouraged to void frequently. The prednisolone dose is 1–1.5 mg/kg on alternate days during CTX therapy; subsequently, steroids are tapered and discontinued over 4–6 weeks. Another alkylating agent, chlorambucil (0.1–0.2 mg/kg/day), may be given for 8 weeks (maximum cumulative dose 11.2 mg/kg) as an alternative to CTX [21].

2. *SRNS:* Two randomized controlled trials (RCTs) comprising 84 children with SRNS demonstrated no significant differences in the number achieving re-

mission with CTX and prednisone compared with prednisone alone, with an increase in adverse effects in the CTX groups providing evidence that CTX should not be used in children with SRNS [22, 23]. Although imprecision may affect these risk estimates, risk ratios and confidence intervals were centered around them. Potential harm from cytotoxic agents far exceeds any evidence of benefit when patients fail CNI therapy.

Mycophenolate Mofetil

1. *FRNS/SDNS:* MMF is generally well tolerated, with small numbers of children developing leukopenia and abdominal pain. In a prospective multicenter clinical trial in China, MMF was used in patients with FRNS or SDNS for a relatively long time and showed more efficacy and better tolerance than pulse intravenous infusion of CTX [24]. MMF has also been used with cyclosporine in children with poorly controlled SDNS and has allowed reduction of the cyclosporine dose. MMF (starting dose 1,200 mg/m^2/day or 20–30 mg/kg/day) should be given in two divided doses for at least 12 months, as most children will relapse when MMF is stopped [4].

2. *SRNS:* There is little evidence to support the use of MMF in patients with SRNS [9]. However, it can be considered in children who fail to achieve complete or partial remission with CNIs or the adverse effects of CNIs are unacceptable. MMF is administered at a dose of 20–30 mg/kg/day for at least 6 months, and taped to 10 mg/kg/day if a consistent remission of 3–6 months can be achieved [25].

Rituximab

1. *FRNS/SDNS:* Rituximab, a monoclonal anti-CD20 antibody, is a new therapy for nephrotic syndrome. Rituximab can be considered only in children with SDNS who have continuing frequent relapses despite optimal combinations of prednisone and steroid-sparing agents, or who have serious adverse effects of other drugs. Prolonged remissions in 80% of children following rituximab have been reported, with doses of 375 mg/m^2 per dose given for up to four weekly doses. Rituximab caused acute reactions, such as fever, vomiting and diarrhea, skin rash, and bronchospasm in about one third of patients in one series [26]. It should be administered under close supervision at a specialized center.

2. *SRNS:* Therapy of rituximab for SRNS is controversial due to the lack of RCTs and the risk for serious adverse events. One study suggested that it was not as effective in SRNS as in SSNS [8]. However, a clinical trial from our center indicated rituximab may have a role in SRNS in reducing exposure to steroids and other immunosuppressants and it may be more effective in children with SRNS

whose pathology was MCD than in those where it was FSGS [27]. This may be an area that will benefit from future RCTs.

Mizoribine
Mizoribine is widely used as a steroid-sparing agent in Japan [4]. A single RCT demonstrated that the relapse rate did not differ significantly between treatment and placebo groups. At present, mizoribine is not recommended as a steroid-sparing agent in FRNS and SDNS [28].

Azathioprine
It is recommended that azathioprine not be used as a steroid-sparing agent in frequent relapses and SDNS, since RCTs have demonstrated no significant difference in the risk of relapse between azathioprine and placebo [21].

Evaluation and Monitoring

Underlying Factors for FRNS/SDNS

Infections
Infections could be a complication of PNS as well as the potential cause of FRNS. The most common forms of infection seen in PNS are cellulitis, peritonitis, and sepsis, which should be treated timely with antibiotics. There are some infections with few symptoms that cannot be detected easily in patients with FRNS/SDNS. Zinc supplements were reported to reduce the risk of infections, especially in patients with FRNS.

Hypothalamic-Pituitary-Adrenal Axis Suppression
A significant proportion of children receiving long-term prednisone for PNS have evidence of adrenal suppression, which is great risk of relapse. Therefore, assessment for the hypothalamic-pituitary-adrenal axis is suggested for patients with FRNS/SDNS receiving long-term steroid treatment [29].

Genetic Test

A genetic test is not a routine evaluation due to the variability in availability, significant cost, current uncertainty about the individual response to therapy and prognosis when a mutation is identified [8]. In children with congenital nephrotic syndrome, a family history of SRNS, or sporadic initial steroid resistance, a mutational analysis should be offered when available. The identification of pathogenic mutations can help to avoid adverse effects of steroid/immunosuppressive treatment and to encourage living donor kidney transplantation. To reduce the time consumed and costs, the method of combining snapshot (NPHS1, NPHS2, APOL1, CD2AP, PLCE1, ACTN4, TRPC6, INF2, MYO1E, MYH9) and direct sequencing (NPHS2, WT1) of exons of candidate genes could be used in screening genetic mutations in children with sporadic SRNS [30].

Monitoring Treatment-Related Side Effects

In addition to complications resulting from nephrotic syndrome, a number of side effects can also result directly or indirectly from the treatment of the disease, including steroids, immunosuppressants, diuretics, etc. Blood pressure, growth curve, bone mineral density and intraocular pressure should be screened in patients given steroids. Baseline assessment of renal function, electrolytes, blood levels of albumin and cholesterol, complete blood counts and urinalysis guides future evaluation of efficacy as well as side effects of therapy. Monitoring for drug levels is necessary when using either cyclosporine or tacrolimus, because individual variations in bioavailability may result in the occurrence of either subtherapeutic or toxic levels. A 12-hour trough level should be estimated about 2 weeks after introduction of therapy, after any dose change, and if drug toxicity or poor compliance is suspected.

Conclusion

The long-term management of nephrotic syndrome requires regular interactions between patients and caregivers to ensure their cooperation in disease control. Patients with both a steroid-sensitive and steroid-resistant disease require frequent monitoring of the clinical course and biochemical response, and individual management of disease in order to optimize patient care and enable satisfactory long-term outcomes.

References

1 Sinha A, Bagga A: Nephrotic syndrome. Indian J Pediatr 2012;79:1045–1055.

2 Ernould S, Godron A, Nelson JR, Rigothier C, Llanas B, Harambat J: Idiopathic nephrotic syndrome in children: incidence, clinical presentation, and outcome in the county of Gironde, France. Arch Pediatr 2011;18:522–528.

3 The Subspecialty Group of Nephrology, Society of Pediatrics, Chinese Medical Association: Evidence-based guidelines on diagnosis and treatment of childhood common renal diseases. 1. Evidence-based guideline on diagnosis and treatment of steroid-dependent nephrotic syndrome. Chin J Pediatr 2009;47:167–170.

4 Lombel RM, Gipson DS, Hodson EM: Treatment of steroid-sensitive nephrotic syndrome: new guidelines from KDIGO. Pediatr Nephrol 2013;28:415–426.

5 Yan Husen M, Kemper MJ: New therapies in steroid-sensitive and steroid-resistant idiopathic nephrotic syndrome. Pediatr Nephrol 2011;26:881–892.
6 Zheng YB, Xu H, Zhou LJ, Cao Q, Sun L, Shen Q: Time to reflect our renal biopsy indications: a thirty-one-year review of renal biopsy histopathology. Chin J Evid Based Pediatr 2011;16:190–198.
7 Yuen LK, Lai WM, Lau SC, Tong PC, Tse KC, Chiu MC: Ten-year review of disease pattern from percutaneous renal biopsy: an experience from a paediatric tertiary renal centre in Hong Kong. Hong Kong Med J 2008;14:348–355.
8 Lombel RM, Hodson EM, Gipson DS: Treatment of steroid-resistant nephrotic syndrome in children: new guidelines from KDIGO. Pediatr Nephrol 2013;28:415–426.
9 Hodson EM, Willis NS, Craig JC: Corticosteroid therapy for nephrotic syndrome in children. Cochrane Database Syst Rev 2007; 4:CD001533.
10 Srivastava RN, Vasudev AS, Bagga A, Sunderam KR: Long-term, low-dose prednisolone therapy in frequently relapsing nephrotic syndrome. Pediatr Nephrol 1992;6: 247–250.
11 Abeyagunawardena AS, Trompeter RS: Increasing the dose of prednisolone during viral infections reduces the risk of relapse in nephrotic syndrome: a randomised controlled trial. Arch Dis Child 2008;93:226–228.
12 Gulati A, Sinha A, Sreenivas V, Math A, Hari P, Bagga A: Daily corticosteroids reduce infection-associated relapses in frequently relapsing nephrotic syndrome: a randomized controlled trial. Clin J Am Soc Nephrol 2011; 6:63–69.
13 Hodson EM, Willis NS, Craig JC: Interventions for idiopathic steroid-resistant nephrotic syndrome in children. Cochrane Database Syst Rev 2010;11:CD003594.
14 Sinha MD, Macleod R, Rigby E, Clark AG: Treatment of severe steroid-dependent nephrotic syndrome in children with tacrolimus. Nephrol Dial Transplant 2006;21:1848–1854.
15 Wang W, Xia Y, Mao J, Chen Y, Wang D, Shen H, Fu H, Du L, Liu A: Treatment of tacrolimus or cyclosporine A in children with idiopathic nephrotic syndrome. Pediatr Nephrol 2012;27:2073–2079.
16 Choudhry S, Bagga A, Hari P, Sharma S, Kalaivani M, Dinda A: Efficacy and safety of tacrolimus versus cyclosporine in children with steroid-resistant nephrotic syndrome: a randomized controlled trial. Am J Kidney Dis 2009;53:760–769.
17 Rao J, Xu H, Cao Q, Huang WY, Zhou LJ: Comparison of cyclophosphamide and cyclosporine in the treatment of steroid-resistant idiopathic nephrotic syndrome in children. J Cent South Univ Med Sci 2007;32:958–963.
18 Malina M, Cinek O, Janda J, Seeman T: Partial remission with cyclosporine A in a patient with nephrotic syndrome due to NPHS2 mutation. Pediatr Nephrol 2009;24:2051–2053.
19 Gellermann J, Stefanidis CJ, Mitsioni A, Querfeld U: Successful treatment of steroid-resistant nephrotic syndrome associated with WT1 mutations. Pediatr Nephrol 2010;25: 1285–1289.
20 Bensman A, Niaudet P: Non-immunologic mechanisms of calcineurin inhibitors explain its antiproteinuric effects in genetic glomerulopathies. Pediatr Nephrol 2010;25:1197–1199.
21 Hodson EM, Willis NS, Craig JC: Non-corticosteroid treatment for nephrotic syndrome in children. Cochrane Database Syst Rev 2008;1:CD002290.
22 International Study of Kidney Disease in Children: Prospective, controlled trial of cyclophosphamide therapy in children with nephrotic syndrome. Report of the ISKDC. Lancet 1974;2:423–427.
23 Tarshish P, Tobin JN, Bernstein J, Edelmann CM Jr: Cyclophosphamide does not benefit patients with focal segmental glomerulosclerosis. A report of the International Study of Kidney Disease in Children. Pediatr Nephrol 1996;10:590–593.
24 Yi ZW, Wu XC, Xu H, Zhou LJ, Wu YB, Feng SP, Zhou JH, Yang Q, Zhu GH, Liu AM, Wei MJ, Yu L, Zeng ZF: A prospective multicenter clinical control trial on treatment of refractory nephrotic syndrome with mycophenolate mofetil in children. Chin J Contemp Pediatr 2008;10:575–578.
25 De Mello VR, Rodrigues MT, Mastrocinque TH, Martins SP, de Andrade OV, Guidoni EB, Scheffer DK, Martini Filho D, Toporovski J, Benini V: Mycophenolate mofetil in children with steroid/cyclophosphamide-resistant nephrotic syndrome. Pediatr Nephrol 2010;25:453–460.

26 Ravani P, Magnasco A, Edefonti A, Murer L, Rossi R, Ghio L, Benetti E, Scozzola F, Pasini A, Dallera N, Sica F, Belingheri M, Scolari F, Ghiggeri GM: Short-term effects of rituximab in children with steroid- and calcineurin-dependent nephrotic syndrome: a randomized controlled trial. Clin J Am Soc Nephrol 2011; 6:1308–1315.
27 Sun L, Xu H, Shen Q, Cao Q, Liu HM: Efficacy of rituximab therapy in children with refractory nephrotic syndrome: a prospective observational study in Shanghai. World J Pediatr 2013 (accepted).
28 Yoshioka K, Ohashi Y, Sakai T, Ito H, Yoshikawa N, Nakamura H, Tanizawa T, Wada H, Maki S: A multicenter trial of mizoribine compared with placebo in children with frequently relapsing nephrotic syndrome. Kidney Int 2000;58:317–324.
29 Abeyagunawardena AS, Hindmarsh P, Trompeter RS: Adrenocortical suppression increases the risk of relapse in nephrotic syndrome. Arch Dis Child 2007;92:585–588.
30 Li GM, Xu H, Shen Q, Cao Q: Research on pathogenic genes and their testing methods in patients with sporadic steroid-resistant nephrotic syndrome (SRNS) Unpublished data.
31 KDIGO: New guidelines for nephrotic syndrome in children. Kidney Int Suppl 2012;2: 143–153.

Hong Xu
Department of Nephrology, Children's Hospital of Fudan University
399 Wanyuan Road, Minhang
Shanghai 200031 (PR China)
E-Mail hxu@shmu.edu.cn

Chen N (ed): New Insights into Glomerulonephritis.
Contrib Nephrol. Basel, Karger, 2013, vol 181, pp 131–142 (DOI: 10.1159/000348472)

Pathogenesis of Membranous Nephropathy: A New Paradigm in Evolution

Richard J. Glassock

Laguna Niguel, Calif., USA

Abstract

Membranous nephropathy (MN) has been recognized as a distinct morphological entity for over 50 years, but it is only recently that the underlying pathogenesis of the primary form of the disorder has been elucidated. This brief overview catalogues recent advances in understanding the pathogenesis of MN, focused mainly on its primary form. These studies have enumerated and identified several autologous podocyte antigens that serve as targets of autoantibody responses in primary MN. The dominant autoantigen is M-type phospholipase A_2 receptor protein (PLA2R1) expressed on the surface of native glomerular podocytes. Autoantibodies to PLA2R1, usually of IgG4 subclass, are found in about 80% of patients with primary MN. These autoantibodies bind to genetically determined, conformational epitopes on PLA2R1, form immune complexes in situ and induce proteinuria, mostly likely via local activation of complement via the mannose-binding lectin pathway. The autoimmune response is modulated by genes at the HLA-DQA1 locus. The level of autoantibody to PLA2R correlates with the severity of the clinical disease and may predict recurrences in renal allografts, at least in some patients. Most forms of secondary MN appear to be due to distinctly different pathogenetic mechanisms.

Copyright © 2013 S. Karger AG, Basel

Almost 60 years has elapsed since membranous nephropathy (MN) was identified as an important cause of nephrotic syndrome and separated morphologically from other forms of primary glomerular disease evoking this stereotypical syndrome [1]. The distinctive lesion of MN has two main clinical presentations: (i) a primary one occurring in the absence of any systemic disease process and without any identifiable inciting causes [1, 2], often also referred to as 'idiopathic' MN,

This paper is a revised and updated version of a previously published article by the author [Curr Opin Nephrol Hypertens 2012;21:235–242].

Table 1. Classification of MN

Primary Membranous Nephropathy – Anti-phospholipase A_2 receptor (PLA2R) autoantibody-positive with IgG4 and PLA2R glomerular deposition (80% or more of cases) – Anti-PLA2R-negative with IgG4 glomerular deposition – Other (anti-superoxide dismutase, aldose reductase, α-enolase, megalin, neutral endopeptidase, anti-bovine serum albumin-related disease)
Secondary Membranous Nephropathy – Systemic autoimmune disease (lupus nephritis, mixed connective tissue disease, rheumatoid arthritis)[1] – Chronic infections (hepatitis B virus, hepatitis C virus, syphilis, malaria, leprosy) – Drug-related (NSAIDs, gold, mercury, penicillamine, probenecid, captopril) – Neoplasia-related (carcinoma, lymphoma, leukemia)[1] – Monoclonal IgG3κ disease with anti-PLA2R antibody activity – Other (sarcoidosis[1], sickle cell disease, Kimura's disease, Castleman's disease, chronic thyroiditis, de novo in renal allografts, graft vs. host disease in bone marrow allografts, allogeneic stem cell transplantation[1])
Recurrent in renal allograft/isograft

[1] Some cases may be anti-PLA2R-positve.

and (ii) a secondary form arising consequent to a systemic disease or a well-recognized etiologic process (such as drug exposure or a chronic viral infection). A remarkably diverse array of disorders can be associated with secondary MN (table 1) [2]. It is also likely that a similar degree of heterogeneity exists for the underlying pathogenesis of primary and secondary MN, since differing pathogenetic events can give rise to similar morphological features in many glomerular diseases. Animal models for human MN have been developed (Heymann nephritis) that clearly show that the in situ formation of immune complexes is largely responsible for the characteristic morphologic feature of MN (subepithelial deposits containing immunoglobulins) [1, 3]. Thus, it seemed likely that an autoantibody to an antigen intrinsic to the glomerular capillary wall, most likely in the vicinity of the podocyte, was the operative vector in primary ('idiopathic') MN and that such MN is largely an autoimmune renal disease. The nature of the autoantibody and its target antigen was unknown until comparatively recently [4]. The discovery of the M-type phospholipase A_2 receptor (PLA2R1) protein constitutively expressed on the surface of podocytes as a major target antigen in human primary ('idiopathic') MN by Beck et al. [4, 5] in 2009 has revolutionized the understanding of human primary MN, even marginalizing the term 'idiopathic' MN to a lesser relevance. Nevertheless, the events responsible for production of autoantibody to the PLA2R1 antigen remain uncertain. Also, circulating autoantibodies to PLA2R1 may not explain the full spectrum of primary MN [6–8].

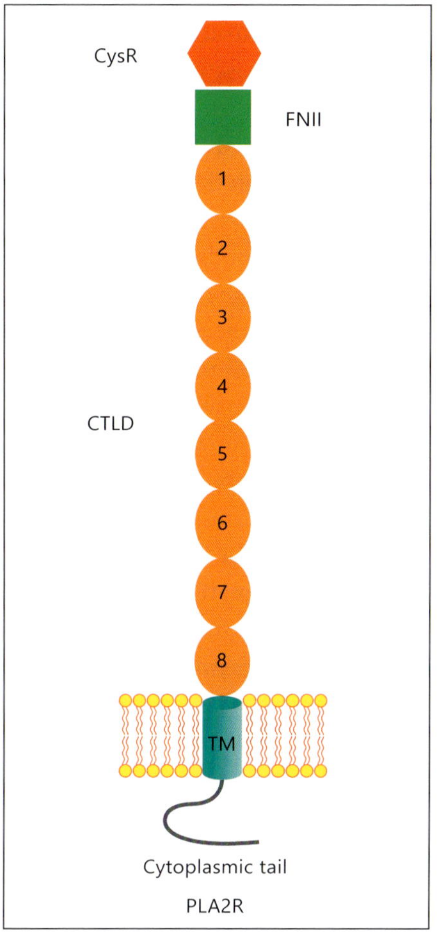

Fig. 1. Molecular depiction of PLA2R1 and its chemical domains. Reproduced by courtesy of Quansheng Zhu, PhD, UCLA School of Medicine.

Primary Membranous Nephropathy Mediated by Anti-PLA2R1 Autoantibody

The dominant nephritogenic antigen (PLA2R1) in human primary MN is synthesized and secreted by podocytes and expressed in the membrane surface [4, 8]. PLA2R1 is a 180-kDa polypeptide with long extracellular domains, consisting of a cysteine-rich head and fibronectin type II-like repeat domain and 8 repeated carbohydrate-recognition domains, and short membrane-spanning intracellular domain [9] (fig. 1). PLA2R1 has been cloned and synthesized in a recombinant form. The epitope (or epitopes) responsible for its nephritogenicity appears to be conformational in nature and dependent on cysteine-rich sequences as its ability to react with autoantibody is destroyed by disruption of cysteine-cysteine disulfide bridges [4, 5]. It is likely that the autoepitopes are located in the extracellular domains, perhaps on one of carbohydrate-recognition domains close to the N-terminus of the protein. The function of PLA2R1 and the autoantigenic epitope specifically within the podocyte is uncertain, but studies have raised the possibility that PLA2R1 may be involved in cellular senescence through the p53 pathway [10] or to its interaction with adjacent integrins associated with glomerular basement membrane [Q. Zhu, pers. commun., 2013]. Autoantibodies (most often of the IgG1 or IgG4 subclasses) to this conformational antigen can be detected by Western blot or by other in vitro assays using recombinant protein in about 80% or more patients with primary MN [5, 11]. An immunofluorescent assay may be clinically useful as well [12]. Such autoantibodies are usually not found in cases with secondary MN (e.g. hepatitis B viral infection), however malignancy-related MN may have circulating anti-PLA2R1 autoantibodies in as many as 30% of cases [11]. The factors responsible for the 10–20% 'anti-PLA2R1-negative' primary MN are not completely understood but might be due to: (i) poor sensitivity of current assays; (ii) heterogeneity of the conformational PLA2R1 epitope among cases of MN; (iii) spontaneous disappearance of the circulating antibody prior to clinical detection of disease, and (iv) involvement of other non-PLA2R1 antigen-antibody systems (e.g. neutral endopeptidase, superoxide dismutase, aldose reductase, α-enolase, cationic serum albumin and renal tubular antigens (megalin) [6, 7, 13–19].

Strong evidence exists for the pathogenicity of the anti-PLA2R1 autoantibodies in human MN. Anti-PLA2R1 autoantibodies can be eluted from diseased kidneys, the anti-PLA2R1 antibody colocalizes with the PLA2R1 antigen in glomeruli (in most but not all cases), recurrences of MN may develop in the transplanted kidney in the presence of pretransplant anti-PLA2R1 antibody in the circulation (but not consistently so), and the levels of anti-PLA2R1 autoantibody highly correlate with the clinical manifestations of disease [20, 21]. Passive transfer of the disease to experimental animals with human anti-PLA2R1 has not been shown, possibly because of low levels of expression of PlA2R1 in non-human species. Circulating levels of anti-PLA2R1 autoantibody decrease or disappear with remission of disease (spontaneous or treatment-induced) [20, 21], often weeks or months prior to the onset of clinical improvement [21]. However, discrepancies may exist between the presence of anti-PLA2R1 autoantibodies in the circulation and the detection of granular deposits of PLA2R1 antigen in glomeruli (presumably as immune complexes) in MN [22, 23]. Debiec and Ronco [22] studied 42 cases of primary MN; 21 had both anti-PLA2R1 antibody in circulation and PLA2R1 antigen in the glomerular deposits; 3 had anti-PLA2R1 autoantibody in the circulation but no PLA2R1 antigen detectable in glomerular deposits; 18 had no circulating anti-PLA2R1 autoantibody but 10 of these had PLA2R1 antigen detectable in the immune deposits. The latter finding could possibly be explained by persistence of immune complexes in subepithelial sites long after the circulating antibody had disappeared and active in situ formation of deposits had long since ceased. Detection of PLA2R1 antigen in glomeruli is a sensitive means of identifying primary MN in renal biopsies [23] and such an antigen deposition shows a high correlation with anti-PLA2R autoantibodies in the circulation and a diagnosis of primary MN [24].

The immune response to the PLA2R1 autoantigen is controlled by single nucleotide polymorphisms (SNPs) at both chromosome 2 (the PLA2R1 gene locus) and chromosome 6 (HLA-DQA1 or other closely contiguous genes at this locus, including DQA1 05:01, DQB1 03:01 or DQB1 02:01) [25]. This di-genic control likely represents the influence of the PLA2R1 locus on generation of immunogenic conformational epitopes and subsequent binding of these PLA2R1-derived immunogens to antigen-presenting cells expressing HLA-DQA1 followed by activation of T cells and augmentation of autoantibody production via T- and B-cell cooperation. Exactly how such SNPs act to define a nephritogenic conformational epitopes on the domains of PLA2R1 is under intense investigation. However, it is tempting to speculate that such polymorphisms help explain the inconsistency of recurrent MN in renal transplants even in patients with positive anti-PLA2R autoantibodies [26, 27]. Donor kidneys lacking the relevant PLA2R1 SNP might not express the conformational PLA2R1 epitopes and thus be resistant to recurrence. Additional research is needed to clarify the role of PLA2R1 SNPs in recurrent primary MN and de novo MN in the transplanted kidney.

Primary MN due to anti-PLA2R1 autoantibodies is characterized by dominancy of an IgG4 subclass immune response both in circulating antibodies and by the immunoglobulin deposits in glomeruli, even though autoantibodies of other classes may also appear (usually IgG1) [1, 28–33]. Similar IgG4 deposits are also seen in recurrences of primary MN in renal allografts [28]. Segawa et al. [32], studying primary MN in children (a rare disease) found more intense deposition of IgG1, IgG2, IgG4, factor B and mannose-binding lectin (MBL) in 'global' MN compared to 'segmental' MN. Collectively, these observations strongly suggest that primary MN should be regarded as an IgG4-associated disease entity. Th2 cytokine activation appears to promote IgG4 synthesis in primary MN [34], and thus MN is also a Th2 activation-associated disease. It is well known that IgG4 fixes complement poorly via the alternative or classical pathway [4, 35]. It also tends to interfere with the formation of an immune complex lattice when IgG1 antibodies are present [36] and can interfere with complement activation [37]. In this sense, IgG4 might be a 'protective' antibody, while other Ig subclasses (e.g. IgG1) might be more pathogenic. However, alterations in the carbohydrate side-chain characteristics, specifically galactose deficiency, of the IgG4 autoantibody may also confer special properties with respect to activation of complement, particularly via the MBL pathway [38]. Whether this abnormal galactosylation is acquired or native remains uncertain. Deposits of C4d in glomeruli are the signposts of MBL activation.

Local activation of the complement cascade and formation of the membrane-attack complex (C5b–C9) is a prerequisite for the development of proteinuria in experimental MN (Heymann nephritis). T-cell activation (CD8+) also participates in the generation of proteinuria, at least in experimental animal studies (Heymann nephritis) [39, 40]. Mannose (mannan)-binding lectin (MBL) is often co-deposited with IgG in 'global' forms of MN [28] and the MBL pathway of complement activation is very important in the pathogenesis of human MN [38]. In the alloantibody form of infantile MN due to neutral endopeptidase (NEP), alloantibodies of the IgG1 subclass rather than IgG4 subclass appear to track more closely with the development of proteinuria [13]. Malignancy-related MN [41] is known to rather consistently show a deficiency of IgG4 subclass in the immune deposits [24, 29, 33], yet the clinical and morphological features of malignancy-related MN are nearly identical to primary MN. Very clearly, the role of the in situ deposition of IgG4 subclass anti-PLA2R1 autoantibodies and the generation of abnormal permselectivity of glomeruli in human MN needs much more research.

The recent observation that a large minority (about 30%) of patients with malignancy-related MN have circulating anti-PLA2R1 autoantibodies (by an improved Western blot assay), raise questions concerning the pathogenesis of primary MN [11]. Are the anti-PLA2R antibody-positive cases of malignancy-associated MN also characterized by IgG4-dominant deposits of IgG? Are some malignancies associated with aberrant expression of PLA2R1 nephritogenic epitopes leading to autoimmunity resembling that seen in primary MN? Resolution of the fine structure of the relevant epitopes on PLA2R1 in various disease states will likely resolve this conundrum.

Finally, the stimuli for autoantibody production in PLA2R1-positive human MN remain largely unknown. Is cross-reacting 'molecular mimicry' evoked by exposure to some ubiquitous environmental antigen (e.g. virus) in the presence of a susceptible genetic milieu (e.g. a PLA2R1 SNP)? If so, the prospect for a preventative strategy for MN looms large on the horizon.

Recurrence of Primary Membranous Nephropathy in Renal Allografts

A recurrence of MN in renal allografts introduced into patients with a primary form of MN in their native kidneys has been repetitively reported [42]. The frequency of recurrent MN varies between 10 and 40% of cases of primary MN requiring renal transplantation (depending on the method of ascertainment) and such recurrences, if severe, can result in graft loss [24, 42]. In addition, some patients with diseases other than primary MN in their native kidneys may develop MN de novo posttransplantation [27], often in association with chronic allograft rejection but usually not accompanied by anti-PLA2R1 autoantibody formation [27]. Patients with detectable circulating anti-PLA2R1 autoantibodies prior to transplantation may be predisposed to such recurrences, but not universally so [24, 25, 43]. Whether 'non-recurrence' of MN in the presence of circulating anti-PLA2R autoantibodies pretransplant is due to the existence of 'non-pathogenetic' anti-PLA2R autoantibodies or to the immunogenetic properties of the transplanted kidney itself (e.g. possession of the relevant conformational PLA2R1 epitopes) cannot be determined at the present time. Recurrences of primary MN are associated with IgG4 deposition while de novo MN is associated with IgG1 deposition in the glomeruli of allografts [28].

A recent study (using protocol or clinically indicated allograft renal biopsies) described the early immunopathologic manifestations of recurrent primary MN in the renal allograft [26]. Twenty-one cases of recurrent MN were studied in biopsies taken 0.3–4 months posttransplantation. The features of early recurrence of MN were granular glomerular deposition of IgG, κ and λ light chains, C4d but usually no C3. Electron-dense subepithelial deposits were observed in 11/19 cases studied by electron microscopy. Studies of IgG subclass deposition or MBL were not performed. Control biopsies lacked IgG, C4d, C3 and electron-dense deposits. This elegant study again raises the issue of a possible role of the MBL pathway of complement activation in the pathogenesis of MN and might encourage exploration of complement-inhibiting therapy for recurrent disease. The lack of C3 deposits in the early recurrence of primary MN is unexplained.

Non-PLA2R1 Antigen-Antibody Systems in Primary Membranous Nephropathy

Although PLA2R1-anti-PLA2R1 appears to be the major antigen-antibody system operative in primary MN, other target antigens and antibodies may also be involved [1, 4, 14, 15]. In 2002, Debiec et al. [13] first described an *alloimmune* form of MN due to antibodies to NEP, mostly in infants and children. Mothers congenitally deficient for NEP pregnant with a fetus expressing NEP develop IgG1 and IgG4 anti-NEP alloantibodies which permeate through the placenta and bind to the native NEP on the fetal podocyte leading to MN, expressed at birth or later. Similar anti-NEP autoantibodies, as well as other podocyte-related cytoplasmic antigens, may participate in some cases of MN of adult onset [44].

In addition, on rare occasions, infants fed cow's milk (containing bovine serum albumin (BSA)) may develop apparently primary MN based on absorption of a cationic component of BSA and its subsequent binding to anionic sites in the glomerular capillary wall where it can serve as a 'planted antigen' target for anti-BSA antibodies to specific epitopes on the BSA molecule [19]. This sequence of events has been elegantly described in a small number of infants/children with apparent primary MN, but has not yet been described in adults with primary MN, despite the relative frequency of anti-BSA antibodies in adults.

Additional cases of apparent primary MN involving superoxide dismutase, aldose reductase and α-enolase have also been reported [15–17, 44]. Antibodies to α-enolase (IgG4 subclass) have been described in MN and eluted from biopsies of patients with MN [16, 17]. Some of these non-anti-PLA2R autoantibodies may represent examples of 'epitope spreading', a common phenomenon in chronic autoimmune diseases. The pathogenetic role of these autoantibodies, that can frequently coexist with anti-PLA2R autoantibodies in primary MN, still remains to be clarified.

Secondary Membranous Nephropathy: Antigen-Antibody Systems

Circulating immune complexes, composed of an exogenous or endogenous (autologous) antigen not native to the kidney and its corresponding antibody, have been clearly shown to be capable of evoking a lesion of MN in experimental models of disease [1]. Classically, chronic serum sickness induced by repeated injections of BSA or bovine γ-globulin is the model of this disease [45]. Possible human examples of this chronic serum sickness model can be cited: (i) malignancy-associated MN where neo-epitopes of tumor-related antigen supply a constant source of antigen; (ii) drug-induced MN, such as mercury or gold, where a neo-antigen is created by the reaction of the metal with autologous proteins and their disulfide residues; (iii) chronic viral diseases, such as hepatitis B virus-related MN, where constantly circulation viral epitopes (HBc or HBe antigens) provide a source for formation of circulating immune complexes; (iv) thyroid-disorders (thyroglobulin as the antigen) and systemic autoimmune diseases, such as lupus erythematosus where dsDNA or nucleosomes serve as the sources for immune complex formation [46]. The exact mechanisms whereby the immune complexes in the circulation are concentrated in the subepithelial space of glomeruli are poorly understood. Theoretically, an oligo-valent antigen combined with a low-affinity antibody could dissociate and re-associate during migration from the circulation to the subepithelial space and be localized underneath slit-pore membrane between podocytes. Alternatively, in a process not requiring deposition of immune complexes from the circulation, a biophysical attraction (e.g. electrostatic charge interaction) between the antigen and constituents of the basement membrane or podocyte cell membrane might allow certain antigens to 'plant' in subepithelial sites to become targets of circulating antibody. Such a sequence has been posited to be, at least in part, responsible for MN in lupus erythematosus [46]. Very rarely, *megalin* (the antigen involved in experimental MN of rats – Heymann nephritis) has been suggested as a target antigen in human secondary MN (such as sickle cell nephropathy, with MN) [18], but *megalin* is not believed to be an antigen participating in cases of primary MN [47]. On occasion, anti-PLA2R1 autoantibodies have been observed in sarcoidosis [48] and lupus nephritis [49] as well as in malignancy-related MN [11] and in postallogeneic stem cell transplantation with a secondary MN [50]. Monoclonal immunoglobulins with unusual charge characteristics may bind to anionic sites in the subepithelial space like cationic BSA) [1, 3, 51]) and evoke a picture of MN [52]. In addition, a remarkable case of an IgG3κ monoclonal antibody that had PLA2R-binding characteristics and provoked MN has been reported [53], thus confirming the pathogenicity of antiPLA2R1 antibodies in man [54].

Concluding Remarks

For primary human MN, the delineation of the role of PLA2R1 and the immunogenetic foundations of susceptibility have transformed its diagnosis, prognostic evaluation and therapy, and raised the tantalizing possibility of avoidance of a recurrence of the disease in renal allografts. Research has now established the in situ mechanism of formation of subepithelial immunoglobulin containing deposits as the main event in primary MN, but other pathogenetic mechanisms may be operative as well, particularly in secondary MN. However, much still remains to be discovered. We still need to understand how the conformational (immunogenic) epitopes on native PLA2R1 become expressed and how do they elicit an autoantibody response? What is the molecular nature of the connection between PLA2R1 SNPs and the susceptibility to MN? Why is the autoantibody response so IgG4 subclass dominant in human primary MN and how do subtle abnormalities of glycosylation of the IgG4 antibodies alter their ability to activate the complement cascade? What are the molecular details that translate the in situ binding of autoantibody to podocytes into a perturbation in glomerular permselectivity? What molecular and cellular entities constitute the mediators of podocyte injury in primary MN? What is the spectrum of antigen-antibody systems that operate in primary MN? What distinct pathogenetic mechanisms are responsible for the development of secondary MN? Much progress has been made in recent years and the breakthroughs in identifying target antigens in human MN are likely to create new opportunities for unraveling the mysteries associated with this disorder.

Disclosure Statement

The author has no conflicts of interest to disclose.

References

1 Glassock RJ: The pathogenesis of idiopathic membranous nephropathy: a 50-year odyssey. Am J Kidney Dis 2010;56:157–167.
2 Glassock RJ: Secondary membranous glomerulonephritis. Nephrol Dial Transplant 1992;7(suppl 1):64–71.
3 Makker SP, Tramontano A: Idiopathic membranous nephropathy: an auto-immune disease. Semin Nephrol 2011;31:333–340.
4 Beck LH Jr, Salant DJ: Membranous nephropathy: recent travels and new roads ahead. Kidney Int 2010;77:765–770.
5 Beck LH Jr, Bonegio RG, Lambeau G, Beck DM, Powell DW, Cummins TD, Klein JB, Salant DJ: M-type phospholipase A_2 receptor as target antigen in idiopathic membranous nephropathy. N Engl J Med 2009;361:11–21.
6 Glassock RJ: Human idiopathic membranous nephropathy – a mystery solved? N Engl J Med 2009;361:81–83.
7 Ronco P, Debiec H: Antigen identification in membranous nephropathy moves toward targeted monitoring and new therapy. J Am Soc Nephrol 2010;21:564–569.
8 Ronco P, Debiec H: Membranous glomerulopathy: the evolving story. Curr Opin Nephrol Hypertens 2010;19:254–259.
9 Lambeau G, Lazdunski M: Receptors for a growing family of secreted phospholipases A_2. Trends Pharmacol Sci 1999;20:162–170.
10 Augert A, Payré C, de Launoit Y, Gil J, Lambeau G, Bernard D: The M-type receptor PLA2R regulates senescence through the p53 pathway. EMBO Rep 2009;10:271–277.
11 Qin W, Beck LH Jr, Zeng C, Chen Z, Li S, Zuo K, Salant DJ, Liu Z: Anti-phospholipase A_2 receptor antibody in membranous nephropathy. J Am Soc Nephrol 2011;22:1137–1143.
12 Hoxha E, Harendza S, Zahner G, Panzer U, Steinmetz O, Fechner K, Helmchen U, Stahl RA: An immunofluorescence test for phospholipase A_2 receptor antibodies and its clinical usefulness in patients with membranous glomerulonephritis. Nephrol Dial Transplant 2011;26:2526–2532.
13 Debiec H, Guigonis V, Mougenot B, Decobert F, Haymann JP, Bensman A, Deschênes G, Ronco PM: Antenatal membranous glomerulonephritis due to anti-neutral endopeptidase antibodies. N Engl J Med 2002;346: 2053–2060.
14 Prunotto M, Carnevali ML, Candiano G, Murtas C, Bruschi M, Corradini E, Trivelli A, Magnasco A, Petretto A, Santucci L, Mattei S, Gatti R, Scolari F, Kador P, Allegri L, Ghiggeri GM: Autoimmunity in membranous nephropathy targets aldose reductase and SOD2. J Am Soc Nephrol 2010;21:507–519.
15 Bruschi M, Carnevali ML, Murtas C, Candiano G, Petretto A, Prunotto M, Gatti R, Argentiero L, Magistroni R, Garibotto G, Scolari F, Ravani P, Gesualdo L, Allegri L, Ghiggeri GM: Direct characterization of target podocyte antigens and autoantibodies in human membranous glomerulonephritis: α-enolase and borderline antigens. J Proteomics 2011; 74:2008–2017.
16 Gitlits VM, Toh BH, Sentry JW: Disease association, origin, and clinical relevance of autoantibodies to the glycolytic enzyme enolase. J Investig Med 2001;49:138–145.
17 Wakui H, Imai H, Komatsuda A, Miura AB: Circulating antibodies against α-enolase in patients with primary membranous nephropathy. Clin Exp Immunol 1999;118:445–450.

18 Strauss J, Pardo V, Koss MN, Griswold W, McIntosh RM: Nephropathy associated with sickle cell anemia: an autologous immune complex nephritis. I. Studies on nature of glomerular-bound antibody and antigen identification in a patient with sickle cell disease and immune deposit glomerulonephritis. Am J Med 1975;58:382–387.
19 Debiec H, Lefeu F, Kemper MJ, Niaudet P, Deschênes G, Remuzzi G, Ulinski T, Ronco P: Early-childhood membranous nephropathy due to cationic bovine serum albumin. N Engl J Med 2011;364:2101–2110.
20 Hofstra JM, Beck LH Jr, Beck DM, Wetzels JF, Salant DJ: Anti-phospholipase A_2 receptor antibodies correlate with clinical status in idiopathic membranous nephropathy. Clin J Am Soc Nephrol 2011;6:1286–1291.
21 Beck LH Jr, Fervenza FC, Beck DM, Bonegio RG, Malik FA, Erickson SB, Cosio FG, Cattran DC, Salant DJ: Rituximab-induced depletion of anti-PLA2R autoantibodies predicts response in membranous nephropathy. J Am Soc Nephrol 2011;22:1543–1550.
22 Debiec H, Ronco P: PLA2R autoantibodies and PLA2R glomerular deposits in membranous nephropathy. N Engl J Med 2011;364: 689–690.
23 Svobodova B, Honsova E, Ronco P, Tesar V, Debeic H: Kidney biopsy is a sensitive tool for retrospective diagnosis of PLA2R-related membranous nephropathy. Nephrol Dial Transplant 2012 (E-pub ahead of print).
24 Hoxha E, Kneibler U, Stege G, Zahner D, Thiele I, Panzer U, Harendza S, Helmchen UM, Stahl RA: Enhanced expression of the M-type phospholipase A_2 receptor in glomeruli correlates with serum receptor antibodies in human membranous nephropathy. Kidney Int 2012;82:797–804.
25 Stanescu HC, Arcos-Burgos M, Medlar A, Bockenhauer D, Kottgen A, Dragomirescu L, Voinescu C, Patel N, Pearce K, Hubank M, Stephens HA, Laundy V, Padmanabhan S, Zawadzka A, Hofstra JM, Coenen MJ, den Heijer M, Kiemeney LA, Bacq-Daian D, Stengel B, Powis SH, Brenchley P, Feehally J, Rees AJ, Debiec H, Wetzels JF, Ronco P, Mathieson PW, Kleta R: Risk HLA-DQA1 and PLA2R1 alleles in idiopathic membranous nephropathy. N Engl J Med 2011;364:616–626.
26 Rodriguez EF, Cosio FG, Nasr SH, Sethi S, Fidler ME, Stegall MD, Grande JP, Fervenza FC, Cornell LD: The pathology and clinical features of early recurrent membranous glomerulonephritis. Am J Transplant 2012 (E-pub ahead of print).
27 Debiec H, Martin L, Jouanneau C, Dautin G, Mesnard L, Rondeau E, Mousson C, Ronco P: Autoantibodies specific for the phospholipase A_2 receptor in recurrent and de novo membranous nephropathy. Am J Transplant 2011; 11:2144–2152.
28 Kearney N, Podolak J, Matsumura L, Houghton D, Troxell M: Patterns of IgG subclass deposits in membranous glomerulonephritis in renal allografts. Transplant Proc 2011;43: 3743–3746.
29 Qu Z, Liu G, Li J, Wu LH, Tan Y, Zheng X, Ao J, Zhao MH: Absence of glomerular IgG4 deposition in patients with membranous nephropathy may indicate malignancy. Nephrol Dial Transplant 2012;27:1931–1937.
30 Omokawa A, Komatsuda A, Nara M, Fujiwara S, Sato R, Togashi M, Okuyama S, Sawada KI, Wakui H: Distribution of glomerular IgG subclass deposits in patients with membranous nephropathy and anti-U1 ribonucleoprotein antibody. Nephrol Dial Transplant 2012;27:1937–1941.
31 Cravedi P, Abbate M, Gagliardini E, Galbusera M, Buelli S, Sabadini E, Marasà M, Beck LH Jr, Salant DJ, Benigni A, D'Agati V, Remuzzi G: Membranous nephropathy associated with IgG4-related disease. Am J Kidney Dis 2011;58:272–275.
32 Segawa Y, Hisano S, Matsushita M, Fujita T, Hirose S, Takeshita M, Iwasaki H: IgG subclasses and complement pathway in segmental and global membranous nephropathy. Pediatr Nephrol 2010;25:1091–1099.
33 Ohtani H, Wakui H, Komatsuda A, Okuyama S, Masai R, Maki N, Kigawa A, Sawada K, Imai H: Distribution of glomerular IgG subclass deposits in malignancy-associated membranous nephropathy. Nephrol Dial Transplant 2004;19:574–579.
34 Kuroki A, Iyoda M, Shibata T, Sugisaki T: Th2 cytokines increase and stimulate B cells to produce IgG4 in idiopathic membranous nephropathy. Kidney Int 2005;68:302–310.
35 Aalberse RC, Stapel SO, Schuurman J, Rispens T: Immunoglobulin G4: an odd antibody. Clin Exp Allergy 2009;39:469–477.

36 Van der Zee JS, van Swieten P, Aalberse RC: Serologic aspects of IgG4 antibodies. II. IgG4 antibodies form small, nonprecipitating immune complexes due to functional monovalency. J Immunol 1986;137:3566–3571.
37 Van der Zee JS, van Swieten P, Aalberse RC: Inhibition of complement activation by IgG4 antibodies. Clin Exp Immunol 1986;64:415–422.
38 Hong M, Beck LH: Salant. Membranous nephropathy-associated anti-phospholipase A_2 receptor IgG4 auto-antibodies activate the lectin complement pathway (abstract). Am Soc Nephrol 2011;22:62A.
39 Cybulsky AV, Quigg RJ, Salant DJ: Experimental membranous nephropathy redux. Am J Physiol Renal Physiol 2005;289:F660–F671.
40 Penny MJ, Boyd RA, Hall BM: Permanent CD8+ T-cell depletion prevents proteinuria in active Heymann nephritis. J Exp Med 1998;188:1775–1784.
41 Beck LH Jr: Membranous nephropathy and malignancy. Semin Nephrol 2010;30:635–644.
42 Ponticelli C, Glassock RJ: Posttransplant recurrence of primary glomerulonephritis. Clin J Am Soc Nephrol 2010;5:2363–2372.
43 Stahl R, Hoxha E, Fechner K: PLA2R autoantibodies and recurrent membranous nephropathy after transplantation. N Engl J Med 2010;363:496–498.
44 Murtas C, Bruschi M, Candiano G, Moroni G, Magistroni R, Magnano A, Brunoi F, Radice F, Furci L, Argentiero L, Carnevali ML, Messa P, Scolari F, Sinico RA, Gesualdo L, Fervenza FC, Allegri L, Ravani P, Ghiggeri GM: Co-existence of different circulating anti-podocye antibodies in membranous nephropathy. Clin J Am Soc Nephrol 2012;7: 1395–1400.
45 Dixon FJ, Feldman JD, Vazquez JJ: Experimental glomerulonephritis. The pathogenesis of a laboratory model resembling the spectrum of human glomerulonephritis. J Exp Med 1961;113:899–920.
46 Van der Vlag, Berden J: Lupus nephritis: role of anti-nucleosome autoantibodies. Semin Nephrol 2011;31:376–389.
47 Whitworth JA, Leibowitz S, Kennedy MC, Cameron JS, Evans DJ, Glassock RJ, Schoenfeld LS: Absence of glomerular renal tubular epithelial antigen in membranous glomerulonephritis. Clin Nephrol 1976;5:159–162.
48 Knehtl M, Debiec H, Kamgang P, Callard P, Cadranel J, Ronco P, Boffa JJ: A case of phospholipase A_2 receptor-positive membranous nephropathy preceding sarcoid-associated granulomatous tubulointerstitial nephritis. Am J Kidney Dis 2011;57:140–143.
49 Gunnarsson I, Schlumberger W, Rönnelid J: Antibodies to M-type phospholipase A_2 receptor (PLA2R) and membranous lupus nephritis. Am J Kidney Dis 2012;59:585–586.
50 Huang X, Qin W, Zhang M, Zheng C, Zeng C, Liu Z: Detection of anti-PLA2R autoantibodies and IgG subclasses in post-allogeneic hematopoietic stem cell transplantation membranous nephropathy. Am J Med Sci 2012 (E-pub ahead of print).
51 Border WA, Ward HJ, Kamil ES, Cohen AH: Induction of membranous nephropathy in rabbits by administration of an exogenous cationic antigen. J Clin Invest 1982;69:451–461.
52 de Seigneux S, Bindi P, Debiec H, Alyanakian MA, Aymard B, Callard P, Ronco P, Aucouturier P: Immunoglobulin deposition disease with a membranous pattern and a circulating monoclonal immunoglobulin G with charge-dependent aggregation properties. Am J Kidney Dis 2010;56:117–121.
53 Recurrent membranous nephropathy in an allograft caused by IgG3κ targeting the PLA2 receptor. J Am Soc Nephrol 2012;23:1949–1954.
54 Beck LH Jr: Monocloanl anti-PLA2R and recurrent membranous nephropathy: another piece of the puzzle. J Am Soc Nephrol 2012; 23:1911–1913.

Richard J. Glassock, MD
8 Bethany
Laguna Niguel
CA 92677 (USA)
E-Mail Glassock@cox.net

Chen N (ed): New Insights into Glomerulonephritis.
Contrib Nephrol. Basel, Karger, 2013, vol 181, pp 143–151 (DOI: 10.1159/000349976)

Short-Term Complications of Membranous Nephropathy

Sean Barbour[a] · Heather Reich[b] · Daniel Cattran[b]

[a]Division of Nephrology, Department of Medicine, University of British Columbia, Vancouver, B.C. and [b]Division of Nephrology, Department of Medicine, University of Toronto, Toronto, Ont., Canada

Abstract

Short-term complications of membranous nephropathy (MN) are important components of the process and their proper management can significantly alter the outcomes of patients with this disease. These complications vary from the non-specific ones associated with many types of chronic kidney disease such as hypertension, proteinuria and the symptoms associated with the nephrotic syndrome including edema and dyslipidemia. These are briefly discussed in the chapter. In addition, there are some specific issues more commonly associated with MN than other types of glomerulonephritis. This includes a hypercoagulable state and the associated thromboembolism. There is more recent information in this domain and this is included in the chapter. The associated increased risk of infection in MN patients is the final chapter component. This section is included to remind us not only of the direct impact of that disease process on reducing resistance to infection, but also to indicate the need to consider prophylactic antibiotics when the addition of potent immunosuppressive drug treatment is required for these patients.

Copyright © 2013 S. Karger AG, Basel

Many of the complications related to membranous nephropathy (MN) are a consequence of the clinical presentation rather than the pattern of injury at the histologic level. Their management should always be considered given they frequently impact on the natural history of the disease. These include measures to treat blood pressure, reduce proteinuria, control edema, and address other metabolic and thrombophilic consequences of the nephrotic syndrome. Many of complications can significantly alter morbidity and even mortality. Successful management may modulate or even negate the need for immunosuppressive drugs. Some of these complications relate to all variants of glomerulonephritis (GN) not just MN, especially if associated with chronic or progressive kidney failure (CKD).

Hypertension

Although hypertension is not generally a problem in the early stage of MN, it can occur especially in the elderly where the hypertension may coexist or when the MN is associated with progressive kidney failure. This will not be discussed aside from emphasizing that blood pressure control is a critical element both in terms of protecting against the accelerated cardiovascular risk of patients with high-grade proteinuria and that its control will help delay progressive loss of renal function. Modifications of other life style habits of the patient also need to be instituted including salt restriction, weight normalization, and smoking cessation all factors considered integral to good blood pressure control. The ideal goal for blood pressure is not firmly established in MN, but current recommendations indicate that on average 130/80 mm Hg should be targeted. There is some data to support a lower target of 125/75 mm Hg if proteinuria exceeds 1 g/day [1]. There is both theoretical and clinical level support for considering angiotensin-converting enzyme inhibitors (ACEi) or angiotensin receptor blockers (ARB) as first-line therapy [2]. Children with MN are rare if present and associated with hypertensive their blood pressure should be reduced to below the 50th percentile for both systolic and diastolic pressure [3].

Dyslipidemia

Treatment of the abnormal lipid profile commonly seen in patients with MN should follow the guidelines that apply to patients at high risk for the development of cardiovascular disease. Such abnormalities in nephrotic patients with MN usually begin early in the disease and persist as long as the proteinuria remains high and thus is considered an early complication. It does abate if the proteinuria is controlled below the nephrotic range or is normalized. The clinical implication of the dyslipidemia is manifested by accelerated vascular disease and cardiovascular complications, which take years to develop but since the natural history of individual patients is unknown the recommendation is when present, treatment should be instituted. Statins (HMG-CoA reductase inhibitors) are well tolerated and effective in correcting the lipid profile, although probably related to the time lag between disease onset and resultant problems, they have not yet been proven to reduce cardiovascular events in patients with MN and the nephrotic syndrome. The use of statins carries a somewhat increased risk of myalgia/myositis when they are combined with calcineurin inhibitors, a drug group commonly used in the treatment of MN patients with severe proteinuria.

Proteinuria

Reduction in proteinuria in MN is important, as it reflects not only control of the primary disease, but potentially a reduction in podocyte damage an important element of subsequent glomerular scarring. Many studies have suggested that the loss of kidney function in MN can largely be prevented if proteinuria is stabilized at values <3–4 g/day. In MN, a reduction of proteinuria to a non-nephrotic range commonly leads to normalization of serum albumin. This, in turn, corrects many of the patients' symptoms and the associated metabolic complications of the nephrotic syndrome leading to improved quality of life. The antiproteinuric agents of choice are as with hypertension the ACEi or ARB, which may reduce proteinuria by 30–40% provided the patient complies with dietary salt restriction. There is little evidence to suggest that ACEi differ from ARBs in this respect. The combination of the two may result in additive antiproteinuric activity but the risks versus benefits of this strategy are unknown, especially if the GFR is low. Since these classes of drugs lower GFR through hemodynamic mechanisms, a 10–20% increase in serum creatinine is often observed and unless serum creatinine continues to rise, this moderate and acute increase does not mean worsening MN disease, and should not prompt stopping or reducing the medication. Adequate dietary protein should be ensured in the proteinuric patient (recommended 0.8–1.0 g/kg daily) plus replacement of their daily urinary protein loss with adequate replacement of calories by carbohydrate intake to allow maximize utilization of that protein. There is evidence that a reduction in proteinuria to subnephrotic range in MN, no matter how achieved (spontaneous or with specific treatment) is associated with both a significant slowing of the rate of progression and improvement in long-term kidney survival [4].

Edema Management

The mainstay of edema control is diuretics accompanied by moderate restriction of dietary sodium (1.5–2 g sodium/day). Patients who are nephrotic secondary to MN are often diuretic-resistant even if the GFR is normal: loop diuretics administered orally as a once- or twice-daily routine is the initial preferred routine. In cases of severe nephrotic syndrome, commonly encountered in MN, gastrointestinal absorption of the diuretic may be variable because of gut wall edema, and intravenous diuretics may be necessary to initiate an effective diuresis. Additional therapy such as combining a loop diuretic with a thiazide diuretic, such as metolazone, can often be an effective oral regimen that may overcome resistance to a single agent. Although intravenous albumin infusions have been used in the past in combination with diuretics to treat diuretic resistance, their value is questionable. Rarely, mechanical ultrafiltration using hemodialysis may be required for resistant edema. Although generally not an issue, rapid depletion of the intravascular volume and subsequent hypovolemia and hypotension should be avoided. This can be a particularly important clinical issue in the pediatric and elderly populations and in patients on high-dose ACEi and/or ARB drugs. This emphasizes the need to take care that the fluid removal is controlled and gradual. The frequency of this complication increases significantly in the elderly and/or if a comorbid condition such as diabetes mellitus and hypertension exists. This scenario has the potential to result in both hypovolemic shock and acute ischemic kidney injury.

Hypercoagulability and Venous Thromboembolism

Venous thromboembolism (VTE) includes deep vein thrombosis (DVT), pulmonary embolism (PE), and renal vein thrombosis (RVT). While RVT can often be asymptomatic, in 20–25% of cases it is complicated by concurrent PE [5]. Patients with MN are thought to be at higher risk of VTE. Further, the mortality of MN patients with VTE is 10% per year, substantially higher than MN patients in general [6].

Risk of VTE in MN

The risk of VTE in patients with the nephrotic syndrome has been variably described between 2 and 60%, with a disproportionately higher risk in patients with MN [7–11]. Although there is substantial variability in the reported prevalence in the literature, there is a trend for more VTE events in patients with MN compared to other causes of nephrotic syndrome. In an early study by Llach et al. [10], amongst 151 nephrotic patients, 30% had a VTE and 22% had a RVT. Of the patients who had a RVT, 88% were asymptomatic and 61% had MN. These findings are representative of the literature on this topic where VTE averaged 44.7% in MN with RVT being particularly common in MN (37%) compared to other glomerular diseases (18.8–26.2% in FSGS, minimal change disease and membranoproliferative GN) [5, 6, 12]. The actual clinical impact in this largely asymptomatic cohort is however harder to assess. Furthermore, these studies did not determine if the increased risk observed in patients with MN was due to more severe proteinuria or hypoalbuminemia, or was specifically related to the underlying histopathology of MN.

Several recent studies addressed the disease-specific risk of clinically relevant venous thrombotic events in MN patients. Mahmoodi et al. [13] followed a cohort of 298 patients with nephrotic syndrome, including 72 with MN, for symptomatic VTE. They found an occurrence rate of 9.7% at a median of 10 months after presentation. More recently, an investigation of an even large cohort of 1,313 patients with idiopathic GN including 385 with MN was carried out to look for such events [14]. Venous thrombosis occurred in 7.9% of patients with MN at a median of 5 months after presentation, compared to only 3.0% of patients with FSGS and 0.4% of those with IgA nephropathy [14]. Although patients with MN tended to have more proteinuria, lower albumin and a more frequent history of malignancy, multivariable analysis suggested that after accounting for these differences, patients with MN maintained an 11-fold increased risk of VTE compared to those with IgA nephropathy and 2-fold increased risk compared to those with FSGS [14]. In contrast to earlier reports dominated by asymptomatic RVT, this study found 68% were symptomatic thrombotic events, i.e. DVT or PE [14]. The current largest review focused solely on MN described the VTE risk in 898 MN patients. 7% had symptomatic VTE at a median of 3.8 months after presentation [15]. In both of these recent studies, the average proteinuria at the time of the thrombotic events was 10 g/day and serum albumin was 22–25 g/l [14, 15]. In summary, recent literature suggests that clinically significant venous thrombosis affects 7–10% of patients with MN, is more common in MN compared to other types of GN, is not explained by differences in proteinuria or hypoalbuminemia, and that this complication most commonly occurs within the first year of the nephrotic phase of the disease.

Hypoalbuminemia and the Risk of VTE in MN

In addition to traditional risk factors for VTE, hypoalbuminemia appears to be a strong predictor of venous thrombosis in patients with MN. In multivariable models, hypoalbuminemia trumped proteinuria as a risk factor for VTE [13, 14]. Lionaki et al. [15] found for each 10 g/l decrease in serum albumin, the risk of VTE in MN doubled. This risk noticeably increased when serum albumin was <28 g/l, where the risk of VTE was 9.4% compared to only 3.2% in those whose albumin was above this level [15].

Pathophysiology of VTE in MN

Hypercoagulability in the nephrotic syndrome may be due to a combination of changes in coagulation factors, venous stasis, hemoconcentration, increased platelet aggregation, and decreased fibrinolysis [5, 8, 16, 17]. Antithrombin III has a similar molecular weight and charge as albumin, with antithrombin III levels declining in parallel with serum albumin due to urinary losses [18, 19]. Levels of coagulation factors in the nephrotic syndrome reflect a balance between increased losses in the urine and increased hepatic synthesis stimulated by hypoalbuminemia, resulting in increased levels of fibrinogen, factor V, and factor VIII [5, 19]. Fibrinolysis may be impaired because albumin is a necessary cofactor for the interaction between plasminogen, fibrin and tissue plasminogen activator [5]. An increased propensity to thrombosis in the renal vein may be due to a combination of upstream losses of coagulation factors in the urine, hemoconcentration, and activation of the coagulation system as blood passes through regions of active inflammation in the glomerular capillary loops [5]. Reasons for a higher risk of VTE in MN are not known but theories include qualitative differences in the molecular weight of proteins lost in the urine, antibodies to α-enolase that may reduce fibrinolysis, and increased glomerular expression of plasminogen activator inhibitor [20–23].

Prophylactic Anticoagulation in Patients with MN

There are no randomized controlled trials to help guide the use of prophylactic anticoagulation for the prevention of VTE in high-risk patients with MN. A decision analysis by Bellomo and Atkins [6] concluded a 32% relative reduction in morbid events by a 1-year period of prophylactic anticoagulation, but this study assumed a VTE risk of 40%. More recent and larger studies suggest a much lower VTE risk of only 7–10% [13–15, 24]. A second-decision analysis assumed a DVT or RVT risk of 18% per year, and concluded a 2.5-month gain in quality-adjusted survival for a 50-year-old male prophylactically anti coagulated with warfarin for 2 years [25]. If VTE risk is reduced to 7.2% per year, compatible with the more recent data, the benefit of prophylactic anticoagulation was lost [25]. Further, risk of serious hemorrhage on anticoagulation has not been studied in nephrotic patients although the risk of bleeding has been noted to increase 1.5-fold even with mild renal insufficiency [26]. On balance, routine prophylactic anticoagulation cannot be recommended in all patients with MN. However, although unproven, the above data supporting a much higher risk of thromboembolism when albumin is <28 g/l early in the course of disease, suggests these specific patients might benefit from short-term anticoagulation [15].

The recent KDIGO clinical practice guidelines for GN have recommended that consideration be given to prophylactic warfarin anticoagulation for the duration of the nephrotic syndrome in patients with high-risk MN (albumin <25 g/l), especially with an additional risk factor for thrombosis, such as a previous history of VTE, immobility, or known thrombophilia [27]. Until such time as an appropriately designed randomized trial addresses this issue, the KDIGO recommendations are reasonable.

Risk of Infection

The concern about bacterial infections in nephrotic patients remains an important and often forgotten risk. Although this is most commonly seen in nephrotic children with ascites, in a population where idiopathic MN is uncommon it can occur in adults even without ascites. In all age groups when ascites does occur, suddenly worsens, or is associated with abdominal pain, the ascitic fluid should be examined microscopically and cultured for spontaneous bacterial peritonitis. Bacteremia can occur even when the symptoms are localized to the abdomen. Intravenous antibiotics should be started following obtaining cultures and the treatment regimen should include penicillin (to treat pneumococcal infection). If repeated infections occur, serum immunoglobulins should be measured and if serum IgG is <6 g/l, intravenous immunoglobulin treatment should be considered. There is low-level evidence to support monthly administration of intravenous immunoglobulin targeting a serum IgG above that level [28]. Those with MN and nephrotic syndrome are at increased risk of invasive pneumococcal infection and should receive pneumococcal vaccination as well as the annual influenza vaccination. The antibody response is impaired by concurrent corticosteroid treatment. As in all patients on immunosuppressive drugs, live vaccines (measles, mumps, rubella, varicella, rotavirus, yellow fever) are contraindicated. These vaccines can be implemented if the prednisone is at a dosage of <20 mg/day and/or the other immunosuppressive agents have been discontinued for a minimum of 1–3 months [29]. The combination antibiotic trimethoprim/sulfamethoxazole (Septra) should be given to protect against *Pneumocystis jirovecii* pneumonia in the scenario where the nephrotic syndrome is severe and cytotoxic or B-cell depletion therapy is in use [30].

References

1 Upadhyay A, Earley A, Haynes SM, Uhlig K: Systematic review: blood pressure target in chronic kidney disease and proteinuria as an effect modifier. Ann Intern Med 2011;154: 541–548.

2 Agarwal A, Haddad N, Hebert LA: Progression of kidney disease: diagnosis and management; in Molony DA, Craig JC (eds): Evidence-Based Nephrology, ed 1. Hoboken/NJ, Wiley, 2008, pp 311–322.

3 National High Blood Pressure Education Program Working Group on High Blood Pressure in Children and Adolescents: The fourth report on the diagnosis, evaluation, and treatment of high blood pressure in children and adolescents. Pediatrics 2004; 114(suppl):555–576.

4 Troyanov S, Wall CA, Miller JA, Scholey JW, Cattran DC: Idiopathic membranous nephropathy: definition and relevance of a partial remission. Kidney Int 2004;66:1199–1205.

5 Singhal R, Brimble KS: Thromboembolic complications in the nephrotic syndrome: pathophysiology and clinical management. Thromb Res 2006;118:397–407.

6 Bellomo R, Atkins RC: Membranous nephropathy and thromboembolism: is prophylactic anticoagulation warranted? Nephron 1993;63:249–254.

7 Kayali F, Najjar R, Aswad F, Matta F, Stein PD: Venous thromboembolism in patients hospitalized with nephrotic syndrome. Am J Med 2008;121:226–230.

8 Llach F: Hypercoagulability, renal vein thrombosis, and other thrombotic complications of nephrotic syndrome. Kidney Int 1985;28:429–439.

9 Llach F, Koffler A, Finck E, Massry SG: On the incidence of renal vein thrombosis in the nephrotic syndrome. Arch Intern Med 1977; 137:333–336.

10 Llach F, Papper S, Massry SG: The clinical spectrum of renal vein thrombosis: acute and chronic. Am J Med 1980;69:819–827.

11 Wagoner RD, Stanson AW, Holley KE, Winter CS: Renal vein thrombosis in idiopathic membranous glomerulopathy and nephrotic syndrome: incidence and significance. Kidney Int 1983;23:368–374.

12 Chugh KS, Malik N, Uberoi HS, Gupta VK, Aggarwal ML, Singhal PC, et al: Renal vein thrombosis in nephrotic syndrome – a prospective study and review. Postgrad Med J 1981;57:566–570.

13 Mahmoodi BK, ten Kate MK, Waanders F, Veeger NJ, Brouwer JL, Vogt L, et al: High absolute risks and predictors of venous and arterial thromboembolic events in patients with nephrotic syndrome: results from a large retrospective cohort study. Circulation 2008; 117:224–230.

14 Barbour SJ, Greenwald A, Djurdjev O, Levin A, Hladunewich MA, Nachman PH, et al: Disease-specific risk of venous thromboembolic events is increased in idiopathic glomerulonephritis. Kidney Int 2012;81:190–195.

15 Lionaki S, Derebail V, Hogan SL, Barbour S, Lee T, Hladunewich M, et al: Venous thromboembolism in patients with membranous nephropathy. Clin J Am Soc Nephrol 2012;7: 43–51.

16 Sirolli V, Ballone E, Garofalo D, Merciaro G, Settefrati N, Di Mascio R, et al: Platelet activation markers in patients with nephrotic syndrome. A comparative study of different platelet function tests. Nephron 2002;91:424–430.

17 Remuzzi G, Mecca G, Marchesi D, Livio M, de Gaetano G, Donati MB, et al: Platelet hyperaggregability and the nephrotic syndrome. Thromb Res 1979;16:345–354.

18 Kauffmann RH, Veltkamp JJ, Van Tilburg NH, Van Es LA: Acquired antithrombin III deficiency and thrombosis in the nephrotic syndrome. Am J Med 1978;65:607–613.

19 Citak A, Emre S, Sairin A, Bilge I, Nayir A: Hemostatic problems and thromboembolic complications in nephrotic children. Pediatr Nephrol 2000;14:138–142.

20 Hamano K, Iwano M, Akai Y, Sato H, Kubo A, Nishitani Y, et al: Expression of glomerular plasminogen activator inhibitor type 1 in glomerulonephritis. Am J Kidney Dis 2002; 39:695–705.

21 Bazzi C, Petrini C, Rizza V, Arrigo G, Beltrame A, D'Amico G: Characterization of proteinuria in primary glomerulonephritides: urinary polymers of albumin. Am J Kidney Dis 1997;30:404–412.

22 Wakui H, Imai H, Komatsuda A, Miura AB: Circulating antibodies against α-enolase in patients with primary membranous nephropathy. Clin Exp Immunol 1999;118:445–450.
23 Lopez-Alemany R, Longstaff C, Hawley S, Mirshahi M, Fabregas P, Jardi M, et al: Inhibition of cell surface mediated plasminogen activation by a monoclonal antibody against α-enolase. Am J Hematol 2003;72:234–242.
24 Glassock RJ: Prophylactic anticoagulation in nephrotic syndrome: a clinical conundrum. J Am Soc Nephrol 2007;18:2221–2225.
25 Sarasin FP, Schifferli JA: Prophylactic oral anticoagulation in nephrotic patients with idiopathic membranous nephropathy. Kidney Int 1994;45:578–585.
26 Fang MC, Go AS, Chang Y, Borowsky LH, Pomernacki NK, Udaltsova N, et al: A new risk scheme to predict warfarin-associated hemorrhage: The ATRIA (Anticoagulation and Risk Factors in Atrial Fibrillation) Study. J Am Coll Cardiol 2011;58:395–401.
27 KDIGO Clinical Practice Guidelines for Glomerulonephritis. Kidney Int 2012;S2:139–274.
28 Ogi M, Yokoyama H, Tomosugi N, Hisada Y, Ohta S, Takaeda M, et al: Risk factors for infection and immunoglobulin replacement therapy in adult nephrotic syndrome. Am J Kidney Dis 1994;24:427–436.
29 Furth SL, Arbus GS, Hogg R, Tarver J, Chan C, Fivush BA: Varicella vaccination in children with nephrotic syndrome: a report of the Southwest Pediatric Nephrology Study Group. J Pediatr 2003;142:145–148.
30 Philibert D, Cattran D: Remission of proteinuria in primary glomerulonephritis: we know the goal but do we know the price? Nat Clin Pract Nephrol 2008;4:550–559.

Daniel Cattran
Suite 1256, 11 CSB, University Health Network
585 University Avenue
Toronto, ONT M5G2N2 (Canada)
E-Mail daniel.cattran@uhn.ca

Chen N (ed): New Insights into Glomerulonephritis.
Contrib Nephrol. Basel, Karger, 2013, vol 181, pp 152–162 (DOI: 10.1159/000348475)

Tacrolimus Combined with Corticosteroids in Idiopathic Membranous Nephropathy: A Randomized, Prospective, Controlled Trial

Jing Xu · Wen Zhang · Yaowen Xu · Pingyan Shen · Hong Ren · Weiming Wang · Xiao Li · Xiaoxia Pan · Nan Chen

Department of Nephrology, Ruijin Hospital, School of Medicine, Shanghai Jiao Tong University, Shanghai, PR China

Abstract

Although idiopathic membranous nephropathy (IMN) is the most common cause of adult-onset nephrotic syndrome, the management of IMN remains controversial. The aim of this prospective study was to compare the efficacy and drug safety of tacrolimus with that of cyclophosphamide (CTX; control group) in IMN patients receiving corticosteroid therapy. A total of 100 IMN patients with nephrotic syndrome were randomly assigned to receive a combination of corticosteroid therapy and either CTX or tacrolimus. During a follow-up period of at least 18 months, the remission rate after 2 months in the tacrolimus group was 65.1%, which was higher than that of the CTX group (44.2%) ($p = 0.02$). The mean time to partial or complete remission was 2.20 months in the tacrolimus group and 3.92 months in the CTX group ($p < 0.001$). We also found significantly greater improvements in the serum albumin levels in the tacrolimus group compared with the CTX group at the 2-month ($p = 0.003$) and 3-month time points ($p = 0.01$). The serum creatinine levels remained stable in both groups. Although remission was quicker and more common in the tacrolimus group (compared with the CTX group) before 3 months, there was no superiority of tacrolimus after 6 months. Glucose intolerance, urinary tract infections, and pneumonia were the major side effects observed in this study. All of the side effects were mild and controlled, and there were fewer side effects in the tacrolimus group compared with the CTX group, indicating a better treatment tolerance in the tacrolimus group.

Copyright © 2013 S. Karger AG, Basel

Idiopathic membranous nephropathy (IMN) is one of the most common causes of nephrotic syndrome (NS) in adults [1]. IMN has been reported in 36.8% of 1,203 primary NS patients in Japan [2]. The proportion of primary glomerular disease patients with IMN is 9.89–13.3% in mainland China [3, 4]. Furthermore, the prevalence of IMN has been progressively increasing over time. Although 30% of patients experience the spontaneous complete (CR) or partial remission (PR) of NS [5], 30–40% of patients progress towards end-stage renal disease (ESRD) within 5–15 years [6]. Kidney Disease: Improving Global Outcomes (KDIGO) recommends that the initial therapy for IMN be started only in patients with NS whose urinary protein excretion persistently exceeds 4 g/day, remains at >50% of the baseline value, and does not show a progressive decline during antihypertensive and antiproteinuric therapy over an observation period of at least 6 months (1B) or in patients with severe, disabling, or life-threatening symptoms related to their NS (1C) [7]. Although a number of randomized, controlled trials of different immunosuppressive agents are underway, the treatment of IMN remains highly controversial, and there is an acute need for agents with a higher response rate and fewer adverse effects.

The immunosuppressant tacrolimus is a calcineurin inhibitor with widespread use for the prevention of graft rejection in transplantation medicine. Tacrolimus inhibits the activation of an essential transcription factor for the expression of cytokine genes in T cells, leading to the decreased production of cytokines such as interleukin-2 and interferon-γ [8]. Because T-cell activation plays a crucial role in the pathogenesis of inflammatory glomerular diseases, there is an increasing number of reports on the use of tacrolimus in idiopathic NS and secondary glomerulonephritis. In a placebo-controlled, randomized trial, tacrolimus monotherapy was recently suggested to be a useful therapeutic option for patients with IMN [9]. Furthermore, several trials treating IMN with tacrolimus and corticosteroids are being performed to compare a tacrolimus-corticosteroid combination with a cyclophosphamide (CTX)-steroid combination. To evaluate the efficacy and safety of tacrolimus, we conducted a prospective, randomized, controlled trial in patients with nephrotic IMN.

Materials and Methods

Patients

This study was a prospective, randomized, controlled trial conducted at Ruijin Hospital from June 2007 to October 2012. The study protocol was approved by the Shanghai Ruijin Hospital Ethics Committee, and all the patients signed an informed consent form. The inclusion criteria were the following: (1) biopsy-proven IMN patients have severe NS or renal dysfunction with moderate proteinuria or severe oedema, (2) age between 18 and 80 years, and (3) no immunosuppressive treatment in the previous 6 months.

Definitions: Severe NS referred to a 24-hour urinary protein (24-hour Upro) excretion ≥5 g or albumin (ALB) ≤25 g/l on admission. NS patients with renal dysfunction were enrolled in our study if they had an estimated glomerular filtration rate (eGFR) of 15–60 ml/min at the time of renal biopsy. Renal dysfunction patients without NS were enrolled if their eGFR had declined to <60 ml/min at the observation 6 months after the renal biopsy. Severe oedema is defined as having severe hydrothorax or pyoperitoneum.

The following exclusion criteria were applied: (1) systemic disease, such as lupus, rheumatoid arthritis, Sjögren's syndrome, hyperthyroidism, or another autoimmune disease; (2) malignancy, including bronchiogenic carcinoma, gastrointestinal tumors, prostate carcinoma, and breast carcinoma; (3) infection, such as hepatitis B or C, tuberculosis, or syphilis; (4) the presence of drugs or toxicants, such as organic gold, mercury, or penicillamine; (5) a history of diabetes with a fasting blood glucose level >6.2 mmol/l on admission; (6) chronic kidney disease stage 4; (7) alanine aminotransferase and/or aspartate aminotransferase levels twice the normal range; (8) pregnancy or inadequate contraception, and (9) life-threatening complications of NS such as a severe infection or heart failure.

The criteria for removal from the study included the following: no response (NR) of the patient after 6 months of treatment, followed by a switch to another immunosuppressive agent; NS relapse; pregnancy, or reaching the secondary endpoint.

The clinical data were collected from the time of the renal biopsy until death, the development of ESRD, or the end of the follow-up period (the last clinical visit before October 2012). The following variables were measured at the follow-up time points: serum creatinine (SCr, μmol/l), eGFR (using an equation from the Modification of Diet in Renal Disease (MDRD) study, ml/min/1.73 m^2), 24-hour Upro excretion rate (g/24 h), serum ALB (g/l), serum triglycerides (mmol/l), serum cholesterol (mmol/l), and serum uric acid (μmol/l). The following complicated MDRD equation was used to calculate the eGFR: eGFR (ml/min/1.73 m^2) = 170 × SCr$^{-0.999}$ × age$^{-0.176}$ (×0.762 if female) × blood urea nitrogen (BUN)$^{-0.170}$ × ALB$^{0.318}$ [10]. Patients' characteristics (including age, sex, height, weight, medical history, complications of NS, and blood pressure) were recorded at the beginning of the study (at the initial renal biopsy) and during the observation period. Administration of immunosuppressive medications, the patients' response to the treatments, and the probable side effects were all recorded. Follow-up visits were scheduled at 1, 2, 3, 6, 12, and 18 months after the renal biopsy.

Treatment Protocols

The patients were randomly assigned to the CTX group or the tacrolimus group, and all the patients were followed up for at least 18 months. The patients in the CTX group received an initial CTX dosage of 0.5–0.75 g/m^2/month, with a maximum dosage of 1.0 g/month. The initial prednisone dosages were 1 mg/kg/day (maximum of 70 mg/day) for CTX group patients <65 years and 0.5 mg/kg/day for patients >65 years. The CTX treatment lasted 9 months, with pulsed intravenous CTX administered once every month initially for 6 months and then once every 2 or 3 months. Some patients were treated with angiotensin-converting enzyme inhibitors (ACEIs) and/or angiotensin receptor blockers (ARBs) to control their blood pressure.

The patients in the tacrolimus group received an initial tacrolimus dosage of 0.1 mg/kg/day, which was then adjusted in accordance to the measured serum concentration. These patients visited the doctor every 2 weeks during the first month of treatment to adjust the tacrolimus dose according to their whole blood trough levels. The tacrolimus group patients also received an initial prednisone dosage of 0.5 mg/kg/day. The tacrolimus was slowly tapered and was maintained longer than the prednisone.

Renal Pathology

The renal specimens were evaluated using direct immunofluorescence, light and electron microscopy. Silver methenamine staining and the presence of typical immunofluorescence findings were used to estimate the IMN stage. Interstitial and tubular lesions were scored semiquantitatively as (–) for 0%, (+) for 1–25%, (++) for 26–50%, and (+++) for >50%.

Endpoints

The primary endpoint was CR or PR. CR was defined as proteinuria <0.5 g/day with normal renal function, and PR was defined as proteinuria of 0.5–3.5 g/day that had declined to ≤50% of the baseline value with stable renal function. If neither CR nor PR was achieved, the outcome was defined as NR. Relapse in patients who had achieved CR or PR was defined as a 24-hour Upro >3.5 g in two consecutive measurements or serum ALB <25 g/l. We compared the remission rates of the two groups at 1, 2, 3, 6, 12, and 18 months.

The secondary endpoint was estimated renal survival, defined as a 50% increase in the baseline SCr concentration.

Statistical Analysis

Continuous variables were compared using the independent samples t test, and the results are expressed as the means (±SD). Missing values were replaced by serial means. Categorical variables were compared using the χ^2 test. A p value <0.05 was considered significant. For the estimation of renal survival, the endpoint was defined as patient death or an abnormal SCr level of at least double of the baseline level. The statistical analysis were performed using SPSS 11.0 software (Chicago, Ill., USA).

Results

A total of 100 patients who had biopsy-proven IMN and fulfilled the trial inclusion criteria were randomly assigned to receive treatment with either tacrolimus (tacrolimus group, n = 48) or CTX (CTX (control) group, n = 52) (fig. 1). The demographic, histological, and biochemical parameters at baseline are listed in table 1. Most of the patients had previously been treated with ACEIs/ARBs. As shown in table 1, the baseline SCr tended to be higher in the CTX group than in the tacrolimus group, although this difference was not significant. During the 18-month study period, 9 patients were removed from the tacrolimus group (including 6 patients with NR to treatment who were switched to other immunosuppressive agents and 3 relapsing patients), whereas 5 patients were removed from the CTX group (including 3 with NR, 1 with NS relapse, and 1 who became pregnant).

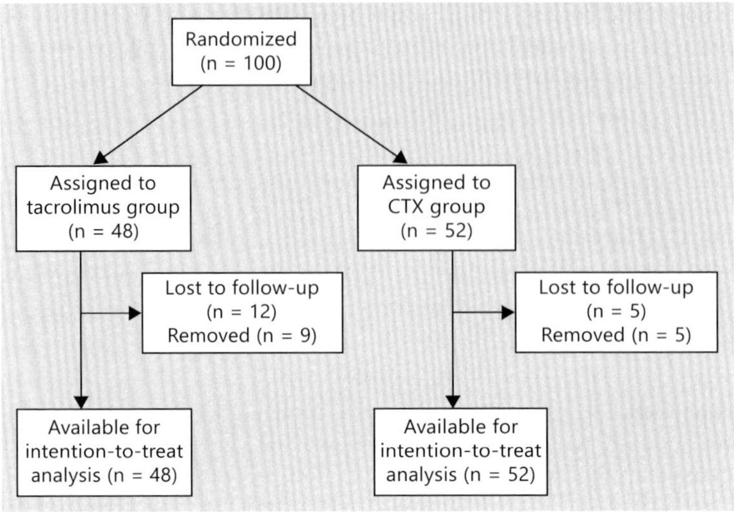

Fig. 1. Trial profile.

Endpoints

The intention-to-treat analysis showed that significantly higher rates of CR and PR were observed in the tacrolimus-treated patients at 2 months, but there were no significant differences at the other follow-up times. The rates of remission (CR plus PR) in the tacrolimus and CTX groups were 39.6 and 36.5%, respectively, at 1 month ($p = 0.16$); 65.1 and 44.2%, respectively, at 2 months ($p = 0.02$); 70.7 and 57.7%, respectively, at 3 months ($p = 0.08$); 91.4 and 83.7%, respectively, at 6 months ($p = 0.16$); 90.6 and 87.5%, respectively, at 1 year ($p = 0.26$), and 87.5 and 92.9%, respectively, at 18 months ($p = 0.26$). After 2 years of follow-up, the percentages of remission were 76.9 and 57.7% in the tacrolimus and CTX groups, respectively ($p = 0.15$). In the tacrolimus group, the remission rate gradually increased for 6 months and decreased moderately thereafter. In the CTX group, the remission rate gradually increased from 1 to 18 months (table 2). The mean time to PR or CR was 2.20 months in the tacrolimus group and 3.92 months in the CTX group ($p < 0.001$).

After 1 month of treatment, the urinary protein excretion decreased from 5.10 ± 2.20 to 3.40 ± 2.04 g/24 h in the CTX group ($p < 0.001$) and from 5.39 ± 2.51 to 4.09 ± 2.71 g/24 h in the tacrolimus group ($p < 0.001$). However, there was no significant difference in the reduction of proteinuria between the two groups ($p = 0.166$). The serum ALB level increased from 19.3 ± 3.7 to 22.6 ± 4.7 g/l in the CTX group ($p < 0.001$) and from 18.4 ± 5.1 to 24.0 ± 5.6 g/l ($p < 0.001$) in the tacrolimus group. The difference between the two groups was also not significant ($p = 0.196$). After 2 months, the urinary protein excretion decreased

Table 1. Demographic, histological, and laboratory characteristics of the patients at baseline and adverse events during the 18-month study period

	Total (n = 100)	CTX group (n = 52)	Tacrolimus group (n = 48)	p value
Male/female	1.56:1	1.36:1	1.82:1	NS
Age, years	57.0±14.0	57.8±14.8	56.3±13.2	NS
Distribution of ages by tertile				NS
(<40/40–60/>60 years)	11/46/43	5/23/24	6/23/19	
SCr, μmol/l	82.8±37.4	87.7±46.8	77.5±22.7	NS
ALB, g/l	18.9±4.4	19.3±3.7	18.4±5.1	NS
Proteinuria, g/24 h	5.24±2.35	5.10±2.20	5.39±2.51	NS
Distribution of proteinuria by tertile				NS
(<5/5–9/>9 g/24 h)	50/44/6	27/22/3	23/22/3	
Distribution of ALB by tertile				NS
(<20/20–25/>25 g/l)	53/40/7	26/24/2	27/16/5	
eGFR (ml/min/1.73 m^2)	92.1±30.9	90.0±35.8	94.4±24.6	NS
Distribution of eGFR by tertile				NS
(>90/60–89/<60 ml/min/1.73 m^2)	47/41/12	21/23/8	26/18/4	
Serum uric acid, μmol/l	364±86	368±94	358±77	NS
Systolic BP, mm Hg	133±21	132±20	134±21	NS
Diastolic BP, mm Hg	80±11	80±11	81±11	NS
Cholesterol, mmol/l	7.90±2.14	7.77±1.75	8.05±2.51	NS
Triglycerides, mmol/l	2.72±1.40	2.66±1.43	2.78±1.37	NS
Histological grade of IMN				
Stage I	14	5	9	NS
Stage II	66	35	31	NS
Stage III	18	11	7	NS
Stage IV	2	1	1	NS
Infection	39	31	8	<0.001
Urinary tract	19	15	4	0.009
Pneumonia	15	11	4	NS
Lower extremities	2	2	0	NS
Other infections	3	3	0	NS
Glucose intolerance	19	12	7	NS
Diarrhea	3	1	2	NS
Hepatic dysfunction	7	6	1	NS
Embolism of deep vein	1	1	0	NS
Gouty arthritis	2	1	1	NS

to 3.13 ± 2.31 g/24 h in the CTX group and to 3.10 ± 2.46 g/24 h in the tacrolimus group (p = 0.952). The serum ALB level increased to 23.5 ± 5.9 g/l in the CTX group and to 27.3 ± 5.9 g/l in the tacrolimus group, with a significant difference between the groups (p = 0.003). After 3 months, the serum ALB level increased to 25.7 ± 5.8 g/l in the CTX group and to 29.1 ± 5.9 g/l in the tacrolimus group, with a significant difference between the groups (p = 0.01). The other follow-up data and comparisons are shown in figure 2.

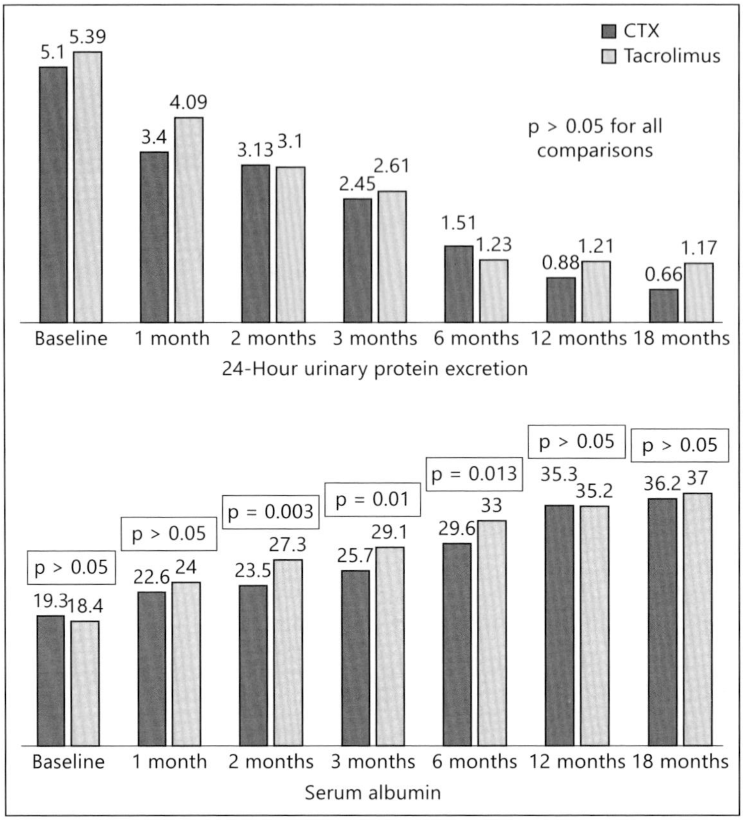

Fig. 2. Follow-up data for the CTX and tacrolimus groups and between-group comparisons.

Table 2. Rates of remission in the tacrolimus and CTX groups

Group	Month					
	1	2	3	6	12	18
Tacrolimus						
CR	2	5	5	16	17	12
PR	17	23	24	16	12	9
Total	48	43	41	35	32	24
Rate of remission (CR+PR), %	39.6	65.1	70.7	91.4	90.6	87.5
CTX						
CR	1	5	4	14	24	30
PR	18	18	26	27	18	9
Total	52	52	52	49	48	42
Rate of remission (CR+PR), %	36.5	44.2	57.7	83.7	87.5	92.9
p value	0.16	0.02	0.08	0.16	0.26	0.26

The SCr remained stable in both groups during the entire follow-up period. None of the patients in either group reached the secondary endpoint of a 50% SCr increase during the study.

Relapse

During the 18-month study period, relapse occurred in 4 patients: 3 in the tacrolimus group and 1 in the CTX group. All the relapse patients had achieved PR or CR during the initial treatment. Two of the relapses in the tacrolimus group occurred at 12 months, and 1 occurred at 7 months. The only relapse in the CTX group occurred at 12 months. The relapse rate in the tacrolimus group was higher than that in the CTX group.

Adverse Events

A list of the side effects observed in the CTX and tacrolimus groups is shown in table 1. The rate of urinary tract infections was higher in the CTX group compared with the tacrolimus group; there were no significant differences between two groups in terms of the incidence of other side effects. Tacrolimus was better tolerated than CTX. None of the patients in either group withdrew from the study because of side effects.

Discussion

The severity of renal dysfunction, the progression in renal function over time, and the level of persistent proteinuria are key factors that determine outcome and affect the decision to initiate therapy in IMN patients [11]. In high-risk patients with IMN, immunosuppressive treatment is effective in inducing remission. Early treatment shortens the duration of the nephrotic phase but does not improve the preservation of renal function. Thus, treatment decisions must be based on a risk-benefit assessment for individual patients [12]. At least five specific treatment options are available: (i) cyclical therapy alternating glucocorticoids and a cytotoxic agent; (ii) calcineurin inhibitors; (iii) a mycophenolate salt; (iv) sACTH, or (v) rituximab [13]. KDIGO has recommend that the initial therapy consist of a 6-month course of alternating monthly cycles of oral and intravenous corticosteroids and oral alkylating agents (1B) and has suggested the use of CTX rather than chlorambucil for the initial therapy (2B). They have also recommended that cyclosporine or tacrolimus be administered for at least 6 months as an initial therapy (1C) [7]. We decided to use tacrolimus, which was preferred over cyclosporin A because it is a more potent inhibitor of antigen-driven T-cell activation, cytokine production, and lymphocyte proliferation in vitro [14] and has demonstrated greater immunosuppressive power with a better adverse effect profile in renal transplantation [15, 16]. Our study was a large, prospective, randomized trial to compare the efficacy and safety of tacrolimus and CTX, with both drugs combined with corticosteroids, in patients with biopsy-proven IMN. The remission rate at 2 months in the tacrolimus group was 65.1%, which was higher than both the remission rate (44.2%) in the CTX group ($p = 0.02$) and the remission achieved with tacrolimus monotherapy (36%) as reported in a previous study [9]. Our data showed that the tacrolimus group achieved a higher remission at 2 months, which has not been found in other studies. At 6 months, the remission rate was 91.4% in the tacrolimus group, which was similar to the remission rate in another study in China (85%) [17] and higher than the remission rates reported for cyclosporine (75%) [18] and tacrolimus monotherapy (56%) [9]. Our results indicate that tacrolimus in combination with corticosteroids induced the remission of nephrotic IMN more quickly than CTX (also in combination with corticosteroids).

We also found a significantly greater improvement in the serum ALB levels in the tacrolimus group compared with the CTX group. We found that at 2 and 3 months, the increase in serum ALB differed between the CTX and tacrolimus groups ($p = 0.003$ at 2 months, and $p = 0.01$ at 3 months), a finding similar to that of another study of Chinese IMN patients [19]. In our study, however, there was no superiority of tacrolimus with respect to the serum ALB and proteinuria levels at 6 months.

Another important finding of our study is that tacrolimus reduced the risk of renal function deterioration. The natural history of untreated IMN includes a considerable rate of spontaneous remission (30%) and a similar percentage of patients who exhibit an aggressive clinical course, with progressively deteriorating renal function (26%), as reported by Praga et al. [9] Other studies have reported similar percentages of patients who experience spontaneous remission or an aggressive clinical course [20–22], including membranous nephropathy patients with chronic renal impairment [23]. In our study, the SCr remained stable in both groups. No patients in either group reached the secondary endpoint of a 50% SCr increase during the study. However, a longer follow-up will be necessary to confirm the safety of tacrolimus in IMN patients in the future.

Tacrolimus was well tolerated in this study, and no severe side effects were detected. Glucose intolerance, urinary tract infections, and pneumonia were the major side effects observed, and all of them were mild and well controlled. The only difference between the groups with respect to side effects was that urinary tract infections occurred less often in the tacrolimus group. Tacrolimus was better tolerated than CTX. We monitored the serum concentration of tacrolimus, and it was considerably lower than that used for immunosuppression in solid organ transplantation, which may explain the absence of severe side effects.

The treatment of IMN remains controversial. The use of immunosuppressive therapies with high-dose corticosteroids and cytotoxic agents has conclusively demonstrated efficacy; however, the high frequency and severity of the side effects of these drugs, especially urinary tract infections, pneumonia, and hepatic dysfunction, are disturbing. Rituximab, an anti-CD20 monoclonal antibody, has also been shown to induce a significant reduction in proteinuria in an uncontrolled trial of IMN patients [24]. As recent prospective trials have demonstrated, tacrolimus can induce the same beneficial effects as corticosteroids and CTX in IMN [17, 19], but there is a lack of prospective, long-term, controlled trials on tacrolimus combined with corticosteroids. As suggested by our study, in patients with persistent nephrotic proteinuria after an observation period of 18 months, tacrolimus combined with corticosteroids was an expedient and well-tolerated therapeutic option. This drug combination induced remission in a great majority of patients before 3 months and significantly reduced the risk of renal function deterioration; however, tacrolimus was not superior to CTX after 6 months.

In conclusion, tacrolimus combined with corticosteroids is an alternative therapeutic regimen for the treatment of patients with nephrotic IMN. Its short-term efficacy may be better than that of CTX combined with corticosteroids, and it is associated with fewer adverse events. However, through the follow-up endpoint of 18 months, tacrolimus demonstrated no superiority with respect to serum ALB and proteinuria levels. Further research is needed to evaluate the long-term efficacy and safety of tacrolimus.

Acknowledgments

This work was supported by grants from the National Basic Research Program of China 973, grant No. 2012CB517600 (grant No. 2012CB517604), the Research on Hypertensive Nephropathy and Ischemic Kidney Diseases National Key Technology R&D Program (12-5), grant No. 2011BAI10B00 (grant No. 2011BAI10B06), and the National Natural Science Foundation of China (grant No. 30871001).

References

1 Glassock RJ: The pathogenesis of idiopathic membranous nephropathy: a 50-year odyssey. Am J Kidney Dis 2010;56:157–167.

2 Hitoshi Y, Takashi T, Hitoshi S, et al: Membranous nephropathy in Japan: analysis of the Japan Renal Biopsy Registry (J-RBR). Clin Exp Nephrol 2012;16:557–563.

3 Zhou FD, Zhao MH, Zou WZ, Liu G, Wang H: The changing spectrum of primary glomerular diseases within 15 years: a survey of 3,331 patients in a single Chinese centre. Nephrol Dial Transplant 2009;24:870–876.

4 Li LS, Liu ZH: Epidemiologic data of renal diseases from a single unit in China: analysis based on 13,519 renal biopsies. Kidney Int 2004;66:920–923.

5 Schieppati A, Mosconi L, Perna A, et al: Prognosis of untreated patients with idiopathic membranous nephropathy. N Engl J Med 1993;329:85–89.
6 Honkanen E, Tornroth T, Gronhagen-Riska C: Natural history, clinical course and morphological evolution of membranous nephropathy. Nephrol Dial Transplant 1992; 7(suppl 1):35–41.
7 KDIGO Clinical Practice Guideline for Glomerulonephritis. Kidney Int Suppl 2012;2: 143–153.
8 Westhoff TH, van der Giet M: Tacrolimus in the treatment of idiopathic nephrotic syndrome. Expert Opin Investig Drugs 2007;16: 1099–1110.
9 Praga M, Barrio V, Juárez GF, et al: Tacrolimus monotherapy in membranous nephropathy: a randomized controlled trial. Kidney Int 2007;71:924–930.
10 Levey AS, Coresh J, Greene T, et al: Using standardized serum creatinine values in the modification of diet in renal disease study equation for estimating glomerular filtration rate. Ann Intern Med 2006;145:247–254.
11 Segal PE, Choi MJ: Recent advances and prognosis in idiopathic membranous nephropathy. Adv Chronic Kidney Dis 2012;19: 114–119.
12 Hofstra JM, Branten AJ, Wirtz JJ, et al: Early versus late start of immunosuppressive therapy in idiopathic membranous nephropathy: a randomized controlled trial. Nephrol Dial Transplant 2010;25:129–136.
13 Ponticelli C, Passerini P: Management of idiopathic membranous nephropathy. Expert Opin Pharmacother 2010;11:2163–2175.
14 Thomson AW: FK506. How much potential. Immunol Today1989;10:6–9.
15 Jardine AG: Assessing the relative risk of cardiovascular disease among renal transplant patients receiving tacrolimus or cyclosporine. Transpl Int 2005;18:379–384.
16 Artz MA, Boots JMM, Ligtenberg G, et al: Conversion from cyclosporine to tacrolimus improves quality of life indices, renal graft function and cardiovascular risk profile. Am J Transplant 2004;4:937–945.
17 Min C, Hang L, Xia-Yu L, et al: Tacrolimus combined with corticosteroids in treatment of nephrotic idiopathic membranous nephropathy: a multicenter randomized controlled trial. Am J Med Sci 2010;339:233–238.
18 Cattran DC, Appel GB, Hebert LA, et al: Cyclosporine in patients with steroid-resistant membranous nephropathy: a randomized trial. Kidney Int 2001;59:1484–1490.
19 He L, Peng Y, Liu H: Treatment of idiopathic membranous nephropathy with combination of low-dose tacrolimus and corticosteroids. J Nephrol 2012 (E-pub ahead of print).
20 Schieppati A, Mosconi L, Perna A, et al: Prognosis of untreated patients with idiopathic membranous nephropathy. N Engl J Med 1993;329:85–89.
21 Donadio JV, Torres VE, Velosa JA, et al: Idiopathic membranous nephropathy: the natural history of untreated patients. Kidney Int 1988;33:708–715.
22 Glassock RJ: The therapy of idiopathic membranous glomerulonephritis. Semin Nephrol 1991;11:138–147.
23 Polanco N, Gutiérrez E, Rivera F, et al: Spontaneous remission of nephrotic syndrome in membranous nephropathy with chronic renal impairment. Nephrol Dial Transplant 2012; 27:231–234.
24 Ruggenenti P, Chiurchiu C, Brusegan V, et al: Rituximab in idiopathic membranous nephropathy: a one-year prospective study. J Am Soc Nephrol 2003;14:1851–1857.

Prof. Nan Chen, MD
Department of Nephrology, Ruijin Hospital
School of Medicine, Shanghai Jiao Tong University
No. 197 Ruijin Er Road, Shanghai 200025 (PR China)
E-Mail chen-nan@medmail.com.cn

Chen N (ed): New Insights into Glomerulonephritis.
Contrib Nephrol. Basel, Karger, 2013, vol 181, pp 163–174 (DOI: 10.1159/000348635)

Membranoprolferative Glomerulonephritis – Mechanisms and Treatment

Gerald B. Appel

Clinical Nephrology and The Glomerular Kidney Center, Columbia University Medical Center, New York, N.Y., USA

Abstract

Membranoproliferative glomerulonephritis (MPGN) is a histologic pattern of glomerular damage with many etiologies. In the past, it was divided into secondary and primary, or idiopathic forms. In turn, the idiopathic forms were labeled MPGN types I, II, and III based on ultrastructural localization of the electron-dense 'deposits'. A recent major change in classification relies on the presence of immunoglobulins and complement to determine etiologies that are more in line with the pathogenesis of the disease. Idiopathic MPGN II, now called dense deposit disease and C3 glomerulonephritis, are now recognized as diseases of the alternate complement pathway. They are caused by a variety of alternate complement pathway defects including genetic forms and those due to autoantibodies that either activate accelerators of the pathway or block inhibitors of the cascade. These patients need to be evaluated in a different fashion and receive different therapy than those patients with MPGN forms associated with immunoglobulin deposition.
Copyright © 2013 S. Karger AG, Basel

Membranoproliferative glomerulonephritis (MPGN), also referred to as mesangiocapillary glomerulonephritis, refers to a light microscopic (LM) histologic pattern of glomerular disease [1, 2]. In some series it comprises between 5 and 10% of biopsied cases of glomerular disease and it often progresses to end-stage renal disease [3–5]. In the past, classification was based on two factors: the presence of an underlying etiology, secondary MPGN, and in those primary or idiopathic cases, the morphologic presence and localization of electron density within the glomerular tuft [6, 7]. Recent classifications for idiopathic MPGN rely upon new understanding of the mechanism(s) behind the glomerular damage [1, 2, 8, 9]. The presence of immunoglobulin and complement component de-

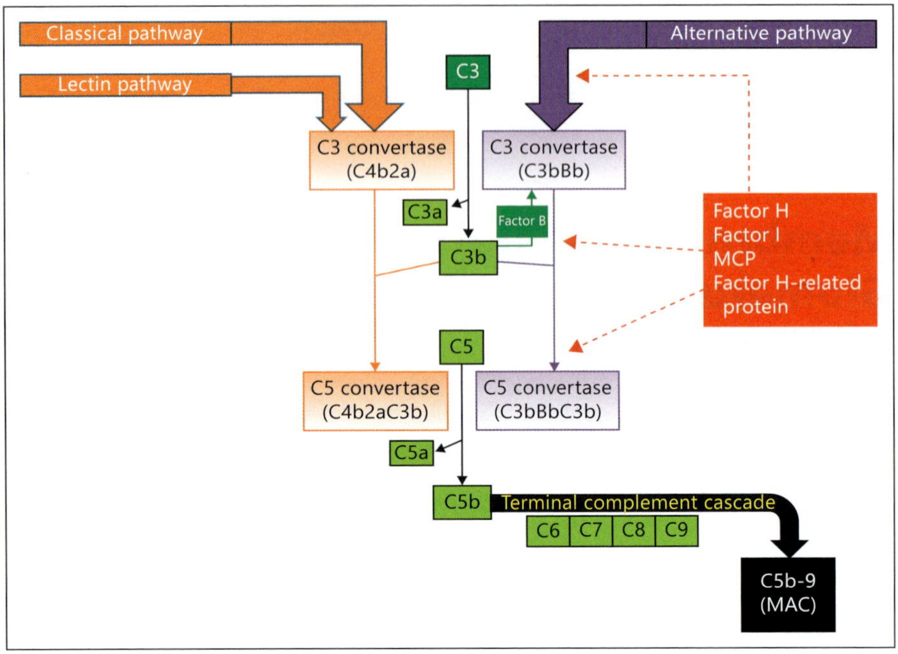

Fig. 1. Three mechanisms of activation of complement cascade and pathways. Reprinted with permission of Bomback and Appel [9].

position by immunofluorescence (IF) microscopy indicates an immune complex etiology for most cases. The presence of isolated C3 deposition by IF indicates an alternate pathway of complement activation and an entirely different pathogenesis. Studies are now focusing on the mechanisms behind the activation of the alternate complement pathway and a mechanistic approach to the treatment of patients with an MPGN pattern of glomerular damage on renal biopsy (fig. 1–3).

Classification of MPGN

MPGN is defined by the LM findings of diffuse mesangial hypercellularity, endocapillary proliferation, a lobulation of the glomerular tuft and splitting of the glomerular capillary wall (so-called tram tracking) [1, 9, 10]. The splitting of the glomerular capillary wall is due to both immune complex and complement component deposition and mesangial cell interposition. While this defines the histologic pattern, there are many etiologies of MPGN including autoimmune disease, infectious etiologies, dysproteinemic diseases, thrombotic microangiopathies, and disorders of the complement system. In the past, MPGN was

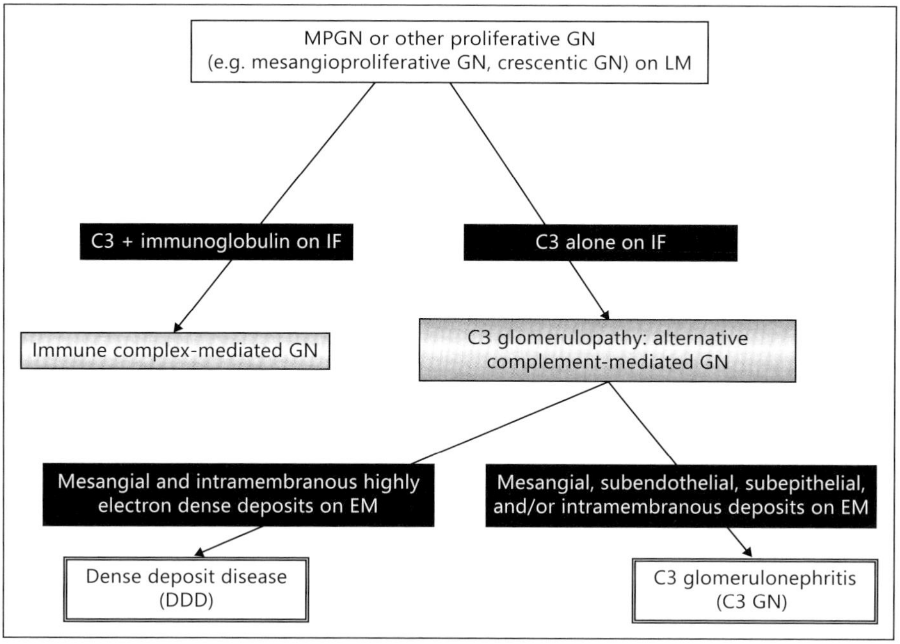

Fig. 2. Classification of MPGN by a mechanistic approach. Reprinted with permission of Bomback and Appel [9].

traditionally classified as being idiopathic when no associated disease entity could be diagnosed. Idiopathic MPGN, in turn was traditionally classified as MPGN I, MPGN II, or MPGN III based on the localization of electron-dense deposits on electron microscopy [6, 7, 9]. In MPGN I the deposits were found in the subendothelial and mesangial location; in MPGN II, also called dense deposit disease (DDD), electron density is found in a ribbon-like pattern along the glomerular basement membrane (GBM), and in MPGN III subepithelial and subendothelial deposits are found [6, 7, 11]. By immunofluorescence microscopy for both MPGN I and III it was clear some patients' biopsies had both immunoglobulin and complement deposition while others had only complement staining. In MPGN II, DDD, only complement deposition was noted.

In this former schema, patients could present with mild asymptomatic urinary findings, full-blown nephrotic syndrome, or a rapidly progressive renal failure. Some patients originally classified as idiopathic MPGN I or III later were found over time to have an autoimmune etiology of their disease such as systemic lupus or a postinfectious etiology such as hepatitis C virus-related glomerulonephritis. Patients with all types of 'idiopathic' MPGN were often included together in both clinical and treatment studies and results were confusing and conflicting [12]. The discovery of hepatitis C virus-related MPGN led to the

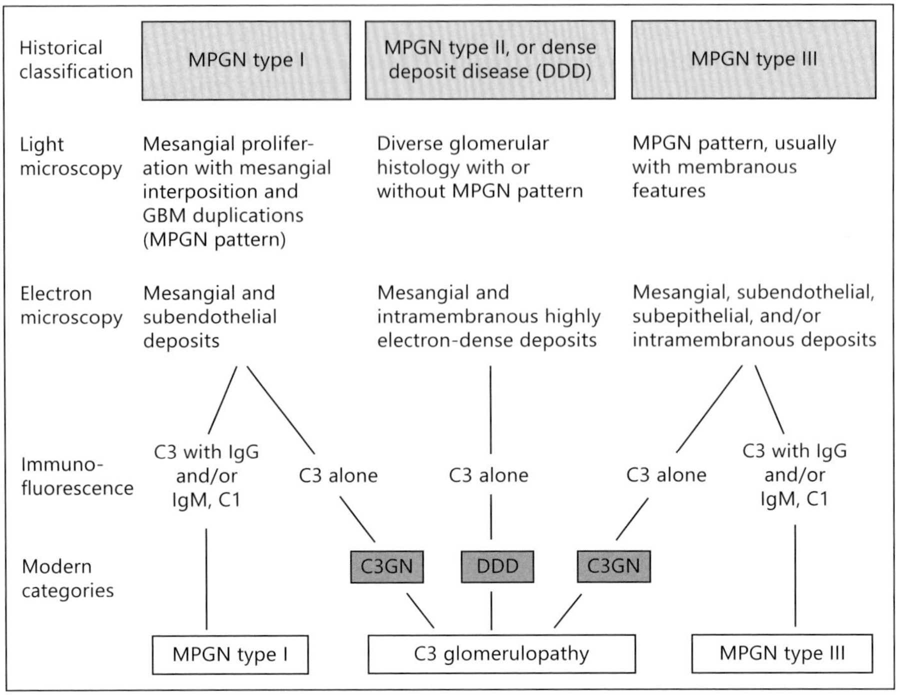

Fig. 3. Reclassification of idiopathic MPGN. Reprinted with permission of D'Agati and Bomback [16].

immediate awareness of the problems of older studies which included this disease with idiopathic and other forms of MPGN [10, 13, 14]. For example, while immunosuppressive agents may have a role in certain forms of idiopathic MPGN, antiviral therapy is clearly indicated in many patients with hepatitis C virus-related MPGN. Recent reclassification of MPGN, based on new knowledge of the pathogenesis of the disease entities, avoids many of the problems of earlier studies.

Current classifications divide MPGN into those diseases with immunoglobulin present by IF and those with only complement deposition by IF [1, 2, 8, 15]. An additional group of patients with an MPGN pattern by LM will have no immunoglobulin or complement deposition, e.g. the thrombotic microangiopathies [10]. Those diseases with immune complex deposition include autoimmune disorders, infection-associated disease, dysproteinemias, as well as a variety of other diseases. Systemic lupus erythematosus, rheumatoid arthritis, Sjögren's syndrome and other mixed connective tissue disorders have all been associated with an MPGN pattern of glomerulonephritis [8, 10]. Clearly many other patterns of glomerular damage can also be associated with these diseases. Among infections associated with an MPGN pattern are a variety of bacterial, viral, parasitic, and other infections. Hepatitis C may be associated with an MPGN pattern either with or without associated cryoglobulins. A host of infections related to streptococci, staphylococci, meningococci, mycobacteria, mycoplasma, *Nocardia*, and *Brucella* in association with bacterial endocarditis, shunt nephritis, glomerulonephritis associated with deep visceral abscess, and as a postinfectious event may cause MPGN. Dysproteinemias, with or without cryoglobulins, may also produce an MPGN pattern of glomerular damage. These may include light chain deposition disease, heavy chain disease, light and heavy chain disease, and cryoglobulinemic GN. Many of these are associated with a monoclonal 'M' spike in the serum, a documented plasma cell dyscrasia, and at times myeloma or a B-cell lymphoma, or chronic lymphocytic leukemia.

There clearly are some patients with an MPGN pattern on LM that have immunoglobulin and complement deposited by IF but have no known associated disease. Many who were young at onset have been followed for decades without another systemic disease entity appearing. These are patients with true idiopathic MPGN [6, 16]. Formerly they were called MPGN I or III depending upon the localization of the electron-dense immune-type deposits along or within the GBM [7]. This is probably less relevant now than as to whether their disease is truly immune complex mediated or actually a *forme fruste* of C3 glomerulonephritis (see below).

Mechanism of Glomerular Damage in MPGN

Damage to the endothelium is a primary inciting factor in all thrombotic microangiopathies [10]. This is true for disease with different pathogeneses leading to the coagulation disorder such as thrombotic thrombocytopenic purpura, hemolytic uremic syndrome, antiphospholipid syndrome, and drug-induced thrombotic microangiopathies. It may result in glomerular mesangiolysis, endothelial swelling, fibrin and fibrin products depositing in the glomeruli and small vessels. With chronicity, glomerular remodeling leads to mesangial expansion and interposition along the glomerular capillary wall [8]. While endothelial damage can cause activation of the complement cascade, in the MPGN LM pattern seen with thrombotic microangiopathies, there is no evidence of immunoglobulin or complement deposition by IF or electron microscopy.

Deposition of immune complexes in the glomeruli may cause an MPGN pattern of glomerular disease. There are three initiating pathways in the activation of the complement system: the classical, mannose-binding lectin, and alternative pathways. Immune complex-mediated glomerular disease is associated with activation of the complement cascade through the classical activation pathway (see Figure 1) [9]. Activation of C1 leads to cleavage of C4 and C2 and generates C4b and C2a leading to C4b2a, the classical pathway convertase. This subsequently activates C3 convertase and the full complement cascade. This is associated with recruitment of leukocytes and macrophages, release of cytokines and enzymes, and capillary wall damage. It is the entrapment of immune complex deposits and cellular and matrix components with new GBM formation, during the reparative processes that account for the double contours of the glomerular capillary wall. Because of the immune complex nature of the disease and the activation of the classical complement cascade, immunoglobulin is virtually always found in the glomeruli of patients with this pattern of active MPGN.

It has been realized for a number of years that one pattern of MPGN, DDD formerly called MPGN II, lacked immunoglobulin deposits on IF examination [7, 9, 17, 18]. Subsequent studies using mass spectrometry have shown that the density along the GBM in DDD is not composed of immune complex deposition, but rather components of the alternative and terminal complement pathways [19]. This includes C5, C6, C7, C8, C9, and a variety of other proteins such as factor H-related protein, apolipoprotien E, and vitronectin. These components are not seen in the glomeruli of normal individuals or to the same extent in glomeruli from patients with other glomerular diseases [19]. Since not all patients with DDD have an MPGN picture on LM (only about 25% in one large series), the name MPGN II is a misnomer and has been discarded in favor of the more appropriate name of DDD [17, 18, 20]. It has also become clear that other patients have the same mechanisms of glomerular complement activation and presumably glomerular damage without having the dense deposits along the GBMs. These patients may have only mesangial or subendothelial electron density and they may or may not show an MPGN pattern by LM. This entity has now been termed C3 glomerulonephritis [8, 9, 15, 16, 21–23]. Nevertheless, in all of these patients, regardless of the LM pattern, IF microscopy shows predominantly strong C3 deposition without immunoglobulin deposits. This implies activation of the alternative complement pathway in these diseases which are now called the C3 glomerulopathies [15, 16, 21–23].

Alternative Pathway of Complement Activation

As previously noted there are three mechanisms for activating the complement cascade: the classical, the mannose-lectin, and the alternative pathway [9]. All converge on C3 to generate C3 convertase an enzyme complex that cleaves C3 into C3a and C3b. This in turn generates C5 convertase, which activates the terminal complement cascade and assembly of the membrane attack complex (C5b-C9) and subsequent cell lysis. Evidence supports the theory that the alternate pathway is always active at a low level, the so-called 'tick-over' activation. Staining patterns for C3 and C5b-9 in glomeruli from healthy controls found mild to moderate C5b-9 staining within the glomerular mesangium and some glomerular capillary walls in the absence of any LM or ultrastructural abnormalities [24]. This suggests C5b-9 is activated and deposited in a constitutive manner at a low level even in normal kidney tissue by this 'tick-over' mechanism, which leads to C5 recruitment in the fluid phase.

The alternative complement pathway is modulated by a number of regulatory proteins. Some increase activation while others dampen the activity of the system [9]. The inhibitory proteins include complement factor H, complement factor I, and membrane cofactor protein. Study of a number of animal models with genetic deficiencies of regulatory proteins has helped clarify the mechanisms of alternative complement pathway activation [17, 25]. Likewise, studies of families with inherited defects of regulatory proteins or individuals with autoantibodies to the regulatory proteins of the system have greatly increased our understanding of the alternative complement pathway [26–28]. Some MPGN patients with either C3 glomerulonephritis or DDD will have an autoantibody C3 nephritic factor which stabilizes the C3 convertase and leads to continuous C3 activation [17, 29]. Some families will have autoantibodies to the inhibitory factors H or I and again have continuous activation of the alternative complement cascade, while others will have genetic abnormalities in the genes producing the quantity or activity of factor H or I. Without these inhibitors the system runs in a hyperdynamic unregulated fashion. All can result in either C3 glomerulonephritis or DDD.

It has been of great interest and clinical significance that at the same time that the C3 glomerulopathies, C3 glomerulonephritis and DDD, were being defined, an understanding of the pathogenesis and treatment of atypical hemolytic uremic syndrome (aHUS) evolved [30]. This disease also exhibits overactivity of the alternative complement pathway. Patients with aHUS have been found to have genetic abnormalities in the same genes encoding complement inhibitory regulators (factor H, factor I, MCP) as well as a gain of function mutations in genes encoding complement activators (C3 and factor B). Up to 10% will have autoantibodies to factor H [15]. However, despite damage to the glomeruli through a microangiopathic thrombotic glomerulopathy in aHUS, there is no glomerular deposition of complement. This is felt to be due to activation of the system in aHUS at the level of the cell membrane. In C3 glomerulopathies the activation of the alternative pathway is in the soluble fluid phase in the plasma [9]. Of note, the efficacy of eculizumab, a monoclonal antibody against C5, in aHUS again supports the role of complement activation and a potential treatment for C3 glomerulopathies [31].

Evaluation of the Patient with MPGN

The evaluation of a patient whose renal biopsy shows an MPGN pattern by LM now clearly depends upon an understanding of the mechanism of disease and their correlation with the biopsy findings [1, 9]. If there is no evidence for immunoglobulin or complement deposition, the clinician must consider a thrombotic microangiopathy. An examination of the platelet count and LDH enzyme level, and potential testing for level of ADAMTS 13 (to exclude thrombotic thrombocytopenic purpura), an anticardiolipin antibody level (to exclude antiphospholipid syndrome), and testing for shiga toxin or complement abnormalities for aHUS may all be indicated. Other considerations will be nephropathy post-bone marrow transplant, malignant hypertension, and smoker's glomerulopathy.

If the biopsy shows by IF microscopy immunoglobulin deposition, the clinician should evaluate the patient for dysproteinemias, infections, and autoimmune disease. Evaluation may include testing for hepatitis C virus, ANA, ENA, cryoglobulins, serum and urine protein immunoelectrophoresis or immunofixation, and free lyte assay for light chains.

If there is no evidence of immunoglobulin deposition but strong C3 staining, disorders of the alternative complement system should be explored [1, 9]. Serum complement levels, which may be low in immune complex-mediated glomerular diseases as well, should be measured. A more detailed evaluation of the alternative complement pathway can include levels of alternate pathway inhibitor proteins (factor H, I, and MCP), an assay for the C3 convertase stabilizing autoantibody (C3 nephritic factor) serum levels of the membrane attack complex, and analysis for genetic defects associated with specific alternate pathway proteins. Genetic testing can check for mutations in genes that encode complement regulators, principally CFH, CFI, and MCP (CD46), as well as complement factor H-related proteins (CFHR1–5). Unfortunately, at present these can only be performed at a few laboratories and are expensive to perform. It is hoped with greater awareness of these diseases and greater usage of these tests, they will become more accessible.

It is unclear whether patients with true idiopathic type I MPGN should undergo a similar series of genetic, autoantibody, and complement testing. However, a recent study analyzed 134 patients with DDD (n = 29), C3GN (n = 56), and idiopathic type I MPGN (n = 49) [22]. Alternative complement pathway abnormalities were found in more than half of their patients with MPGN type I: mutations in CFH (n = 5), mutations in CFI (n = 3), and positive C3NeF activity (n = 18), and C3 nephritic factor was found as often in MPGN type I as in C3GN. It is possible that some patients were misclassified as MPGN 1 rather than C3GN with only trace immunoglobulin staining. Misclassification was a factor in a small study of C3GN from the United States, in which 9 of the 12 patients with C3GN were initially misclassified as MPGN type I (n = 4), MPGN type III (n = 1), and postinfectious glomerulonephritis (n = 4) [23]. Nonetheless, some of these patients may actually have a dysregulated alternative complement pathway system as the cause of their 'idiopathic' MPGN. If so, this can have major implications for therapy.

Treatment of MPGN in 2013

Due to prior lumping of all forms of idiopathic MPGN together, early treatment studies are of little value in these diseases. Indeed, many early studies are contaminated with patients with hepatitis C virus patients [13, 14]. Thus, even in children with idiopathic MPGN early studies suggesting a benefit from alternate-day corticosteroid treatment must be reconsidered [12].

For MPGN with no immunoglobulin or complement deposition, a search for an underlying disease entity is important [1, 9, 10]. The finding of a low level of ADAMTS 13 and a diagnosis of TTP would lead to plasmapheresis and perhaps steroids and rituximab therapy. The finding of a thrombotic microangiopathy with anticardiolipin antibodies might lead to anticoagulation, while the finding of a genetic abnormality of the alternate complement cascade would suggest aHUS and a role for eculizumab treatment. In immune complex-mediated MPGN, the underlying disease again will dictate the therapy. If the patient has lupus nephritis, hepatitis C virus glomerular disease, or cryoglobulins or a plasma cell dyscrasia, each disease will be treated accordingly.

For the C3 glomerulopathies, treatment at present remains uncertain [9]. There are still too few studies that link specific therapy with genetic, autoantibody, and complement functional screening. It has been proposed, but not yet proven, that despite similar defects in the alternative complement pathway, C3 glomerulonephritis may have a more benign course than DDD. In proteinuric patients the use of renin-angiotensin-aldosterone system (RAAS) blockade has been widely utilized. At what stage of proteinuria this should be used or even whether use in a prophylactic fashion is beneficial is unknown. Immunosuppressive and immunomodulatory medications have been used in small numbers of patients with C3 glomerulopathies mixed results [9, 22, 23]. KDIGO (Kidney Disease: Improving Global Outcomes) clinical practice guidelines do recommend immunosuppressive therapy for adults or children with idiopathic type I MPGN accompanied by nephrotic syndrome and/or deteriorating kidney function, despite low levels of evidence supporting this suggestion [32].

Despite the evidence for alternate complement pathway defects the C3 glomerulopathies are a very heterogeneous group pathogenetically. Some have autoantibodies, others genetic defects, and others abnormal levels of functional proteins of the complement system. Moreover, dual defects are not rare. A patient may have a genetic defect in inhibitory factor H levels and C3 nephritic factor as well. Thus, targeted therapy will be a problem in these relatively rare diseases [9].

Replacement of factor H, which currently is only available via plasma exchange, has been used in a small number of patients successfully as a targeted therapy [9]. Clearly, plasma exchange would need to be lifelong therapy for these patients as for other patients with genetic defects of other inhibitory proteins of the alternate pathway of complement (factor I and MCP). Immunosuppressive agents would seem a logical course to decrease antibody production for patients with an acquired antibody to an inhibitory protein of the alternate pathway [9, 29]. This treatment has only been studied in a small number of patients and preliminary data suggest both mycophenolate mofetil and rituximab are effective in some patients but clearly not all with C3 glomerulonephritis. Neither the dose, duration of treatment nor combination of agents has been defined and some patients have clearly failed to respond to intensive immunomodulatory therapy [9].

Eculizumab is a humanized monoclonal antibody directed against C5. It is FDA-approved for use in paroxysmal nocturnal hemoglobinuria and in aHUS [29]. It prevents both the formation of C5a, a potent anaphylatoxin, and the generation of membrane attack complex, C5b-C9. It has been used in a small number of patients with DDD, C3 glomerulonephritis, and even in MPGN I both in native kidneys and in recurrences in the allograft. An open-label study of 6 patients with C3 glomerulopathies (3 DDD and 3 C3GN) used eculizumab for 1 year with the same dosing schedule used for aHUS. Four patients had clinical benefit (2 subjects significantly reduced serum creatinine, 1 subject remission of nephrotic syndrome, and 1 subject histopathologic improvements on repeat biopsy). Circulating serum levels of the membrane attack complex normalized with treatment and paralleled clinical improvements. While eculizumab may be appropriate treatment for some patients with DDD and C3GN, larger, controlled trials are needed to define the role of specific complement blockers.

References

1 Sethi S, Fervenza FC: Membranoproliferative glomerulonephritis – a new look at an old entity. N Engl J Med 2012;366:1119–1131.

2 Sethi S, Fervenza FC: Membranoproliferative glomerulonephritis: pathogenetic heterogeneity and proposal for a new classification. Semin Nephrol 2011;31:341–348.

3 Swaminathan S, Leung N, Lager DJ, et al: Changing incidence of glomerular disease in Olmstead County, Minnesota: a 30-year renal biopsy study. Clin J Am Soc Nephrol 2006;1: 483–487.

4 Braden GL, Mulhern JG, O'Shea MH, et al: Changing incidence of glomerular disease in adults. Am J Kidney Dis 2000;35:878–883.

5 Maisonneuve P, Agadoa L, Gellert R, et al: Distribution of primary renal diseases leading to ESRF in the United States, Europe, and Australia/New Zealand. Am J Kidney Dis 2000:35:157–165.

6 Levy M, Gubler MC, Sich M, et al: Immunopathology of membranoproliferative glomerulonephritis with subendothelial deposits (type I MPGN). Clin Immunol Immunopathol 1978;10:477–492.

7 Habib R, Gubler MC, Loirat C, et al: Dense deposit disease: a variant of membranoproliferative glomerulonephritis. Kidney Int 1975; 7:204–215.

8 Fakhouri F, Fremeaux-Bacchi V, Noel L-H, et al: C3 glomerulopathy: a new classification. Nat Rev Nephrol 2010;6:494–499.

9 Bomback AS, Appel GB: Pathogenesis of the C3 glomerulopathies and reclassification of MPGN. Nat Rev Nephrol 2012;8:634–642.

10 Rennke HG: Secondary membranoproliferative glomerulonephritis. Kidney Int 1995;47: 643–656.

11 Strife CF, McEnery PT, McAdams AJ, West CD: Membranoproliferative glomerulonephritis with disruption of the glomerular basement membrane. Clin Nephrol 1977;7:65–72.

12 Tarshih P, Bernstein J, Tobin JN, Edelman CN Jr: Treatment of mesangiocapillary glomerulonephritis with alternate-day prednisone – a report of the International Study of Kidney Disease in Children. Pediatr Nephrol 1992;6:123–130.

13 Johnson RJ, Gretch DR, Yamabe H, et al: Membranoproliferative glomerulonephritis associated with hepatitis C virus infection. N Engl J Med 1993;328:465–470.

14 Rocatello D, Fornasierri A, Gianchino O, et al: Multicenter study on hepatitis C virus-related cryoglobulinemic glomerulonephrtiis. Am J Kidney Dis 2007;49:69–82.

15 Sethi S, Nester CM, Smith RJ: Membranoproliferative glomerulonephritis and C3 glomerulopathy: resolving the confusion. Kidney Int 2012;81:434–441.

16 D'Agati VD, Bomback AS: C3 Glomerulopathy: what's in a name? Kidney Int 2012;82: 379–381.

17 Appel GB, Cook HT, Hageman G, et al: Membranoproliferative glomerulonephritis type II (dense deposit disease): an update. J Am Soc Nephrol 2005;16:1392–1403.

18 Nasr SH, Valeri AM, Appel GB, et al: Dense deposit disease: clinicopathologic study of 32 pediatric and adult patients. Clin J Am Soc Nephrol 2009;4:22–32.

19 Sethi S, Gamez JD, Vrana JA, et al: Glomeruli of DDD contain components of the alternative and terminal complement pathway. Kidney Int 2009;75:852–960.

20 Walker PD, Ferrario F, Joh K, Bonsib SM: Dense deposit disease is not a membranoproliferative glomerulonephritis. Mod Pathol 2007;20:605–616.

21 Servais A, Fremeaux-Bacchi V, Lequintrec M, et al: Primary glomerulonephritis with isolated C3 deposits: a new entity which shares common genetic risk factors with haemolytic uraemic syndrome. J Med Genet 2007;44: 193–199.

22 Servais A, Noel L-H, Roumenina LT, et al: Acquired and genetic complement abnormalities play a critical role in dense deposit disease and other C3 glomerulopathies. Kidney Int 2012;82:454–463.

23 Sethi S, Fervenza F, Zhang Y, et al: C3 glomerulonephritis: clinicopathological findings, complement abnormalities, glomerular proteomic profile, treatment and follow-up. Kidney Int 2012;82:465–473.

24 Herlitz LC, Bomback AS, Markowitz GS, et al: Renal biopsy findings after eculizumab treatment in dense deposit disease and C3 glomerulonephritis. J Am Soc Nephrol 2012; 23:1229–1237.

25 Pickering M, Cook HT: Complement and glomerular disease: new insights. Curr Opin Nephrol Hypertens 2011;20:271–277.

26 Gale DP, de Jorge EG, Cook HT, et al: Identification of a mutation in complement factor H-related protein 5 in patients of Cypriot origin with glomerulonephritis. Lancet 2010; 376:794–801.
27 Habbig S, Mihatsch MJ, Heinen S, et al: C3 deposition glomerulopathy due to a functional factor H defect. Kidney Int 2009;75:1230–1234.
28 Zhang Y, Meyer NC, Wang K, et al: Causes of alternative pathway dysregulation in dense deposit disease. Clin J Am Soc Nephrol 2012; 7:265–274.
29 Bomback AS, Smith RJ, Barile GR, et al: Eculizumab for dense deposit disease and C3 glomerulonephritis. Clin J Am Soc Nephrol 2012;7:748–756.
30 Noris M, Remuzzi G: Atypical hemolytic uremic syndrome. N Engl J Med 2009;361:1676–1687.
31 Maga TK, Nishimura CJ, Weaver AE, et al: Mutations in alternative pathway complement proteins in American patients with atypical hemolytic uremic syndrome. Hum Mutat 2010;31:E1445–E1460.
32 KDIGO Clinical Practice Guidelines for Glomerulonephritis. Idiopathic MPGN. Kidney Int 2012;(suppl 2):198–199.

Gerald B. Appel, MD
Clinical Nephrology and The Glomerular Kidney Center
Columbia University Medical Center
New York, NY 10032 (USA)
E-Mail Gba2@columbia.edu

Chen N (ed): New Insights into Glomerulonephritis.
Contrib Nephrol. Basel, Karger, 2013, vol 181, pp 175–184 (DOI: 10.1159/000348478)

Re-Evaluation of the Classification System for Membranoproliferative Glomerulonephritis

Yimin Lu · Pingyan Shen · Xiao Li · Yaowen Xu · Xiaoxia Pan · Weiming Wang · Xiaonong Chen · Wen Zhang · Hong Ren · Nan Chen

Department of Nephrology, Ruijin Hospital, School of Medicine, Shanghai Jiao Tong University, Shanghai, PR China

Abstract

In recent years, significant advances have been made in understanding the pathogenesis and etiology of membranoproliferative glomerulonephritis (MPGN). A new classification system based on pathological immunofluorescence findings has been proposed to replace the traditional clinical classification system in order to better identify the underlying causes of MPGN and to provide guidance for more individualized treatment. We conducted a retrospective survey of the MPGN patients treated in our hospital from 2000 to 2012 and report here the validation of this new classification system in this cohort. A total of 34 patients were diagnosed with MPGN, including 25 males and 9 females. There were 3 cases of secondary MPGN, including 1 case due to monoclonal gammopathy of undetermined significance (MGUS) and 2 cases related to hepatitis B virus (HBV) infection. Clinical presentations included nephrotic syndrome (76.5%), microscopic hematuria (79.4%), hypocomplementemia (58.8%), renal insufficiency (82.4%), hypertension (100%), and peripheral edema (100%). All patients were treated with prednisone and immunosuppressive agents, mainly cyclophosphamide. During follow-up (median 6 months, range 3–47 months), 4 patients were lost to follow-up and 2 patients progressed to end-stage renal disease. In Western countries the main cause of secondary MPGN was hepatitis C virus or HBV infection, here however we report 2 cases related to HBV infection. MGUS-associated MPGN was less frequent in the Chinese cohort. Future studies should be designed to evaluate the association of the new classification system and clinical outcomes of MPGN.

Copyright © 2013 S. Karger AG, Basel

Membranoproliferative glomerulonephritis (MPGN) denotes a general pattern of glomerular injury characterized by mesangial cell proliferation and structural changes in the glomerular capillary walls. This condition exists in two forms: primary (idiopathic), when no etiology is identified, and secondary, when an underlying etiology is present, such as systemic lupus erythematosus, hepatitis C virus (HCV) infection, or cryoglobulinemia. On the basis of pathological features, MPGN is classified as type I and III, which are considered immune complex-mediated glomerulonephritis, and type II, which is also known as dense-deposit disease (DDD) [1]. This classification system was established prior to an understanding of the pathogenesis of the MPGN process was obtained, and relies on electron microscopy (EM) findings of renal biopsies; unfortunately, EM evaluation is severely limited in its ability to provide differential diagnosis of the underlying causes of MPGN (i.e. dysregulation of alternative complement pathway in the MPGN type I and III). To overcome this limitation, Sethi and Fervenza [2] proposed a new classification system for MPGN based on immunofluorescence microscopy (IF) staining findings rather than EM findings, in order to better understand the underlying pathogenesis and to help guide more individualized treatment strategies. In this study, we aimed to evaluate this new classification system using a single-center cohort of MPGN patients.

Materials and Methods

Patient Selection and Renal Biopsy Evaluation

This study was conducted using a protocol approved by the Institutional Review Board and Ethics Committee of Ruijin Hospital (Shanghai, PR China). To be eligible for this study, patients had to be seen at the Ruijin Hospital between 2000 and 2012 and have undergone a renal biopsy which subsequently showed MPGN. Each biopsy sample was assessed by light microscopy (LM), EM, and IF. LM examination included analysis of sections stained with hematoxylin and eosin, trichrome, periodic acid-Schiff, silver stain, and Congo red. The IF analyses were carried out with antibodies directed against immunoglobulin (Ig)G, IgA, IgM, C3, C1q, albumin, fibrinogen, and κ and λ light chains. The EM analysis was used to reveal the presence of glomerular dense deposits.

Pertinent clinical and laboratory data were extracted from the hospital's electronic databases and from the patient's paper medical records. Cases of MPGN related to lupus and other autoimmune diseases were excluded from the study, as were cases of thrombotic angiopathy with an MPGN pattern of injury.

Statistical Analyses

Differences of quantitative parameters were assessed using the Student's t test (for data that were normally distributed) or the nonparametric test (for data that were not normally distributed). Differences of qualitative data were compared using the χ^2 test or Fisher's exact test. A difference was considered significant if the p value was <0.05. All statistical analyses were performed by the SPSS statistical software package (version 19.0).

Table 1. Demographic, clinical, and etiologies of 34 cases of MPGN according to the new classification system

	Ig-mediated (n = 27) n (%)	Complement-mediated (n = 7) n (%)	p
Male/female	19/8	6/1	NS
Age, years	53.9±15.3	49.4±15.0	NS
Microscopic hematuria	21 (77.8)	6 (85.7)	NS
Urinary protein, g/24 h*	4.8 (0.8–12.4)	4.0 (1.2–5.6)	NS
Nephrotic syndrome	22 (81.5)	4 (57.1)	0.32
Serum albumin, g/dl	2.4±0.7	2.9±0.4	<0.05
Serum creatine at presentation, mg/dl*	1.6 (0.6–8.0)	1.8 (1.1–3.8)	NS
Complement 3, mg/dl*	65.5 (49–117)	67.0 (54–82)	NS
Complement 4, mg/dl*	18.0 (3–40)	19.5 (12–21)	NS
Treatments			
Corticosteroid+CTX	21 (77.8)	4 (57.1)	NS
Corticosteroid+FK506	3 (11.1)	0 (0)	
Corticosteroid alone	1 (3.7)	3 (42.9)	
Antiviral HBV	2 (7.4)	0 (0)	NS
Complete remission	3 (11.1)	1 (14.3)	NS
Partial remission	5 (18.5)	1 (14.3)	
No remission	19 (70.4)	5 (71.4)	
ESRD	2 (7.4)	0 (0)	NS

* Median (range). NS = Not significant.

Results

Clinical and Pathological Characteristics of MPGN (table 1)

During the period 2000–2012, a total of 34 patients were diagnosed with MPGN at Shanghai Ruijin Hospital, including 25 males and 9 females. The age of the patients ranged from 17 to 75 years (average 53.0). Clinical presentations included nephrotic syndrome (76.5%, 26/34), microscopic hematuria (79.4%, 27/34), hypertension (100%, 34/34), and edema (100%, 34/34). More than one half of the patients (58.8%, 20/34) had hypocomplementemia; the mean C3 level was 65.5 mg/dl (normal 79–152) and the mean C4 level was 19 mg/dl (normal 16–38). Renal insufficiency was present at diagnosis in 82.4% (28/34) of patients, and the median serum creatinine level was 1.7 mg/dl (range 0.6–8.0).

Using the new classification system, the cohort was divided into Ig-mediated MPGN (79.4%, 27/34) and complement-mediated MPGN (20.6%, 7/34). There was no significant difference between the two groups for age, sex, microscopic hematuria, excretion of urinary protein, or levels of serum creatinine, serum C3, or serum C4 at MPGN onset. The Ig-mediated MPGN group had more cases of nephrotic syndrome than the complement-mediated MPGN group, but the difference did not reach statistical significance (81.5 vs. 57.1%, $p > 0.05$). On the other hand, the serum albumin level was significantly lower in the Ig-mediated MPGN group than in the complement-mediated MPGN group (2.4 ± 0.7 vs. 2.9 ± 0.4, $p < 0.05$).

LM analysis revealed consistent findings of mesangial hypercellularity, endocapillary proliferation, and capillary wall remodeling with double contour formation, all of which resulted in a lobular accentuation of the glomerular tufts. IF showed positive staining for IgG, IgM, C3, and C4. In the context of the new classification system, all of the 27 patients with Ig-mediated MPGN were positive for IgG, IgM, and both complement factors. The 7 patients with complement-mediated MPGN showed only C3 immunostaining in the mesangium, and no Ig staining. EM showed mesangial and subendothelial deposits, and less commonly, intramembranous and subepithelial deposits. Of the 27 patients with Ig-mediated MPGN, 18 (66.7%) were classified as MPGN type I. Of the 7 patients with complement-mediated MPGN, 4 (57.1%) were classified as MPGN type I. There were no cases of MPGN type II in the cohort.

There were 3 secondary cases of MPGN, including 1 case due to monoclonal gammopathy of undetermined significance (MGUS) and 2 cases related to hepatitis B virus (HBV) infection. There were no cases of MPGN secondary to HCV infection. The 2 cases of secondary MPGN due to hepatitis were in the Ig-mediated MPGN group and the one secondary form of MPGN due to MGUS was in the complement-mediated MPGN group.

Treatment and Outcomes

Steroids were commonly prescribed, with 5.9% (2/34) of patients receiving prednisone alone and 82.4% (28/34) administered prednisone in combination with immunosuppressive agents. The immunosuppressive agents included cyclophosphamide (CTX) and tacrolimus (FK506). Therapy with FK506 was initiated at a dose of 0.1 mg/kg/day with a trough concentration between 5 and 10 ng/ml, and accompanied by oral prednisone initiated at 0.5 mg/kg/day. Therapy with intravenous CTX (IVC) was administered on a monthly basis at 0.5–0.75 g/m^2 and accompanied by oral prednisone initiated at 1.0 mg/kg/day. IVC was administered on a monthly basis for 6 months as induction therapy, with mycophenolate mofetil or azathioprine used as maintenance therapy. Of the 27 patients in the Ig-mediated MPGN group, 21 were treated with prednisone plus IVC, 3 were treated with prednisone plus FK506, 1 was treated with prednisone alone, and the 2 patients with secondary MPGN due to hepatitis were treated with antiviral agents. Of the 7 patients in the complement-mediated MPGN group, 4 were treated with prednisone plus IVC, including the patient with MPGN secondary to MGUS. All patients in this group were also treated with angiotensin-converting enzyme inhibitor and/or angiotensin receptor blockers.

During follow-up (median 6 months, range 3–47 months), 29.4% (10/34) of patients showed remission, including 4 who achieved complete remission (CR, defined as a decrease in proteinuria to a level ≤0.15 g/day) and 6 who achieved partial remission (PR, defined as a 50% decrease in proteinuria with stable creatinine level). In the Ig-mediated MPGN group, the remission rate was 3.0% (8/27) and was based on 3 CR cases and 5 PR cases. In the complement-mediated MPGN group, the remission rate was 2.9% (2/7) which was based on 1 CR case and 1 PR case. The difference in remission rates between the two groups did not reach statistical significance ($p > 0.05$). In the entire cohort, 82.4% (28/34) of patients presented with renal insufficiency at diagnosis and 5.9% (2/34) of patients progressed to end-stage renal disease (ESRD). Although the Ig-mediated MPGN group had fewer patients with renal insufficiency at diagnosis than the complement-mediated MPGN group (81.5 vs. 85.7%), more patients in the Ig-mediated MPGN group (7.4%, 2/27) progressed to ESRD during follow-up than in the complement-mediated MPGN group (0%, 0/7).

Discussion

The current study is, to our knowledge, the first to evaluate the recently proposed classification system for MPGN using a single-institute cohort. MPGN is considered a 'lesion', and not a 'disease'. Over the past few years, nephrologists and pathologists have both emphasized that MPGN is not a diagnosis per se, but rather a histopathologic pattern of injury that requires the identification of its underlying cause(s) [2–5]. According to epidemiological studies [6–8], the incidence of MPGN ranges from 1.4 to 9.3 cases per million population (pmp) per year [6], accounting for 7–10% of all cases of biopsy-confirmed glomerulonephritis [2]. Of note, the incidence has decreased over time (from ~7 pmp/year in the 1970s to ~2 pmp/year in the 1990s) [9–11]. The declining incidence of MPGN may be explained by the concomitant decrease in infection rates, which previously accounted for a major part of the MPGN cases [12, 13]. Nonetheless, MPGN continues to rank as the third or fourth leading cause of ESRD among the primary glomerulonephritides [2]. In this retrospective study of 34 Chinese patients, we determined the clinical and pathological characteristics of patients with MPGN and the underlying causes.

MPGN can occur at any age. As the current study population was composed solely of adult patients, the incidence, clinical manifestation, and prognosis may differ from those of children. It is well recognized that the clinical presentation and course of MPGN are extremely variable, ranging from benign and slowly progressive to rapidly progressive [2]. In the current cohort, the patients presented largely with nephrotic syndrome, microscopic hematuria, and hypocomplementemia, but hypertension and peripheral edema were present in all patients. There were no cases of gross hematuria or rapidly progressive glomerulonephritis. The varied clinical presentation is believed to reflect differences in the pathogenic mechanisms of the disorder and in the timing of the diagnostic biopsy relative to the clinical course.

In the current cohort, of the 7 patients with complement-mediated MPGN (as defined by new classification proposed by Sethi and Fervenza [2]), 4 were classified as type I MPGN based on EM. These patients showed only C3-positive staining by IF. It is possible that some of these patients may have been misclassified as MPGN type I. Such misclassification was previously demonstrated in a smaller cohort of patients from the USA for which 4 of the 12 patients with C3 glomerulonephritis (C3GN) were initially misclassified as MPGN type I [14]. The major shortcoming of the traditional clinical classification system is that many cases classified as MPGN types I and III show only C3 deposits and no Ig deposits on IF. It has been known for a long time that the alternative pathway of complement is abnormal in many patients with MPGN types I and III [1]. Thus, the traditional classification system fails to distinguish between the two major pathophysiologic factors leading to the development of a MPGN (i.e. abnormalities resulting in Ig deposition and abnormalities resulting in complement deposition). There were no cases of DDD in our cohort. In the new classification system of MPGN, DDD and C3GN share the same aspect on IF, namely positive C3 staining. DDD and C3GN are distinguishable by the appearance and location of deposits on EM, but these diseases should be considered part of a related spectrum of glomerulonephritis caused by alternative pathway of complement dysregulation [1].

Compared with a previous report from Mayo Clinic [15], the causes of MPGN differ largely between China and other parts of the world. During the period of 2001 through 2006, a total of 126 cases of MPGN were diagnosed at Mayo Clinic [15]. Of the 106 patients with evaluation for possible hepatitis virus infection, 25 (23.6%) were positive for HBV (n = 12), HCV (n = 13), or both (n = 2). In the current cohort, all 34 patients were evaluated for potential hepatitis viral infection, and only 2 (5.9%) were positive for HBV and none were positive for HCV. Thus, the positive rate of hepatitis was significantly lower for our study cohort than the Mayo Clinic cohort (5.9 vs. 23.6%, $p = 0.02$) (table 2). In Western coun-

Table 2. Comparison of the current Chinese cohort with the Western cohort of MPGN

	Current study (n = 34), n (%)	Mayo Clinic study (n = 126), n (%)	p
Screening for hepatitis, n	34	106	
HBV-positive	2 (5.9)	12 (11.3)	0.52
HCV-positive	0 (0)	13 (12.3)	NS
Negative	32 (94.1)	81 (76.4)	0.02
Screening for MGUS, n	10	68	
Abnormal	1 (10.0)	28 (41.2)	0.08
Normal	9 (90.0)	40 (58.8)	
Undetermined cause	31 (91.2)	73 (57.9)	<0.05

NS = Not significant.

tries, in addition to HBV, chronic HCV infection is the main cause of immune complex-mediated MPGN, and it is commonly associated with cryoglobulinemia [16]. In China, however, HBV remains the main cause of secondary MPGN. The lower positive rate of hepatitis in the current study cohort, compared to that of the Mayo Clinic cohort, could be partly explained by the high incidence of HCV infections detected in the Mayo Clinic cohort, as HCV infection is more common in Western countries.

In addition to chronic infections, autoimmune diseases, such as systemic lupus erythematosus, lupus nephritis [17, 18], Sjögren's syndrome [19], rheumatoid arthritis, mixed connective tissue disorders and thrombotic thrombocytopenic purpura [20], are also associated with persistent circulating immune complexes that can lead to the development of an MPGN lesion.

In the study from the Mayo Clinic [15], of the 68 patients evaluated for abnormal serum or urine protein electrophoresis, 28 (41.2%) were positive for monoclonal/biclonal Igs and 8 were positive for polyclonal Igs. Of the 28 patients with monoclonal and/or biclonal gammopathy, 24 were positive for monoclonal gammopathies and 4 were positive for biclonal gammopathies. Of the 24 monoclonal gammopathies, 10 were positive for IgM-κ, 9 were positive for IgG-λ, 4 were positive for IgG-κ, and 1 showed positivity for only λ light chains. None of their patients showed positivity for IgM-λ. In the current Chinese cohort, of the 10 patients evaluated for abnormal serum or urine protein electrophoresis, only 1 showed positivity for monoclonal IgG-λ. Thus, the diagnostic rate of MGUS in our study was significantly lower than that of the Mayo Clinic study (10.0 vs. 41.2%, p = 0.08) (table 2). As previously demonstrated by the Mayo Clinic study [15], there is a strong association of MPGN with MGUS. The diagnosis of MGUS requires a serum monoclonal paraprotein band of <30 g/l, a bone marrow biopsy showing <10% plasma cells, absence of lytic lesions, anemia and hypercalcemia, and absence of end-organ damage. As the most common plasma cell disorder recognized, it is a potential precursor for multiple myeloma [21]. In the current study cohort, only 1 (3.0%) patient with MPGN was diagnosed with monoclonal IgG-λ. Sethi et al. [15] reported that 41% of MPGN patients without an active autoimmune process or chronic infection had results from serum and/or urine electrophoresis studies that were positive for a monoclonal gammopathy. While underdiagnosis and misdiagnosis of MGUS exists, the positive screening rate for MGUS was significantly higher in the Mayo Clinic study than in our current study (41.2 vs. 10.0%, $p = 0.08$); we believe this was partly due to the higher incidence and prevalence of MGUS in Western countries [21]. Furthermore, correct and timely diagnosis of MGUS should be recommended, especially in patients with MPGN.

Undetermined causes of MPGN in the current study cohort accounted for 91.2% of all cases, which is significantly higher than the 57.9% that were reported in the Mayo Clinic study (table 2). In our study, patients with lupus and other autoimmune diseases were excluded from enrolment; therefore, of the 34 patients with MPGN, only 3 cases of secondary MPGN were identified, which were represented by 2 patients with MPGN secondary to HBV and 1 patient with MPGN secondary to MGUS. There were also significantly more cases without evident underlying causes of MPGN in our study than in the Mayo Clinic study (91.2 vs. 57.9%, $p < 0.05$).

With the aid of the new classification system, 4 patients that had been formerly classified as MPGN type I were reclassified as complement-mediated MPGN, providing a further opportunity to research the potential contribution of AP dysregulation. Thus, our study suggested that this new classification system may promote efforts to identify more underlying causes of MPGN.

Immunosuppressive agents are common in the treatment of MPGN and outcome is variable. To date, there is still controversy regarding the prognosis of MPGN. Nonetheless, in our study, 2 (5.9%) patients progressed to ESRD despite vigorous immunosuppression therapy. Unfortunately, there was no difference observed in the prognosis of the two MPGN groups, Ig-mediated MPGN and complement-mediated MPGN. Due to the limitation of this retrospective study being based on a cohort from a single center, it is difficult to evaluate the effect of the new classification system on prognosis. With improved understanding of the role of complement in the pathogenesis of MPGN, eculizumab [22–25] (a humanized monoclonal antibody to C5 that prevents generation of the membrane attack complex) has been proposed as a potentially effective therapy for complement-mediated MPGN. Nonetheless, large multicenter studies are still needed to gain a better understanding of MPGN and to design effective targeted therapies.

Conclusion

Our understanding of the pathogenesis of MPGN has evolved considerably in recent years. As a diagnosis of exclusion, truly idiopathic MPGN is substantially decreasing in frequency because of a better appreciation of its etiology and pathogenesis. The new classification system aids in identifying the underlying causes of MPGN based on IF findings. Immune complex-mediated MPGN includes MPGN resulting from chronic infections, autoimmune diseases, and dysproteinemias. On the other hand, complement-mediated MPGN results from dysregulation of the alternative pathway of complement. Future studies should focus on the underlying mechanism of MPGN rather than the 'lesion' itself.

Acknowledgments

This work was supported by grants from the National Basic Research Program of China 973 Program (Nos. 2012CB517600 and 2012CB517604), the National Key Technology R&D Program (12-5) (Nos. 2011BAI10B00 and 2011BAI10B06), the National Nature Science Foundation of China (No. 81000295), and the Key Program of Shanghai Science and Technology Commission (No. 08dz1900502).

References

1 Bomback AS, Appel GB: Pathogenesis of the C3 glomerulopathies and reclassification of MPGN. Nat Rev Nephrol 2012;8:634–642.
2 Sethi S, Fervenza FC: Membranoproliferative glomerulonephritis – a new look at an old entity. N Engl J Med 2012;36:1119–1131.
3 Fervenza FC, Sethi S, Glassock RJ: Idiopathic membranoproliferative glomerulonephritis: does it exist? Nephrol Dial Transplant 2012; 27:4288–4294.
4 Sethi S, Fervenza FC: Membranoproliferative glomerulonephritis: pathogenetic heterogeneity and proposal for a new classification. Semin Nephrol 2011;31:341–348.
5 Sethi S, Nester CM, Smith RJ: Membranoproliferative glomerulonephritis and C3 glomerulopathy: resolving the confusion. Kidney Int 2012;81:434–441.
6 Briganti EM, Dowling J, Finlay M, Hill PA, Jones CL, Kincaid-Smith PS, Sinclair R, McNeil JJ, Atkins RC: The incidence of biopsy-proven glomerulonephritis in Australia. Nephrol Dial Transplant 2001;16:1364–1367.
7 Maisonneuve P, Agodoa L, Gellert R, Stewart JH, Buccianti G, Lowenfels AB, Wolfe RA, Jones E, Disney AP, Briggs D, et al: Distribution of primary renal diseases leading to end-stage renal failure in the United States, Europe, and Australia/New Zealand: results from an international comparative study. Am J Kidney Dis 2000;35:157–165.
8 McGrogan A, Franssen CF, de Vries CS: The incidence of primary glomerulonephritis worldwide: a systematic review of the literature. Nephrol Dial Transplant 2011;26:414–430.
9 Swaminathan S, Leung N, Lager DJ, Melton LJ 3rd, Bergstralh EJ, Rohlinger A, Fervenza FC: Changing incidence of glomerular disease in Olmsted County, Minnesota: a 30-year renal biopsy study. Clin J Am Soc Nephrol 2006;1:483–487.
10 Braden GL, Mulhern JG, O'Shea MH, Nash SV, Ucci AA Jr, Germain MJ: Changing incidence of glomerular diseases in adults. Am J Kidney Dis 2000;35:878–883.

11 Zhou FD, Zhao MH, Zou WZ, Liu G, Wang H: The changing spectrum of primary glomerular diseases within 15 years: a survey of 3,331 patients in a single Chinese centre. Nephrol Dial Transplant 2009;24:870–876.
12 Ernst A, Radonic M, Belicza M: The syndrome of long-standing *Staphylococcus albus* bacteremia, bacterial endocarditis and diffuse membranoproliferative glomerulonephritis complicating ventriculoatrial shunt infection – case report and review of literature. Acta Med Iugosl 1980;34:137–152.
13 Michael AF, Herdman RC, Fish AJ, Pickering RJ, Vernier RL: Chronic membranoproliferative glomerulonephritis with hypocomplementemia. Transplant Proc 1969;1:925–932.
14 Sethi S, Fervenza FC, Zhang Y, Zand L, Vrana JA, Nasr SH, Theis JD, Dogan A, Smith RJ: C3 glomerulonephritis: clinicopathological findings, complement abnormalities, glomerular proteomic profile, treatment, and follow-up. Kidney Int 2012;82:465–473.
15 Sethi S, Zand L, Leung N, Smith RJ, Jevremonic D, Herrmann SS, Fervenza FC: Membranoproliferative glomerulonephritis secondary to monoclonal gammopathy. Clin J Am Soc Nephrol 2010;5:770–782.
16 Prakash S, Dash SC, Kumar A, Dinda AK, Agarwal SK, Acharya SK: Frequency and role of hepatitis-C virus and type II cryoglobulinemia in membranoproliferative glomerulonephritis. J Assoc Physicians India 2004;52: 451–453.
17 Zhu B, Chen N, Lin Y, Ren H, Zhang W, Wang W, Pan X, Yu H: Mycophenolate mofetil in induction and maintenance therapy of severe lupus nephritis: a meta-analysis of randomized controlled trials. Nephrol Dial Transplant 2007;22:1933–1942.
18 Li X, Ren H, Zhang Q, Zhang W, Wu X, Xu Y, Shen P, Chen N: Mycophenolate mofetil or tacrolimus compared with intravenous cyclophosphamide in the induction treatment for active lupus nephritis. Nephrol Dial Transplant 2012;27:1467–1472.
19 Ren H, Chen N, Chen X, Fu X, Jiang Y, Hao C, Dong D: Clinical and pathologic analysis of Sjögren's syndrome with renal impairment: a report of 84 cases (in Chinese). Zhonghua Nei Ke Za Zhi 2001;40:367–369.
20 Zhang W, Shi H, Ren H, Shen PY, Pan XX, Li X, Chen YX, Xu YW, Chen XN, Zhu P, et al: Clinicopathological characteristics and outcome of Chinese patients with thrombotic thrombocytopenic purpura-hemolytic uremic syndrome: a 9-year retrospective study. Nephron Clin Pract 2009;112:c177–c183.
21 Kyle RA, Rajkumar SV: Monoclonal gammopathy of undetermined significance. Br J Haematol 2006;134:573–589.
22 Radhakrishnan S, Lunn A, Kirschfink M, Thorner P, Hebert D, Langlois V, Pluthero F, Licht C: Eculizumab and refractory membranoproliferative glomerulonephritis. N Engl J Med 2012;366:1165–1166.
23 Vivarelli M, Pasini A, Emma F: Eculizumab for the treatment of dense-deposit disease. N Engl J Med 2012;366:1163–1165.
24 Daina E, Noris M, Remuzzi G: Eculizumab in a patient with dense-deposit disease. N Engl J Med 2012;366:1161–1163.
25 Bomback AS, Smith RJ, Barile GR, Zhang Y, Heher EC, Herlitz L, Stokes MB, Markowitz GS, D'Agati VD, Canetta PA, et al: Eculizumab for dense deposit disease and C3 glomerulonephritis. Clin J Am Soc Nephrol 2012;7: 748–756.

Prof. Nan Chen, MD
Department of Nephrology, Ruijin Hospital
School of Medicine, Shanghai Jiao Tong University
No. 197 Ruijin Er Road, Shanghai 200025 (PR China)
E-Mail chen-nan@medmail.com.cn

Chen N (ed): New Insights into Glomerulonephritis.
Contrib Nephrol. Basel, Karger, 2013, vol 181, pp 185–193 (DOI: 10.1159/000348654)

C3 Glomerulopathy

Aude Servais[a, d] · Laure-Hélène Noël[b, c] · Véronique Frémeaux-Bacchi[e, f] · Philippe Lesavre[a, d]

Departments of [a]Nephrology and [b]Pathology, and [c]INSERM U 845, Hôpital Necker-Enfants Malades AP-HP, [d]Université Paris Descartes, [e]Cordeliers Research Center, INSERM UMRS 872, [f]Assistance Publique - Hopitaux de Paris, Department of Immunology, Hôpital Européen Georges Pompidou, Paris, France

Abstract

C3 glomerulopathy is a recent disease classification comprising several rare types of glomerulonephritis, including dense deposit disease (DDD) and C3 glomerulonephritis (C3GN). The most common histological feature in these diseases is the presence of glomerular deposition of C3 within the mesangium and along the glomerular basement membrane (GBM) in the subendothelial area or within the GBM. The key role of complement alternative pathway (AP) in these disorders has been recently shown with the identification of acquired or genetic abnormalities. Low serum C3 level but normal C4 is a common finding. The acquired AP dysregulation in DDD and C3GN may be first induced by C3 nephritic factor (C3NeF). C3NeF activity is found in approximately 80% of patients with DDD and in 45% of patients with C3GN. The correlation with the complement is further supported by the detection of homozygous or heterozygous mutations in the regulatory complement proteins factor H (CFH), factor I (CFI), or C3. The genetic background of the patients may also influence the disease manifestation since common genetic variants including single nucleotide polymorphisms in the CFH, C3 and CFHR5 genes are associated with DDD and one at-risk MCP haplotype have been found to be significantly increased in C3GN. C3 glomerulopathies can present over a broad age range. DDD is more often diagnosed in children and age at diagnosis is significantly higher for patients with C3GN. Presenting features comprise any of proteinuria, hematuria, hypertension and renal failure. These glomerulonephritides are associated with chronic deterioration of renal function, leading to ESRD within 10 years of the diagnosis in 36–50% of patients. Outcomes of renal transplantation are characterized by histological recurrence which may contribute to increased rates of allograft failure. Administration of recombinant FH if it becomes available or replacement of FH via plasma exchange may be efficacious in the cases of FH deficiency. However, therapeutic inhibition of complement C3 and C5 is the main perspective.

Copyright © 2013 S. Karger AG, Basel

C3 glomerulopathy is a recent disease classification comprising several rare types of glomerulonephritis, including dense deposit disease (DDD) and C3 glomerulonephritis (C3GN). The most common histological feature in these diseases is the presence of glomerular deposition of C3 within the mesangium and along the glomerular basement membrane (GBM) in the subendothelial area or within the GBM. The key role of complement alternative pathway (AP) in these disorders has been recently shown with the identification of acquired or genetic abnormalities.

Histology

Until recently, the classification for membranoproliferative glomerulonephritis (MPGN) was based on a histopathologic pattern. Clinically, patients were divided into secondary cases when an etiology could be identified, including autoimmune disease, hepatitis C infection, cryoglobulinemia, and primary cases, when no such etiology was identified. The membranoproliferative pattern is characterized by mesangial hypercellularity, endocapillary proliferation, and capillary wall remodeling with mesangial interposition and duplication of GBM. These changes result in a lobular accentuation of the glomerular tufts and are associated with deposits of complement components associated or not with immunoglobulins.

MPGN type I is defined as mesangial cellular proliferation, increase in the mesangial matrix, a diffuse 'double contours' aspect and subendothelial, and mesangial deposits (light microscopy), C3 and significant granular Ig deposits in mesangial area, mainly IgG and often C1q, but also in some cases IgM and/or IgA (immunofluorescence study), and subendothelial, mesangial and some epimembranous deposits (electron microscopy) (fig. 1a–d).

MPGN type II, more accurately called DDD, is defined as thickening of capillary walls due to both GBM deposits and double contours (light microscopy), deposits (mainly C3) in the GBM (immunofluorescence study), and thickening of the GBM due to electron-dense deposits within the lamina densa (electron microscopy) (fig. 1e–h). The essential feature of DDD is that presence of electron-dense transformation of the GBM, and not the membranoproliferative pattern [1]. Infiltrates of neutrophils and cellular crescents may also be observed.

In MPGN type III, deposits can be both subendothelial and subepithelial (Burkholder variant) or produce complex intramembranous, subendothelial and subepithelial formations with fraying of the lamina densa (Strife and Anders variant).

C3GN is a recently described entity defined as deposits containing only C3 without immunoglobulin deposits (immunofluorescence study) with or without

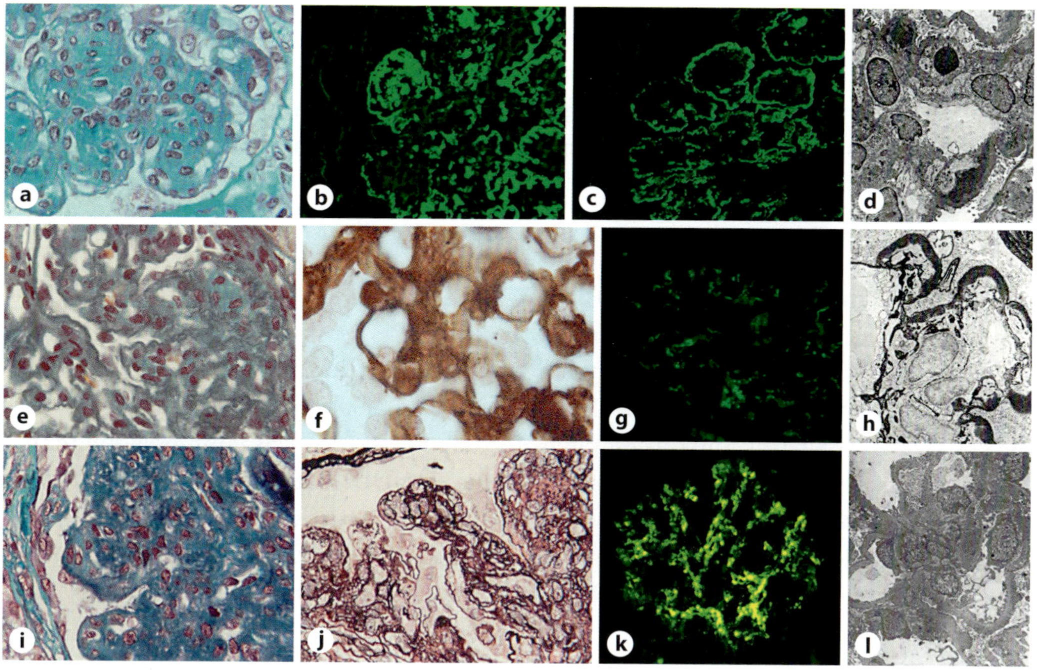

Fig. 1. Illustrative cases of DDD, MPGN type I (MPGN I) and glomerulonephritis with isolated C3 deposits (C3GN) by light, immunofluorescence and electron microscopy. **a** DDD with mesangial proliferation and ribbon-like aspect of the GBMs. Light green trichrome. ×250. **b** DDD with mesangial deposits and characteristic aspect of the GBMs. Jone's staining. ×1,000. **c** DDD with mesangial C3 deposits and a mild fixation along the GBMs by immunofluorescence study. ×250. **d** DDD with electron-dense transformation of the lamina densa by electron microscopy. Uranyl acetate and lead citrate. ×2,500. **e** MPGN I with mesangial proliferation and endomembranous deposits. Light green trichrome. ×250. **f** MPGN I with IgG deposits along the GBMs and in the mesangial space by immunofluorescence study. ×250. **g** MPGN I with C3 deposits in the same localization than IgG by immunofluorescence study. ×250. **h** MPGN I with subendothelial deposits by electron microscopy (uranyl acetate and lead citrate. ×2,500. **i** C3GN with endomembranous deposits and granular deposits in the mesangial space. Light green trichrome. ×250. **j** C3GN with double contours and complex changes of the GBMs. Jone's staining. ×250. **k** C3GN with C3 deposits (and without immunoglobulin) by immunofluorescence study. ×250. **l** C3GN with subendothelial, mesangial, and some epimembranous deposits without transformation of the lamina densa by electron microscopy. Uranyl acetate and lead citrate. ×2,500.

membranoproliferative pattern, and without dense deposits within the GBM (electron microscopy) (fig. 1i–l) [2–4]. In 70% of the C3GN cases, renal biopsy shows a membranoproliferative pattern with mesangial proliferation, subendothelial, mesangial, and less frequently epimembranous deposits, diffuse 'double contours' aspect and accumulation of mesangial matrix. In 30% of cases, renal biopsy discloses mesangial and epimembranous deposits without subendothelial deposits or mesangial proliferation. Crescents may be observed. In all cases, immunofluorescence study reveals isolated C3 deposits without dense intramembranous deposits. This pattern is clearly not acute poststreptococcal glomerulonephritis because of the absence of exudative lesions and the presence of mesangial deposits.

Composition of the Deposits

The most common histological feature in all these patients is the presence of glomerular deposition of C3. The deposits in biopsy samples from patients with MPGN type I contain complement components as well as immunoglobulins. By contrast, biopsy samples from patients with DDD and C3GN show deposits of C3 only, without any immunoglobulin deposition. The MPGN type III Burkholder variant usually shows deposits of immunoglobulin and complement, while the MPGN type III Strife and Anders variant shows deposits of C3 either alone or in combination with immunoglobulins.

Sethi et al. [5] used mass spectrometry to show that deposits in DDD and C3GN contain components of the AP including C3b and its breakdown products iC3b, C3dg or C3c and terminal complement complex. Furthermore, biopsy samples from patients with DDD and C3GN demonstrated strong and equal C3 and C5b-9 staining in the glomerulus providing evidence of activation of C3 leading to C5 recruitment in the fluid phase and assembly of the membrane attack complex [6].

The disorders that share the key histological feature of isolated C3 deposits in the glomerulus, with no immunoglobulin, reflecting complement C3 convertase activation, have been recently classified as C3 glomerulopathy. This entity includes DDD and C3GN and, in turn, signals dysregulation of the AP. A renal biopsy diagnosis of C3 glomerulopathy should prompt investigation for complement abnormalities including protein levels, gene mutations and autoantibodies.

Role of Complement Alternative Pathway in C3 Glomerulopathy

DDD and C3GN are associated with acquired or genetic dysregulation of the complement AP [2, 3, 7]. Low serum C3 level but normal C4 is a common finding, suggesting an AP consumption. Indeed, in our series, 60 and 40% of the DDD and C3GN patients, respectively, had a low C3, and 20% of DDD and C3GN patients presented with less than 20% of normal C3 [3].

The acquired AP dysregulation in DDD and C3GN may be first induced by C3 nephritic factor (C3NeF) [7]. First described in DDD, but not specific for this disease, C3NeF binds and stabilizes the fluid phase and the membrane-bound C3 convertases. C3NeF activity is found in approximately 80% of patients with DDD and in 45% of patients with C3GN [3, 7–10]. The frequency may be variable since a fluctuation of the C3NeF activity may be observed in one third of patients during the follow-up [3]. In most reported DDD patients, C3NeF is typically associated with low C3 levels in plasma [10, 11]. However, C3NeF is not always associated with evidence of complement consumption in plasma. Indeed, in our series, C3 remains in the normal range in approximately 40% of patients, suggesting the capacity of complement proteins to regulate the C3NeF-stabilized C3 convertase [3]. As C3NeF has also been identified in healthy individuals, the exact role of C3NeF in C3 glomerulopathies remains undefined. Furthermore, fluctuating levels do not correlate with the course of nephritis.

Additional serologic autoantibodies directed at a complement regulator protein, such as autoantibodies to factor H [5, 12], or autoantibodies directed at the individual components of the C3 convertase, such as autoantibodies to C3b and/or CFB [13, 14], have been reported.

The strong correlation between the complement and these glomerulopathies is further supported by the detection of homozygous or heterozygous mutations in the regulatory complement proteins CFH, CFI, or C3 [2, 3, 8, 15–19] which can lead to hyperactivity of the AP. We identified a genetic abnormality in complement genes in 17% of DDD patients and 20% of C3GN patients, half these patients also being C3NeF positive [3]. The absence of FH in plasma as a cause of C3 glomerulopathies has been observed in humans, pigs, and mice [20, 21]. However, complete CFH deficiency accounts for only a small minority of cases [3, 15, 22, 23]. Heterozygous mutations in CFH and CFI associated with low protein levels were identified [3], suggestive of a quantitative deficiency (type I mutations). Other mutations were found in patients with normal CFH levels (type II mutations). For some of them a functional deficiency was suggested since they mapped near the C3b binding sites and may affect the generation of the C3b metabolites. Interestingly, CFH deficiency (complete and quantitative) is also frequently found in patients with atypical hemolytic uremic syndrome (aHUS). In our series, we found mutations in CFH and in CFI, which have been previously identified in patients with aHUS [24, 25].

Gale et al. [26] reported mutation in the gene encoding CFHR5 in Cypriot families presenting with microscopic hematuria with autosomal dominant inheritance and C3GN. An internal duplication of exons 2–3 of the CFHR5 gene resulted in a novel CFHR5 that is less effective than native CFHR5 in association with surface-bound C3b to effectively regulate the activity of C3 and C5 convertases. Recently, Malik et al. [27] also reported an autosomal dominant MPGN III associated with abnormal increases in copy number across the CFHR3 and CFHR1 loci. In addition to normal copies of these genes, affected individuals carry a unique hybrid CFHR3-1 gene.

However, the mutations in the complement AP genes by themselves may not be sufficient to determine the disease phenotype and the genetic background of the patients may influence the disease manifestation. Common genetic variants including single nucleotide polymorphisms in the CFH, C3 and CFHR5 genes have been recognized as modifying risk of DDD [28]. CFH Y402H variant is significantly increased in DDD patients [3, 28, 29]. The CFH Y402H variant is also highly associated with an increased risk for development of age-related macular degeneration. DDD glomeruli contain breakdown products of C3b without factor B, suggesting that the major site of AP activity is the fluid phase [10]. The functional difference between Y and H at position 402 of CFH remains unclear. The at-risk Y402H does not affect the fluid phase complement control, but seems to confer a poorer regulation of the C3 convertase on cell surface or in the GBM. The increasing generation of iC3b and C3d in the circulation, due to the presence of C3Nef or CFH mutations, may lead to their direct accumulation in the GBM. It is worth noting that none of the patients with DDD presented with a genetic abnormality in CFI gene. This may suggest a role of CFI in the accumulation of C3 degradation products in the GBM and a key role for iC3b in the pathogenesis of DDD. Indeed, it has been demonstrated that the subendothelial C3 deposits do not develop in mice with combined FH and FI deficiency [30]. Therefore, FI-mediated generation of activated C3 fragments in the circulation under the influence of FH is a critical determinant of the location of the deposits.

One at-risk MCP haplotype (aaggt) was also found to be significantly increased in C3GN [3]. This haplotype is different from the ones identified in aHUS. Therefore, although genetic abnormalities are similar in C3 glomerulopathies and aHUS, the genetic background of the patients seems to play an important role in the disease phenotype. These results highlight that MCP, which is expressed on endothelial and mesangial cells and participates to the generation of the C3 metabolite iC3b, is implicated in the glomerular changes in patients presenting fluid phase activation of complement.

Clinical Features

C3 glomerulopathies can present over a broad age range. DDD is more often diagnosed in children and age at diagnosis is significantly higher for patients with C3GN. Presenting features comprise any of proteinuria (sometimes with the nephrotic syndrome), hematuria, hypertension and renal failure. Individuals with DDD may have acquired partial lipodystrophy and/or ocular drusen.

These glomerulonephritides are associated with chronic deterioration of renal function, leading to ESRD within 10 years of the diagnosis in 36–50% of patients [3, 10, 11]. Renal survival is similar in DDD and C3GN. However, in adults, renal survival is worse in patients with DDD than in patients with C3GN. Furthermore, adults are more likely to suffer from a rapid decline in renal function than children [3]. Renal failure at diagnosis is associated with an adverse prognosis [3, 10]. Renal survival does not differ between patients with normal or low C3 concentration.

Outcomes of renal transplantation are characterized by histological recurrence which may contribute to increased rates of allograft failure [7, 31]. Some patients may develop thrombotic microangiopathy before or after C3 glomerulopathy onset or after transplantation [3].

Treatment

The optimal treatment remains undefined partly because there have been no appropriate trials. Nasr et al. [10] reported that combined immunosuppressive/renin angiotensin blockade was better than either agent alone in DDD. In our large retrospective study, we also found that renin angiotensin blockade was associated with a better renal survival. In MPGN, the KDIGO recommend the use of immunosuppressive therapy in case of nephrotic syndrome and/or deteriorating kidney function which could include cyclophosphamide or MMF and corticosteroids. In patients with an acquired antibody to an inhibitory protein of the AP, immunosuppressive agents might decrease the production of the antibody and lead to clinical improvement. However, this approach has not proven its efficiency and the most appropriate immunosuppressive regimen has to be defined (corticosteroids, mycophenolate mofetil or rituximab).

Administration of recombinant FH if it becomes available or replacement of FH via plasma exchange may be efficacious in the cases of FH deficiency. However, therapeutic inhibition of complement C3 and C5 is the main perspective. Indeed, the inflammatory glomerular lesions, most likely triggered by C5a release, have a major role in the progression towards renal fibrosis. Eculizumab is a recombinant humanized monoclonal antibody directed against human complement component C5. Reports of 10 cases of C3 glomerulopathies treated with eculizumab are available [32], including 6 cases of C3 glomerulopathy affecting native kidney [19] and 4 cases recurring in the renal graft [19]. The use of eculizumab led to an improvement of renal function in 6 out of 10 cases and of proteinuria only with a slight degradation of renal function in an additional patient. An increased level of circulating sC5b-9 and the presence of endocapillary or extracapillary proliferation correlated with the beneficial effect of the treatment. Limited available data suggest that eculizumab is efficient in treating rapidly progressive and/or crescentic forms of C3 glomerulopathy recurring in the renal graft. Whether preventive treatment with eculizumab is warranted after renal transplantation in patients with a history of rapidly progressive C3 glomerulopathy in their native kidney remains unclear [32].

References

1 Walker PD, Ferrario F, Joh K, et al: Dense deposit disease is not a membranoproliferative glomerulonephritis. Mod Pathol 2007;20: 605–616.

2 Servais A, Fremeaux-Bacchi V, Lequintrec M, et al: Primary glomerulonephritis with isolated C3 deposits: a new entity which shares common genetic risk factors with haemolytic uraemic syndrome. J Med Genet 2007;44: 193–199.

3 Servais A, Noel L-H, Roumenina LT, et al: Acquired and genetic complement abnormalities play a critical role in dense deposit disease and other C3 glomerulopathies. Kidney Int 2012;82:454–464.

4 Berger J, Noel L-H, Yaneva H: Complement deposition in the kidney; in Hamburger J, Crosnier J, Maxwell MH (eds): Advances in Nephrology. Chicago, Year Book Medical Publishers, 1974, vol 4, pp 37–48.

5 Sethi S, Fervenza FC, Zhang Y, et al: Proliferative glomerulonephritis secondary to dysfunction of the alternative pathway of complement. Clin J Am Soc Nephrol 2011;6: 1009–1017.

6 Herlitz LC, Bomback AS, Markowitz GS, et al: Pathology after eculizumab in dense deposit disease and C3 GN. J Am Soc Nephrol 2012;23:1229–1237.

7 Appel GB, Cook HT, Hageman G, et al: Membranoproliferative glomerulonephritis type II (dense deposit disease): an update. J Am Soc Nephrol 2005;16:1392–1403.

8 Sethi S, Fervenza FC, Zhang Y, et al: C3 glomerulonephritis: clinicopathological findings, complement abnormalities, glomerular proteomic profile, treatment, and follow-up. Kidney Int 2012;82:465–473.

9 Zhang Y, Meyer NC, Wang K, et al: Causes of alternative pathway dysregulation in dense deposit disease. Clin J Am Soc Nephrol 2012; 7:265–274.

10 Nasr SH, Valeri AM, Appel GB, et al: Dense deposit disease: clinicopathologic study of 32 pediatric and adult patients. Clin J Am Soc Nephrol 2009;4:22–32.

11 Little MA, Dupont P, Campbell E, et al: Severity of primary MPGN, rather than MPGN type, determines renal survival and post-transplantation recurrence risk. Kidney Int 2006;69:504–511.

12 Goodship TH, Pappworth IY, Toth T, et al: Factor H autoantibodies in membranoproliferative glomerulonephritis. Mol Immunol 2012;52:200–206.

13 Strobel S, Zimmering M, Papp K, et al: Anti-factor B autoantibody in dense deposit disease. Mol Immunol 2010;47:1476–1483.

14 Chen Q, Muller D, Rudolph B, et al: Combined C3b and factor B autoantibodies and MPGN type II. N Engl J Med 2011;365:2340–2342.

15 Servais A, Noel L-H, Dragon-Durey MA, et al: Heterogeneous pattern of renal disease associated with homozygous factor H deficiency. Hum Pathol 2011;42:1305–1311.

16 Martinez-Barricarte R, Heurich M, Valdes-Canedo F, et al: Human C3 mutation reveals a mechanism of dense deposit disease pathogenesis and provides insights into complement activation and regulation. J Clin Invest 2010;120:3702–3712.

17 Montes T, de Jorge EG, Ramos R, et al: Genetic deficiency of complement factor H in a patient with age-related macular degeneration and membranoproliferative glomerulonephritis. Mol Immunol 2008;45:2897–2904.
18 Dragon-Durey MA, Fremeaux-Bacchi V, Loirat C, et al: Heterozygous and homozygous factor h deficiencies associated with hemolytic uremic syndrome or membranoproliferative glomerulonephritis: report and genetic analysis of 16 cases. J Am Soc Nephrol 2004;15:787–795.
19 Bomback AS, Smith RJ, Barile GR, et al: Eculizumab for dense deposit disease and C3 glomerulonephritis. Clin J Am Soc Nephrol 2012;7:748–756.
20 Hogasen K, Jansen JH, Mollnes TE, et al: Hereditary porcine membranoproliferative glomerulonephritis type II is caused by factor H deficiency. J Clin Invest 1995;95:1054–1061.
21 Pickering MC, Cook HT, Warren J, et al: Uncontrolled C3 activation causes membranoproliferative glomerulonephritis in mice deficient in complement factor H. Nat Genet 2002;31:424–428.
22 Habbig S, Mihatsch MJ, Heinen S, et al: C3 deposition glomerulopathy due to a functional factor H defect. Kidney Int 2009;75:1230–1234.
23 Schejbel L, Schmidt IM, Kirchhoff M, et al: Complement factor H deficiency and endocapillary glomerulonephritis due to paternal isodisomy and a novel factor H mutation. Genes Immun 2011;12:90–99.
24 Saunders RE, Abarrategui-Garrido C, Fremeaux-Bacchi V, et al: The interactive factor H-atypical hemolytic uremic syndrome mutation database and website: update and integration of membrane cofactor protein and factor I mutations with structural models. Hum Mutat 2007;28:222–234.
25 Nilsson SC, Kalchishkova N, Trouw LA, et al: Mutations in complement factor I as found in atypical hemolytic uremic syndrome lead to either altered secretion or altered function of factor I. Eur J Immunol 2010;40:172–185.
26 Gale DP, de Jorge EG, Cook HT, et al: Identification of a mutation in complement factor H-related protein 5 in patients of Cypriot origin with glomerulonephritis. Lancet 2010;376:794–801.
27 Malik TH, Lavin PJ, de Jorge EG, et al: A hybrid CFHR3-1 gene causes familial C3 glomerulopathy. J Am Soc Nephrol 2012;23:1122.
28 Abrera-Abeleda MA, Nishimura C, Smith JL, et al: Variations in the complement regulatory genes factor H (CFH) and factor H related 5 (CFHR5) are associated with membranoproliferative glomerulonephritis type II (dense deposit disease). J Med Genet 2006;43:582–589.
29 Sethi S, Fervenza FC, Zhang Y, et al: Proliferative glomerulonephritis secondary to dysfunction of the alternative pathway of complement. Clin J Am Soc Nephrol 2011;5:1009–1017.
30 Rose KL, Paixao-Cavalcante D, Fish J, et al: Factor I is required for the development of membranoproliferative glomerulonephritis in factor H-deficient mice. J Clin Invest 2008;118:608–618.
31 Angelo JR, Bell CS, Braun MC: Allograft failure in kidney transplant recipients with membranoproliferative glomerulonephritis. Am J Kidney Dis 2012;57:291–299.
32 Zuber J, Fakhouri F, Roumenina LT, et al: Use of eculizumab for atypical haemolytic uraemic syndrome and C3 glomerulopathies. Nat Rev Nephrol 2012;8:643–657.

Dr. Aude Servais
Service de Néphrologie adultes
Hôpital Necker, 149, rue de Sèvres
FR–75015 Paris (France)
E-Mail aude.servais@nck.aphp.fr

Chen N (ed): New Insights into Glomerulonephritis.
Contrib Nephrol. Basel, Karger, 2013, vol 181, pp 194–206 (DOI: 10.1159/000348477)

Hepatitis C Virus-Associated Glomerulonephritis

Sydney C.W. Tang[a] · Kar Neng Lai[a, b]

[a]Division of Nephrology, Department of Medicine, University of Hong Kong, Queen Mary Hospital, and [b]Nephrology Center, Hong Kong Sanatorium and Hospital, Hong Kong SAR, China

Abstract

Hepatitis C virus (HCV) infection remains a major global health burden. In addition to liver-related morbidity and mortality due to hepatic decompensation and development of hepatocellular carcinoma, extrahepatic manifestations of hepatitis C are frequent. Type I membranoproliferative glomerulonephritis associated with type II cryoglobulinemia is the most frequent association. The pathogenesis of the renal lesions is related to glomerular deposition of immune complexes. International guidelines recommend that patients with HCV-related glomerulopathies should be treated with antiviral therapy in the form of standard or pegylated interferon-α and ribavirin. The recent development of direct-acting antiviral agents that inhibit the various steps of the viral life cycle represents a major milestone for the treatment of chronic HCV infection. Initial reports suggest that they are safe in ESRD subjects. However, data is lacking on their efficacy and safety in HCV-related glomerulonephritis.
Copyright © 2013 S. Karger AG, Basel

Discovery of the HCV

In the early 1970s, when specific diagnostics for hepatitis A and B were developed, it became clear that most cases of hepatitis arising from blood transfusions at the time were not caused by these viral agents. It was only after over a decade of further research that the agent responsible for this so-called non-A, non-B hepatitis (NANBH) was identified in 1989 [1]. It was since referred to as the hepatitis C virus (HCV). Very soon, people began to realize the association between HCV and the previous mysterious nature of 'mixed' cryoglobulinemia [2] that often manifested as a triad of palpable purpura, arthralgias and weakness plus variable degrees of glomerulonephritis (GN), lymphadenopathy, and hepatosplenomegaly. Cryoglobulinemia in these patients had a 'mixed' composition of IgG and IgM rheumatoid factor (RF). It is now apparent that the classical clinical triad occurs in a minority of patients and that the main characteristic of the disease is the markedly heterogeneous manifestations of a systemic vasculitis, with purpuric skin lesions that show leukocytoclastic vasculitis on biopsy being an almost constant and predominant feature.

HCV Virology and Replication

HCV is a small RNA virus that is included in the Flaviviridae family and has been recently classified as the sole member of the genus hepacivirus [3]. HCV is a small double-shelled virus consisting of a lipid envelope (E) with virally encoded glycoproteins (E1 and E2) and an inner nucleocapsid (core) that contains a positive-sense single-stranded RNA genome of 9,500 nucleotides. It has well-defined structural (core, E1, and E2) as well as several nonstructural (from NS2 to NS5) proteins. The nonstructural proteins encode several proteases, a virus-specific helicase, and an RNA-dependent RNA polymerase responsible for replication of the genome. Chronic HCV infection is also known for its high degree of genetic heterogeneity, with six major genotypes identified [4].

The multiple steps of the HCV life cycle have recently been unraveled, as a result of the development of a number of cell-free or cell culture-based model systems. The HCV life cycle comprising early steps, polyprotein synthesis, maturation of viral proteins, replication, and finally late steps of viral particle formation, offers a large number of potential targets of intervention for the development of specific antiviral drugs, including direct-acting antiviral drugs and host-targeted agents [5].

Epidemiology

Chronic HCV infection is a major cause of mortality and morbidity throughout the world, with an estimated 170 million people infected worldwide, amounting to 3.1% of the world's population [6]. Some 350,000 patients die every year from HCV infection. Its prevalence varies widely among countries, being highest in several African and Eastern Mediterranean countries. Most studies report findings of genotypes 1–3, though a large number of patients worldwide have genotypes 4–6. HCV genotype 6 is mostly found in South-East Asia, ranging from

18% in Thailand, 23.6% in Hong Kong, to 49% in Myanmar. It is the second most common genotype in southern China after genotype 1. HCV genotype 1 patients are largely infected through blood transfusion, while HCV genotype 6 is commonly found in intravenous drug users [7].

Natural History of Acute and Chronic HCV Infection

HCV is acquired mainly via blood contact. Thus, the main risk groups are intravenous drug users, recipients of blood or blood product transfusions before systematic screening in 1992 and healthcare workers. Sexual transmission accounts only for a minority of cases, however acute HCV has become a significant problem in HIV-infected men having sex with men (MSM) in recent years. HCV RNA levels rise rapidly after infection followed by a delayed increase of serum alanine aminotransferases (ALT) 4–12 weeks after infection indicating hepatic injury. ALT levels frequently reach values of over 10 times the upper limit of normal with concomitant rises in serum bilirubin. Still, the majority of patients remain asymptomatic during the acute phase and most of the infected individuals do not become aware of their disease. Even in symptomatic patients, most of the clinical signs are nonspecific. Fulminant hepatitis has only been rarely reported.

Once chronic HCV infection is established, spontaneous HCV clearance rarely occurs. Chronic hepatitis C can cause continuous liver damage resulting in cirrhosis and subsequently hepatocellular carcinoma (HCC). The individual course of liver disease is highly variable. The median time for development of cirrhosis is estimated to be 20 years. For patients with established cirrhosis, the annual mortality is 4%. Markov modeling estimates that 1 in 4 patients with chronic HCV will develop cirrhosis, and about 50% of those with cirrhosis will die from liver disease [8]. The cumulative probability of developing HCC in HCV carriers is estimated at 18% in 10 years.

Extrahepatic Manifestation

Although HCV infection appears to be primarily a disease that is almost exclusively confined to the liver, a wide variety of extrahepatic manifestations (EHMs) have been reported [9]. Up to 40–74% of patients with HCV infection develop at least one EHM during their lifetime. Occasionally, EHMs might be the leading clinical manifestation of HCV infection and can determine the overall prognosis of the disease.

The disorder with the strongest link to HCV infection is mixed cryoglobulinemia (MC). MC is defined by the presence of cryoglobulins in the blood, which are immunoglobulins that precipitate at temperatures lower than 37°C. Overall, cryoglobulins can be found in 19–55% of HCV-infected individuals. There are three different types of cryoglobulinemia. In type I immunoglobulins are of monoclonal origin. This variant is mainly found in B-cell malignancies. HCV infection is associated in 60–96% of cases with type II and III cryoglobulinemia which both contain at least one polyclonal 'mixed' component in the precipitating immunoglobulins. Therefore they are called MCs. In type II a polyclonal IgG and a monoclonal IgM element with RF activity can be found whereas in type III all cryoglobulins are of polyclonal origin.

Up to 30% of patients with detectable cryoglobulins develop MC-related symptoms collectively called MC syndrome. Patients with MC syndrome may develop systemic vasculitis. Other common symptoms are peripheral polyneuropathy, Raynaud's syndrome and Sicca-like symptoms. Around 10% of patients develop a B-cell lymphoma at some stage. Clinically, the most relevant manifestation of MC is a membranoproliferative glomerulonephritis (MPGN), which appears in 30–36% of cases and is associated with significantly increased morbidity and mortality. An impaired renal function also complicates antiviral therapy, for instance the management of anemia can be challenging.

In addition, HCV seems to influence several arms of metabolism. The overall prevalence of insulin resistance and type 2 diabetes is increased in patients with chronic HCV infection. Of note, insulin resistance improves after successful eradication of HCV. Insulin resistance has a considerable impact on the natural course of chronic HCV infection. It leads to faster progression of liver fibrosis and a higher prevalence of HCC. HCV seems to affect also the serum lipid proteins. Patients with HCV infection have lower serum levels of cholesterol and triglycerides.

Another important EHM of HCV infection is the involvement of the central nervous system. About 20–80% of the patients with chronic HCV infection develop fatigue at some point independent of the severity of hepatitis. Fatigue often is the predominant complaint that reduces quality of life. Patients may also develop depression or a general cognitive impairment irrespective of the stage of liver disease which may be linked to HCV-induced neuroinflammation and brain dysfunction.

Pathogenesis of HCV-Associated of GN

Two immunologic features of HCV may predispose to chronic infection. First, HCV is known to evade immune elimination, and lead to chronic infection and accumulation of circulating immune complexes. MPGN associated with HCV infection may be the result of this phenomenon. The second feature is the ability of HCV to induce production of monoclonal RFs (mRF). The pathogenesis of cryoglobulinemic GN associated with HCV infection is related to the deposition of immune complexes containing HCV proteins, anti-HCV antibodies and RF IgM with activity anti-IgG [10]. Recently, it has been described that toll-like receptor 3 (TLR3) messenger RNA expression was elevated in mesangial cells in HCV-associated GN and was associated with enhanced proinflammatory cytokines [11]. Therefore, TLR3 could also have a pathogenic role establishing a link between viral infection and GN.

Clinical Findings in HCV-Related Cryoglobulinemia

Full-blown symptomatic cryoglobulinemia occurs infrequently, and the typical symptoms are fatigue and palpable purpura, which histologically consists of a leukocytoclastic vasculitis (with complexes of anti-HCV and HCV in injured tissue). These lesions are usually found on the lower limbs, although they can occur anywhere, and represent small vessel vasculitis. A smaller proportion of patients have fever, arthritis, Raynaud's phenomenon, and neuropathy. Peripheral neuropathy is usually characterized by paresthesia and variable degree of motor deficit. Abdominal pain arises from mesenteric vasculitis, and may mimic an acute abdomen during disease flare. Hepatosplenomegaly is due to chronic liver disease as a result of HCV. Cryoglobulinemia is more common in women than men and typically occurs after a prolonged period, typically years or decades, of HCV infection. Although the course of illness tends to wax and wane, occasionally the systemic illness can be severe or even fulminant.

Cryoglobulinemic Glomerulonephritis

The principal renal manifestation of HCV infection is MPGN type I, usually in the context of type II MC [12]. The prevalence of MPGN in HCV-type II cryoglobulinemia is approximately 30%. A cross-sectional analysis of the NHANES III data demonstrated an age-dependent association between HCV seropositivity and albuminuria (adjusted odds ratio of 1.84 for ages 40–59 years and 1.27 for ages ≥60 years) [13]. In the ≥60 year age group, 46% of HCV-positive individuals had albuminuria compared with 24% of those who were HCV antibody-negative. A similar study from Taiwan found that nondiabetic patients who were anti-HCV positive had a high prevalence of proteinuria (>1+ dipstick) than the seronegative group (8.3 vs. 5.1%, $p = 0.002$) [14]. Both studies found no association with reduced eGFR. In an Italian study of 320 predialysis patients, the incidence of HCV infection was 6.25%, which was much higher than that expected for Europe, i.e. 0.2–3.5% [15]. These data suggest that HCV infection is a risk factor for proteinuria and/or impaired renal function. On the other hand, the prevalence of anti-HCV among patients with MPGN is much lower among Chinese [16].

Renal disease is rare in children and the typical age of disease onset is in the fifth or sixth decade of life after long-standing infection, often in association with mild subclinical liver disease. Clinically, patients may have other symptoms of cryoglobulinemia, such as palpable purpura, and arthralgias. Renal manifestations include nephrotic (20%) or nonnephrotic proteinuria and microscopic hematuria [17]. Acute nephritic syndrome is the presenting feature in about a quarter of cases. Over 80% of patients have refractory hypertension upon presentation which may be responsible for a considerable number of cardiovascular deaths.

As the main clinical manifestations of glomerular disease in HCV-positive patients are proteinuria and microscopic hematuria with or without impaired kidney function, KDIGO Clinical Practice Guidelines recommend that HCV-infected patients be tested at least annually for proteinuria, hematuria and eGFR [18]. This is especially important in patients with cryoglobulinemia. The guidelines also recommend that a kidney biopsy be performed in HCV-infected patients with clinical evidence of GN [18]. Early diagnosis and therapy of HCV-associated GN may improve clinical outcomes.

Laboratory and Pathologic Findings

Laboratory testing establishes the diagnosis of HCV-related MPGN. Most patients will have anti-HCV, as well as HCV RNA, in serum. Serum transaminase levels are elevated in 70% of patients. Cryoglobulins are detected in 50–70% of patients. Serum electrophoresis and immunofixation reveals type II mixed cryoglobulins (fig. 1), in which the mRF, almost invariably an IgMκ, is a distinguishing feature of cryoglobulinemic GN. Their amount, usually measured as cryocrit, varies from one patient to another, and varies from time to time in a given patient (ranging between 2 and 70%). Urine κ light chains are also commonly present. The serum complement pattern, which does not change much with clinical activity, is also discriminative. Characteristically, the early components (C4 and C1q) and CH50 are at very low, or even undetectable, levels, while C3 levels are normal or only slightly depressed.

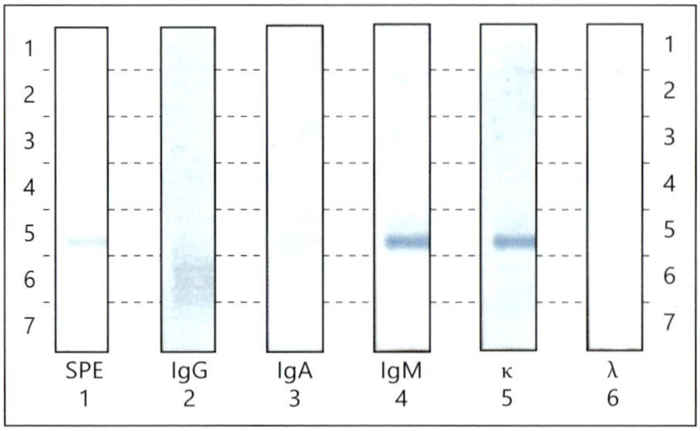

Fig. 1. Immunofixation of the washed, dissolved cryoprecipitate typed a monoclonal IgMκ with polyclonal IgG. By definition, these are type II cryoglobulins. Adapted from Lai and Tang [36].

Renal histological evaluation typically shows evidence of immune complex deposition in glomeruli and changes of MPGN [12]. MPGN refers to a pattern of glomerular injury characterized by diffuse mesangial proliferation and thickening of the capillary wall, hence the synonym of mesangiocapillary GN. In cryoglobulinemic MPGN, light microscopy reveals an increased number of mesangial cells, expansion of the mesangial matrix, and diffuse accentuation of glomerular tufts, which gives a lobular appearance to the glomeruli (fig. 2a). Glomerular capillary walls appear thickened because of the interposition of mesangial matrix between the GBM and the endothelium. Staining of the GBM with periodic acid-Schiff or methenamine silver shows splitting ('double contour') or 'tram-tracking' due to insertion of the mesangial matrix (fig. 2b). Immunofluorescence reveals granular deposits of C3 and IgG in the mesangium and in peripheral capillary loops. A similar morphological appearance may be seen with infective endocarditis and infected ventriculoatrial shunts (shunt nephritis). In addition, glomerular capillaries may have marked inflammatory cell infiltrates with both mononuclear cells and polymorphonuclear leukocytes, a distinguishing feature from noncryoglobulinemic MPGN. Intracapillary globular accumulations of eosinophilic material representing precipitated immune complexes or cryoglobulins may also be present. Viral HCV-containing antigens had been previously detected in glomerular structures using a three-stage indirect immunohistochemical monoclonal antibody technique [19] but not confirmed by subsequent studies. Electron microscopy shows subendothelial deposits which may have a finely fibrillar or tactoid pattern suggestive of cryoglobulin deposition.

Other forms of glomerular injury have been associated with HCV infection in individual case reports and small series, including membranous GN, IgA ne-

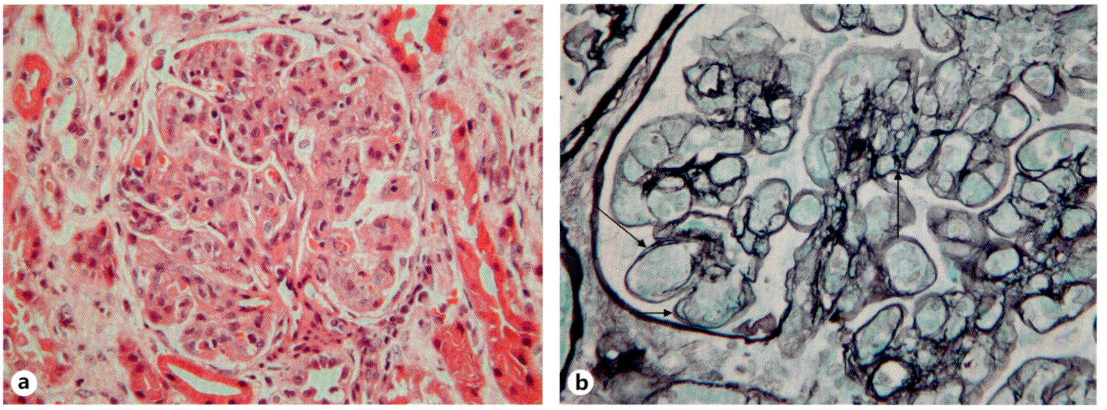

Fig. 2. Pathology of MPGN type I associated with cryoglobulinemia. **a** Glomerulus exhibits diffuse increase in mesangial cellularity and matrix with accentuation of lobulation of tuft architecture, obliteration of capillary lumens, and leukocytic infiltrate. HE. ×200. **b** Periodic acid-Schiff and methenamine silver staining reveals prominent double contours or tram-tracking (arrows) of glomerular basement membrane. ×400. Adapted from Lai and Tang [36].

phropathy, focal and segmental glomerulosclerosis, fibrillary GN, immunotactoid glomerulopathy, rapidly progressive GN, exudative-proliferative GN, and lupus nephritis [20]. Membranous nephropathy (MGN) in HCV carriers is characterized by the absence of cryoglobulins and male predominance. The clinical manifestations, outcome and pathology are similar to those of idiopathic MGN. All patients have detectable serum HCV RNA accompanied by normal serum complement and negative cryoglobulins and RF.

Treatment

As with any glomerulopathy, blood pressure control, blockade of the renin-angiotensin system, as well as the treatment of hyperlipidemia should form an integral part of the therapeutic approach.

Specific treatment of HCV infection and its associated morbidities including GN has evolved over time. For the past decade the standard of care for patients with chronic HCV infection has been pegylated interferon-α (peg-IFN-α) in combination with ribavirin (RBV). Improved knowledge of the viral life cycle and the crystal structure of several important viral proteins [21] have led to the recent development and approval of direct-acting antiviral agents (DAAs) that have revolutionized therapy for chronic HCV infection. The evolution of HCV treatment from the IFN era to the DAA era will be reviewed here.

Combination therapy with IFN-α2b + RBV, reported in 1998, represented the first important milestone for chronic HCV treatment [22], and also of acute hepatitis C after renal transplantation [23]. Such cocktail therapy has also produced favorable initial results in MC, although nonresponses and relapses after initial improvements still occur [24]. The introduction of pegylated forms of IFN in the year 2000 represented another breakthrough [25]. Pegylation refers to the covalent attachment of a large inert molecule of polyethylene glycol to a protein to yield a molecule that retains biological activity, but has delayed absorption and clearance, allowing for weekly rather than daily or three times weekly administration. Delayed clearance also led to greater, more potent, and longer lasting antiviral effects. To date, two molecularly different peg-IFNs are available: peg-α2a-IFN (Pegasys; Hoffmann-La Roche, Basel, Switzerland) and peg-α2b-IFN (Peg-Intron; Schering-Plough, Berlin Germany).

The American Association for the Study of Liver Diseases has published guidelines for the CKD population stating that 'when HCV infection is identified in persons with CKD, IFN-based antiviral treatment must be considered, but the regimen will vary depending on the kidney disease. The decision to treat must take into account the competing severities of the CKD and the chronic liver disease, the risks of the treatment itself, and whether there are comorbid conditions that may affect morbidity and mortality, such as cardiovascular disease [26]. The kidneys play a major role in the catabolism and excretion of both IFN and RBV, hence their clearances may be affected in subjects with CKD. The clearance of peg-IFN is also affected in those with CKD, although hemodialysis does not affect its clearance.

For treatment of HCV-related GN with peg-IFN and RBV, published case reports and uncontrolled studies have only included small numbers of patients. In a prospective uncontrolled study, 20 patients presenting with MPGN (n = 17), MGN (n = 2), and mesangioproliferative GN (n = 1) were treated with IFN-α and either with or without RBV [27]. Both HCV RNA and proteinuria levels decreased significantly. But no data are provided regarding the outcome of renal disease after cessation of anti-HCV therapy. In another study, 18 patients who had HCV-related cryoglobulinemic MPGN were treated with a combined therapy of standard or peg-IFN-α and RBV [28]. The mean duration of anti-HCV therapy was 18 ± 10 months (range 6–24). The mean duration of follow-up after completing anti-HCV treatment was 16.7 ± 17.7 months (range 6–30). A sustained virological response was observed in 67% of patients, along with reduction of proteinuria and cryoglobulin levels. A suggested management algorithm, based on international guidelines, is shown in figure 3.

In severe acute flares of cryoglobulinemia with GN or vasculitis and nephrotic range proteinuria, an appropriate approach is to include high-dose corticoste-

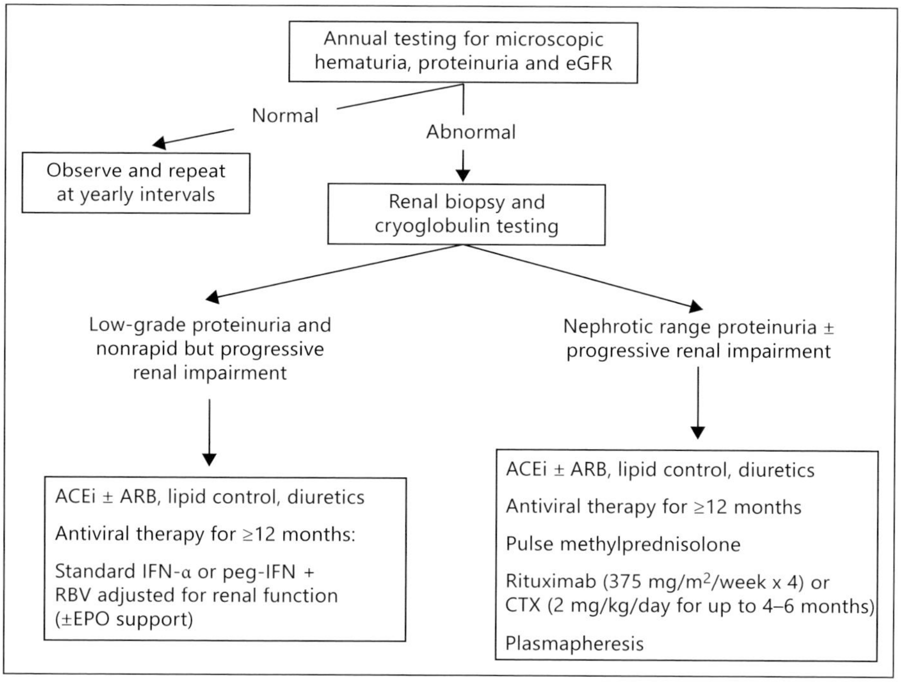

Fig. 3. Screening for kidney disease in HCV-positive patients and management algorithm for HCV-associated GN. ACEi = Angiotensin-converting enzyme inhibitor; ARB = angiotensin II receptor blocker; EPO = erythropoietin; CTX = cyclophosphamide.

roids and cyclophosphamide to control severe cryoglobulinemic symptoms in addition to combination antiviral therapy (fig. 3). In the most severe cases, plasmapheresis (3–4 times weekly exchanges of 3 liters of plasma for 2–3 weeks) might be used to remove circulating cryoglobulins. For refractory cases, the human-mouse chimeric monoclonal antibody against the B-cell surface antigen CD20, rituximab, is preferred to cyclophosphamide due to its efficacy and more favorable side-effect profile [29, 30]. More recently, it has been found to be effective for the treatment of HCV-related type II cryoglobulinemic MPGN when used as a first-line therapy [31]. Further randomized controlled studies are required to define the exact indications and dose of rituximab, as well as its long-term effect on the liver in this context.

A major drawback of peg-IFN + RBV is the high treatment failure rate among HCV carriers with genotype 1. Sustained virologic response, defined as negative HCV RNA 24 weeks after treatment completion, is observed in only about 40–50% of patients [32]. The recent development of small molecule compounds that directly inhibit the viral life cycle represents a further milestone for the treatment of chronic HCV infection. Unlike the nonspecific antiviral activity of peg-

IFN and RBV, DAAs directly inhibit viral proteins that are part of the viral life cycle. Theoretically, each step of the viral life cycle is a potential target for direct intervention. The furthest along in clinical development are direct inhibitors of the nonstructural (NS) 3/4A protease, nucleos(t)ide and nonnucleoside inhibitors of the HCV-RNA-dependent RNA polymerase NS5B and NS5A replication complex inhibitors [33]. Among these, two NS3/4A protease inhibitors are approved for the treatment of HCV genotype 1 infection: boceprevir and telaprevir. Taken together, these two DAAs administered together with peg-IFN and RBV as triple therapy have increased overall cure rates by 20–30% when compared to the former dual combination therapy. Furthermore, treatment duration could be reduced to 24–28 weeks in the majority of treatment-naive patients [34]. However, the majority of DAAs is still in early clinical development. The rapid replication rate of HCV, along with the error-prone polymerase activity, leads to a high genetic diversity among HCV virions that include mutants with reduced susceptibility to DAA therapy. Data on the safety and efficacy of DAAs in HCV-associated GN is lacking. In a small series of 4 cases of HCV-infected hemodialyzed patients who were unresponsive to previous peg-IFN and RBV, the addition of telaprevir has resulted in clearance of HCV RNA in 3 subjects [35]. Anemia requiring increased doses of erythropoietin appeared to be the only observable side effect. The potential of triple therapy with a first-generation protease inhibitor for the treatment of HCV patients with ESRD needs to be confirmed in larger series of patients.

Acknowledgments

This study is supported by the L&T Charitable Foundation and the House of Indocafe and the Hong Kong Concrete and the Continental Cement and Gypsum Co. Ltd.

References

1 Choo QL, Kuo G, Weiner AJ, Overby LR, Bradley DW, Houghton M: Isolation of a cDNA clone derived from a blood-borne non-A, non-B viral hepatitis genome. Science 1989;244:359–362.

2 Meltzer M, Franklin EC, Elias K, McCluskey RT, Cooper N: Cryoglobulinemia – a clinical and laboratory study. II. Cryoglobulins with rheumatoid factor activity. Am J Med 1966; 40:837–856.

3 Robertson B, Myers G, Howard C, Brettin T, Bukh J, Gaschen B, Gojobori T, Maertens G, Mizokami M, Nainan O, Netesov S, Nishioka K, Shin-i T, Simmonds P, Smith D, Stuyver L, Weiner A: Classification, nomenclature, and database development for hepatitis C virus and related viruses: proposals for standardization. International Committee on Virus Taxonomy. Arch Virol 1998;143:2493–2503.

4 Simmonds P, Bukh J, Combet C, Deleage G, Enomoto N, Feinstone S, Halfon P, Inchauspe G, Kuiken C, Maertens G, Mizokami M, Murphy DG, Okamoto H, Pawlotsky JM, Penin F, Sablon E, Shin I, Stuyver LJ, Thiel HJ, Viazov S, Weiner AJ, Widell A: Consensus proposals for a unified system of nomenclature of hepatitis C virus genotypes. Hepatology 2005;42:962–973.
5 Chevaliez S, Pawlotsky JM: Virology of hepatitis C virus infection. Best Pract Res Clin Gastroenterol 2012;26:381–389.
6 Poynard T, Yuen MF, Ratziu V, Lai CL: Viral hepatitis C. Lancet 2003;362:2095–2100.
7 Seto WK, Lai CL, Fung J, Hung I, Yuen J, Young J, Wong DK, Yuen MF: Natural history of chronic hepatitis C: genotype 1 versus genotype 6. J Hepatol 2010;53:444–448.
8 Krahn M, Wong JB, Heathcote J, Scully L, Seeff L: Estimating the prognosis of hepatitis C patients infected by transfusion in Canada between 1986 and 1990. Med Decis Making 2004;24:20–29.
9 Maasoumy B, Wedemeyer H: Natural history of acute and chronic hepatitis C. Best Pract Res Clin Gastroenterol 2012;26:401–412.
10 Perico N, Cattaneo D, Bikbov B, Remuzzi G: Hepatitis C infection and chronic renal diseases. Clin J Am Soc Nephrol 2009;4:207–220.
11 Wornle M, Schmid H, Banas B, Merkle M, Henger A, Roeder M, Blattner S, Bock E, Kretzler M, Grone HJ, Schlondorff D: Novel role of toll-like receptor 3 in hepatitis C-associated glomerulonephritis. Am J Pathol 2006;168:370–385.
12 Johnson RJ, Gretch DR, Yamabe H, Hart J, Bacchi CE, Hartwell P, Couser WG, Corey L, Wener MH, Alpers CE: Membranoproliferative glomerulonephritis associated with hepatitis C virus infection. N Engl J Med 1993;328:465–470.
13 Tsui JI, Vittinghoff E, Shlipak MG, O'Hare AM: Relationship between hepatitis C and chronic kidney disease: results from the Third National Health and Nutrition Examination Survey. J Am Soc Nephrol 2006;17:1168–1174.
14 Huang JF, Chuang WL, Dai CY, Ho CK, Hwang SJ, Chen SC, Lin ZY, Wang LY, Chang WY, Yu ML: Viral hepatitis and proteinuria in an area endemic for hepatitis B and C infections: another chain of link? J Intern Med 2006;260:255–262.
15 Li CG, Ferrantelli A, Bono L, Tortorici C, Giammarresi C, Zagarrigo C, Schillaci O, Tralongo A, Soresi M, Rotolo U: Incidence of hepatitis C virus infection in patients with chronic kidney disease on conservative therapy. Int J Infect Dis 2011;15:e514–e516.
16 Lai FM, Tam JS, Liew CT, Ip M, Lai KN: Low prevalence of hepatitis C virus antibodies with primary membranous nephropathy and membranoproliferative glomerulonephritis in Hong Kong. Nephron 1995;70:367–368.
17 D'Amico G: Renal involvement in hepatitis C infection: cryoglobulinemic glomerulonephritis. Kidney Int 1998;54:650–671.
18 KDIGO Clinical Practice Guidelines for the Prevention, Diagnosis, Evaluation and Treatment of Hepatitis C in Chronic Kidney Disease. Kidney Int 2008;73(suppl 109):S69–S77.
19 Sansonno D, Gesualdo L, Manno C, Schena FP, Dammacco F: Hepatitis C virus-related proteins in kidney tissue from hepatitis C virus-infected patients with cryoglobulinemic membranoproliferative glomerulonephritis. Hepatology 1997;25:1237–1244.
20 Meyers CM, Seeff LB, Stehman-Breen CO, Hoofnagle JH: Hepatitis C and renal disease: an update. Am J Kidney Dis 2003;42:631–657.
21 Bartenschlager R, Cosset FL, Lohmann V: Hepatitis C virus replication cycle. J Hepatol 2010;53:583–585.
22 Poynard T, Marcellin P, Lee SS, Niederau C, Minuk GS, Ideo G, Bain V, Heathcote J, Zeuzem S, Trepo C, Albrecht J: Randomised trial of interferon-α2b plus ribavirin for 48 weeks or for 24 weeks versus interferon-α2b plus placebo for 48 weeks for treatment of chronic infection with hepatitis C virus. International Hepatitis Interventional Therapy Group (IHIT). Lancet 1998;352:1426–1432.
23 Tang S, Cheng IK, Leung VK, Kuok UI, Tang AW, Ho YW, Lai KN, Chan TM: Successful treatment of hepatitis C after kidney transplantation with combined interferon-α2b and ribavirin. J Hepatol 2003;39:875–878.
24 Zuckerman E, Keren D, Slobodin G, Rosner I, Rozenbaum M, Toubi E, Sabo E, Tsykounov I, Naschitz JE, Yeshurun D: Treatment of refractory, symptomatic, hepatitis C virus related mixed cryoglobulinemia with ribavirin and interferon-α. J Rheumatol 2000;27:2172–2178.

25 Zeuzem S, Feinman SV, Rasenack J, Heathcote EJ, Lai MY, Gane E, O'Grady J, Reichen J, Diago M, Lin A, Hoffman J, Brunda MJ: Peginterferon-α2a in patients with chronic hepatitis C. N Engl J Med 2000;343:1666–1672.

26 Ghany MG, Nelson DR, Strader DB, Thomas DL, Seeff LB: An update on treatment of genotype 1 chronic hepatitis C virus infection: 2011 practice guideline by the American Association for the Study of Liver Diseases. Hepatology 2011;54:1433–1444.

27 Sabry AA, Sobh MA, Sheaashaa HA, Kudesia G, Wild G, Fox S, Wagner BE, Irving WL, Grabowska A, El-Nahas AM: Effect of combination therapy (ribavirin and interferon) in HCV-related glomerulopathy. Nephrol Dial Transplant 2002;17:1924–1930.

28 Alric L, Plaisier E, Thebault S, Peron JM, Rostaing L, Pourrat J, Ronco P, Piette JC, Cacoub P: Influence of antiviral therapy in hepatitis C virus-associated cryoglobulinemic MPGN. Am J Kidney Dis 2004;43:617–623.

29 Sansonno D, De Re V, Lauletta G, Tucci FA, Boiocchi M, Dammacco F: Monoclonal antibody treatment of mixed cryoglobulinemia resistant to interferon-α with an anti-CD20. Blood 2003;101:3818–3826.

30 Saadoun D, Resche RM, Sene D, Terrier B, Karras A, Perard L, Schoindre Y, Coppere B, Blanc F, Musset L, Piette JC, Rosenzwajg M, Cacoub P: Rituximab plus Peg-interferon-α/ribavirin compared with Peg-interferon-α/ribavirin in hepatitis C-related mixed cryoglobulinemia. Blood 2010;116:326–334.

31 Quartuccio L, Soardo G, Romano G, Zaja F, Scott CA, De Marchi G, Fabris M, Ferraccioli G, De Vita S: Rituximab treatment for glomerulonephritis in HCV-associated mixed cryoglobulinaemia: efficacy and safety in the absence of steroids. Rheumatology (Oxford) 2006;45:842–846.

32 Fried MW, Shiffman ML, Reddy KR, Smith C, Marinos G, Goncales FL Jr, Haussinger D, Diago M, Carosi G, Dhumeaux D, Craxi A, Lin A, Hoffman J, Yu J: Peginterferon-α2a plus ribavirin for chronic hepatitis C virus infection. N Engl J Med 2002;347:975–982.

33 Vermehren J, Sarrazin C: New HCV therapies on the horizon. Clin Microbiol Infect 2011;17:122–134.

34 Jacobson IM, McHutchison JG, Dusheiko G, Di Bisceglie AM, Reddy KR, Bzowej NH, Marcellin P, Muir AJ, Ferenci P, Flisiak R, George J, Rizzetto M, Shouval D, Sola R, Terg RA, Yoshida EM, Adda N, Bengtsson L, Sankoh AJ, Kieffer TL, George S, Kauffman RS, Zeuzem S: Telaprevir for previously untreated chronic hepatitis C virus infection. N Engl J Med 2011;364:2405–2416.

35 Dumortier J, Guillaud O, Gagnieu MC, Janbon B, Juillard L, Morelon E, Leroy V: Antiviral triple therapy with telaprevir in haemodialysed HCV patients: Is it feasible? J Clin Virol 2013;56:146–149.

36 Lai KN, Tang SC: Hepatitis-associated glomerulonephritis; in Lerma EV, Berns JS, Nissenson AR (eds): Current Diagnosis and Treatment in Nephrology and Hypertension. New York, McGraw-Hill, 2009, p 302.

Prof. Kar Neng Lai
Division of Nephrology, Department of Medicine
University of Hong Kong, Queen Mary Hospital
Hong Kong, SAR (China)
E-Mail knlai@hkucc.hku.hk

Chen N (ed): New Insights into Glomerulonephritis.
Contrib Nephrol. Basel, Karger, 2013, vol 181, pp 207–215 (DOI: 10.1159/000348633)

Pathogenesis of Rapidly Progressive Glomerulonephritis: What Do We Learn?

Yong-Xi Chen · Nan Chen

Department of Nephrology, Ruijin Hospital, School of Medicine, Shanghai Jiao Tong University, Shanghai, PR China

Abstract

Rapidly progressive glomerulonephritis (RPGN) is a life-threatening disease with a poor prognosis. In this review, the pathogenesis of RPGN owing to antineutrophil cytoplasmic antibody-associated crescentic glomerulonephritis and anti-GBM diseases is discussed. By the model of nephrotoxic nephritis, T cells, dendritic cells and toll-like receptors are involved in podocyte activation and parietal epithelial cell proliferation which contribute to the crescent formation and glomerular injury. Furthermore, growth factors and Goodpasture autoantigen are also involved in the onset of the disease. In the study of ANCA-associated glomerulonephritis, the role of lysosome-associated membrane protein (LAMP)-2 and neutrophil extracellular traps is well studied. However, the role of LAMP-2 in the disease pathogenesis remains uncertain. We hope this review can help us to further understand the pathogenesis of the disease. Copyright © 2013 S. Karger AG, Basel

Rapidly progressive glomerulonephritis (RPGN) is a life-threatening disease which is characterized as rapid loss of kidney function clinically and crescentic formation pathologically. RPGN is classified into three main categories according to direct immunofluorescence microscopy: (1) antiglomerular basement membrane (GBM) crescentic glomerulonephritis (CrGN), (2) immune-complex CrGN, and (3) pauci-immune CrGN [1].

Approximately 80% of the patients with pauci-immune CrGN could be positive for antineutrophil cytoplasmic antibody (ANCA) and thus be called ANCA-associated CrGN. One half of the patients with anti-GBM glomerulonephritis can present with Goodpasture's syndrome and the other half with ANCA-associated glomerulonephritis can have ANCA-associated small-vessel vasculitis (AASV), including microscopic polyangiitis (MPA), granulomatosis with polyangiitis (GPA, formally known as Wegener's granulomatosis), and Churg-Strauss syndrome (CSS) [2, 3]. Furthermore, an overlap between different immunopathological phenotypes of RPGN exists in patients as some could be positive for both ANCA and GBM.

Despite aggressive immunosuppressive therapies, the prognosis of RPGN remains poor. Although patients with ANCA-associated glomerulonephritis have a better outcome in comparison with patients with other types of RPGN, our own experience suggested the 5-year survival of ANCA-associated CrGN was <70% and renal involvement at disease onset affected the outcome [4, 5]. Therefore, further understanding the pathogenesis of the disease would help us to develop effective therapeutic strategies to treat the patients with RPGN and improve the prognosis. In this review the pathogenesis of RPGN owing to AASV and anti-GBM diseases is discussed and immune-complex CrGN (which is commonly caused by membranoproliferative nephritis, IgA nephropathy, etc.) will be discussed in other chapters.

Anti-GBM Nephritis

Anti-GBM nephritis is characterized by focal necrotizing glomerulonephritis with crescents and linear deposition of IgG and C3 along the GBM. The form of disease with pulmonary alveolar hemorrhage is called Goodpasture's syndrome. Though anti-GBM nephritis is not a usual cause for end-stage renal diseases, the history of pulmonary alveolar hemorrhage could be associated with an increased mortality risk on dialysis [6]. The anti-GBM disease is caused by autoantibodies to the noncollagenous domain of the α3 chain of type IV collagen which is present in lung and the kidney. The roles of anti-GBM antibody deposition and cellular events that lead to glomerular injury were described by a nephrotoxic nephritis (NTN) model and the mechanisms were also investigated thereafter. In this part, we will discuss recent advances in the pathogenesis of the disease.

Toll-Like Receptors (TLRs)

TLRs are pattern recognition receptors that recognize conserved immunostimulatory molecular patterns including pathogen-associated molecular patterns (PAMPs) and endogenous cell-derived patterns (danger-associated molecular patterns, DAMPs). In the study by Giorgini et al. [7], the results showed that renal cell TLR4 stimulation could trigger crescentic disease in the NTN model. Though TLR4 stimulation could augment the nephritogenic immune response, such stimulation could not affect the response after disease induction but trigger crescentic disease. These results demonstrated an integration of TLR4 stimulation on both adaptive and innate immune systems in a CrGN model. Furthermore, studies showed that TLR4 and other pattern recognition receptors on glomerular immune and nonimmune cells could be activated by innate immune response which was activated by glomerular cell injury and extracellular matrix modification [8]. These studies in whole highlight the important role of TLRs in the adaptive and innate immune response in the onset of the diseases.

T Cells, Dendritic Cells (DCs)

Some CD4+ T cells are known to be responsible for antibody-independent tissue injury in glomeruli [9]. By using the experimental autoimmune glomerulonephritis (EAG) animal model, Hopfer et al. [10] showed that glomerular damage in EAG could result in activation of CD4+ T cells specific for glomerular antigens by tubulointerstitial DCs which highlighted the role of Th1 effector cells in the autoimmunity of the disease. These T cells then recruited further T cells as well as macrophages which escalated the inflammatory reaction and led to CrGN. Paust et al. [11] identified a cytokine-chemokine-driven feedback loop which involved Th1, Th17 and chemokines that exerted pathogenic effects in CrGN. In the study, the spatiotemporal pattern was found of early Th17 and late Th1 immune response which implied that interaction of Th1 and Th17 cells existed in the differentiation and effector phase. Furthermore, the results showed that early recruitment of IL-17-producing Th17 cells triggered the expression of CXCL9 in the kidney and drove the infiltration of CXCR3-bearing Th1 cells. Later, Th1 cell-derived IFN-γ might inhibit local CCL20 expression which limited the trafficking of Th17 cells into the kidney and inhibited Th17 immune response. In the study by Paust et al. [12], depletion of Treg cells followed by enhancement of Th1 response was associated with exacerbation of glomerular injury which was mediated by IFN-γ. The results support the concept that Tregs suppress adaptive T-cell response, however as Tregs also suppress innate immune response [13], the immunosuppressive role of Tregs in glomerulonephritis needs further study. In the study by Nozaki et al. [14], endogenous T-cell immunoglobulin mucin 1 (Tim-1) promoted crescent formation and renal injury in a rat model of NTN. By using anti-Tim-1 antibodies as anti-Tim-1 treatment the results demonstrated that anti-Tim-1 selectively diminished injurious T-cell subsets (mainly Th1 and Th17 response) which promoted renal injury in the model. The study thus suggested that the Th1 and Th17 nephritogenic immune response as well as Tim-1 are involved in the inflammation in the anti-GBM nephritis animal model.

Apart from antigen-specific CD4+ T cells, Turner et al. [15] demonstrated in their excellent investigation that non-CD4+ T cells, namely renal γδ T cells, were also responsible for immune-mediated CrGN. By using mice with NTN, the study showed that γδ T cells were a major source of IL-17A in the early phase of the disease before CD4+ Th17 cells arrived. Furthermore, depletion of γδ T cells as well as IL-17A production in γδ T cells would reduce neutrophil infiltration and glomerular crescent formation. The study identified the pathogenic role of non-CD4+ T cells in the early inflammatory response which contribute to renal tissue injury in experimental glomerulonephritis.

In addition to T cells, DCs were reported to be involved in the renal injury of the NTN model. In the study by Hochheiser et al. [16], kidney DC (KDC) acquired far more filtrated antigens in the NTN and captured an antigen too large to be filtered in a healthy glomerulus under homeostatic conditions and used filterable antigens for driving intrarenal Th1 and Th17 responses. Also, KDC became immunogenic and drove harmful Th responses during the course of nephritis. The authors implied that KDC may act in a harmful manner when they were matured. In the study by Riedel et al. [17], the protective role of immature renal DCs was illustrated. Invariant natural killer T (iNKT) cells constituted a distinct population of mature T lymphocytes which could attenuate glomerular injury. By using in vivo and in vitro experiments, the authors demonstrated that renal DC-derived CXCL16 might attract protective CXCR6+ iNKT cells. The authors thus suggested that iNKT cells which were under the control of the CXCR6-CXCL16 axis as well as immature DCs were involved in the immunoprotective mechanism of CrGN.

Growth Factor Receptors

In a recent study by Bollee et al. [18], the role of heparin-binding epidermal growth factor-like growth factor (HB-EGF) in RPGN was well investigated and discussed. By using an anti-GBM NTN serum (NTS)-induced mouse model, the results demonstrated that HB-EGF deficiency and blockage of epidermal growth factor receptor (EGFR) both prevented the development of renal leukocytic infiltrates before the appearance of crescentic formation and interstitial injury, which implied that the HB-EGF–EGFR pathway promotes renal damages in very early stages. Furthermore, the protective effects of deletion of the *Egfr* gene in vitro and pharmacological blockade of EGFR in vivo on the severity of RPGN suggested that recruitment of the HB-EGF–EGFR pathway was involved during the effector phase of the disease.

In the study focusing on molecular details in such a process, van Roeyen et al. [19] showed that platelet-derived growth factor-D (PDGF-D) was involved in the extracapillary proliferation of CrGN with fibrin/fibrinogen deposition. Podocyte-specific overexpression of PDGF-D caused CrGN and other glomerulonephritides. Therefore, the antiproliferative effects on parietal epithelial cells (PECs) activated by blockade of PDGF signaling using the inhibitor imatinib further proved the role of PDGF in crescentic formation [20].

Parietal Epithelial Cells and Podocytes

In response to the immune stimuli, glomerular PECs and podocyte activation play an important role in the pathogenesis of RPGN. Smeets et al. [21] used an NTS-induced mice model to induce CrGN. They observed the cellular bridges between Bowman's capsule and the glomerular tuft composed of PECs. These novel findings firmly proved that PECs were the major cell population within cellular extracapillary lesions in the crescentic nephritis model. Similarly, PECs constituted proliferative lesions in patients with CrGN in human biopsy tissues [22].

Apart from PECs, in another glomerular epithelial cell population podocytes were also involved in the crescentic formation and a study demonstrated that proliferation of PEC to form a crescent was initiated by podocytes. In the study by Le Hir et al. [23], activated podocytes were observed to adhere to both the GBM and the parietal basement membrane which forms bridges between the tuft and Bowman's capsule; the podocyte bridges were interposed between PECs and the cells of a crescent. The authors thus proposed that spreading of podocytes on the parietal basement membrane initiated proliferation of PEC which then formed the crescent. Similar results were reported by Ohse et al. [24] who showed in their study that cells in the glomerular tuft coexpressed both podocytes and PEC proteins in the anti-GBM model. The authors hypothesized that PECs coexpressed podocyte proteins and moved from their original location to the edge of the glomerular tuft. These cells function as progenitors of podocytes.

Goodpasture Autoantigen

Circulating autoantibodies bind to the noncollagenous-1 (NC1) domain of type IV collagen in the GBM and lead to the glomerular and pulmonary injury. By using enzyme-linked immunosorbent assay, Pedchenko et al. [25] identified that α3NC1 and α5NC1 monomers acted as the autoantigens in the disease. The authors thus proposed that Goodpasture's disease involves conformational transition between a nonpathogenic conformer within the hexamer and a dissociated pathogenic conformer that elicits an autoimmune response.

ANCA-Associated Glomerulonephritis

The pathogenicity of ANCA is well studied, however the question why the ANCA appears has not been well answered. Apart from environmental factors, drugs (most common of which is propylthiouracil), infections, neoplasm and other factors might contribute to the appearance of ANCA and onset of the diseases [26–28]. The pathogenesis and treatment of AASV are discussed in another article by Dr. Tesar and colleagues, we therefore highlight on some research hotspots in this area.

Lysosome-Associated Membrane Protein-2 (LAMP-2)
LAMP-2 is a heavily glycosylated membrane protein which colocalizes to myeloperoxidase (MPO) and proteinase 3 (PR3) in neutrophils as well as in cellular membrane and lysosomes. In the study by Kain et al. [29], molecular mimicry of LAMP-2 by bacterial adhesion protein (FimH) could induce circulating anti-FimH antibodies which cross-reacted with LAMP-2 and caused pauci-immne glomerulonephritis and vasculitis. Rats immunized with FimH developed antibodies to rat as well as human LAMP-2 (hLAMP-2) and developed pauci-immune necrotizing glomerulonephritis. The authors thus proposed that infection and molecular mimicry trigger autoimmunity by inducing antibodies directed against FimH that cross-react with hLAMP-2 and contribute to pauci-immune glomerulonephritis. However, in the study by Roth et al. [30], a low percentage of anti-hLAMP 2 reactivity was found by ELISA in their ANCA-associated patients. Also, the positivity could not be validated by Western blotting or immunofluorescence microscopy. Furthermore, the titer of anti-LAMP-2 was much lower than that of anti-MPO or anti-PR3, and Wistar-Kyoto rats injected with anti-LAMP-2 antibodies could not develop glomerulonephritis. Based on these conflicting results, more work needs to be done to solve the controversy of the role of hLAMP-2 in ANCA-associated glomerulonephritis.

Neutrophil Extracellular Traps (NETs)
Activated neutrophils would release NETs, which are chromatin fibers that trap and kill invading microbes extracellularly. In the study by Kessenbrock et al. [31], ANCA-stimulated neutrophils released NETs which contained targeted autoantigens PR3 and MPO. Extracellular NETs in conjunction with LL37 is an antimicrobial peptide made by neutrophils to convert self-DNA into an activator of plasmacytoid DCs. Hence, NETs not only cause damage to endothelial cells, but get involved in perpetuating the ANCA autoimmune response by presenting ANCA antigens to the immune system as well. These results suggest NETs formation not only triggers vasculitis but also promotes the autoimmune response against neutrophil components in ANCA-associated patients.

Conclusion

Significant advances have been made to understand the pathogenesis of RPGN recently. As summarized in figure 1, the immune response as well as T cells, DCs, and TLRs are involved in podocyte activation and PEC proliferation which contribute to the crescent formation and glomerular injury in anti-GBM nephritis

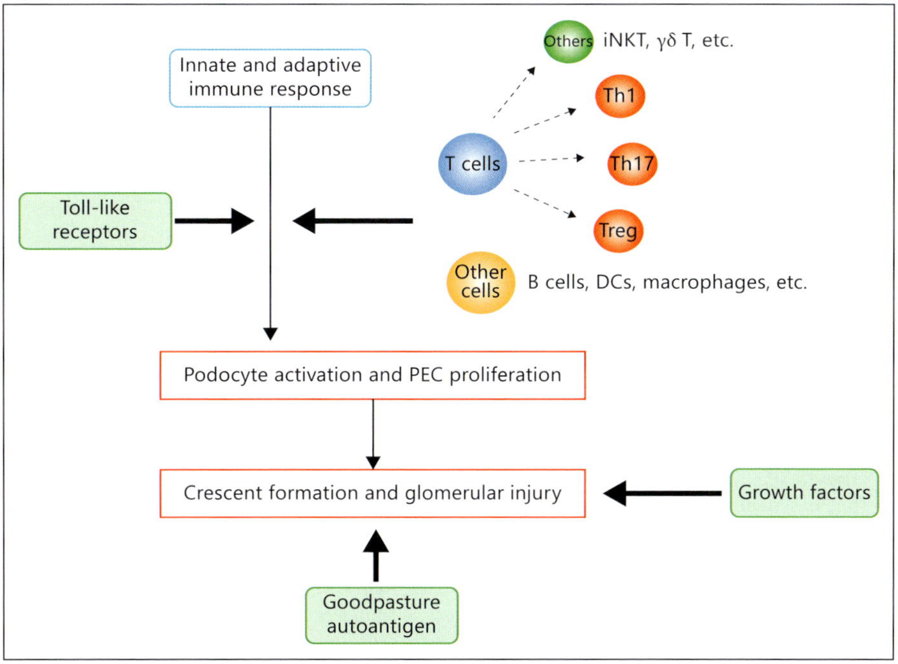

Fig. 1. Brief overview of the mechanism in anti-GBM nephritis.

models. Furthermore, growth factors and Goodpasture autoantigen also lead to glomerular injury in the disease. In the study of ANCA-associated glomerulonephritis the role of LAMP-2 and NETs are demonstrated. However, many questions regarding the details of the disease remain elusive. Further studies are still necessary to clarify the pathogenesis of the disease and develop more specific therapies to improve prognosis and reduce mortality.

Acknowledgements

We apologize to the many colleagues whose original works cannot be cited in this review due to limitations of the references listed. This work is supported by grants from the National Natural Science Foundation (No. 81000285), National Basic Research Program of China 973 (No. 2012CB517600, No. 2012CB517604), National Key Technology R&D Program (12-5) (No. 2011BAI10B00, 2011BAI10B06), Key Program of Shanghai Science and Technology Commission (08dz1900502) and a grant from Ruijin Hospital and Shanghai Educational Committee for young investigators.

References

1 Jennette JC: Rapidly progressive crescentic glomerulonephritis. Kidney Int 2003;63: 1164–1177.

2 Falk RJ, Gross WL, Guillevin L, Hoffman GS, Jayne DR, Jennette JC, Kallenberg CG, Luqmani R, Mahr AD, Matteson EL, Merkel PA, Specks U, Watts RA; American College of Rheumatology; American Society of Nephrology; European League Against Rheumatism: Granulomatosis with polyangiitis (Wegener's): an alternative name for Wegener's granulomatosis. Arthritis Rheum 2011;63: 863–864.

3 Jennette JC, Falk RJ, Andrassy K, Bacon PA, Churg J, Gross WL, Hagen EC, Hoffman GS, Hunder GG, Kallenberg CG, et al: Nomenclature of systemic vasculitides. Proposal of an international consensus conference. Arthritis Rheum 1994;37:187–192.

4 Chen YX, Yu HJ, Zhang W, Ren H, Chen XN, Shen PY, Xu YW, Li X, Pan XX, Ni LY, Wang WM, Chen N: Analyzing fatal cases of Chinese patients with primary antineutrophil cytoplasmic antibodies-associated renal vasculitis: a 10-year retrospective study. Kidney Blood Press Res 2008;31:343–349.

5 Chen YX, Zhang W, Chen XN, Shen PY, Shi H, Xu YW, Ren H, Chen N: Application of rifle criteria in Chinese patients with ANCA-associated renal vasculitis. Clin Exp Rheumatol 2011;29:951–957.

6 Tang W, McDonald SP, Hawley CM, Badve SV, Boudville NC, Brown FG, Clayton PA, Campbell SB, de Zoysa JR, Johnson DW: Anti-glomerular basement membrane antibody disease is an uncommon cause of end-stage renal disease. Kidney Int 2013 (E-pub ahead of print).

7 Giorgini A, Brown HJ, Sacks SH, Robson MG: Toll-like receptor-4 stimulation triggers crescentic glomerulonephritis by multiple mechanisms including a direct effect on renal cells. Am J Pathol 2010;177:644–653.

8 Anders HJ, Schlondorff DO: Innate immune receptors and autophagy: implications for autoimmune kidney injury. Kidney Int 2010; 78:29–37.

9 Couser WG: Basic and translational concepts of immune-mediated glomerular diseases. J Am Soc Nephrol 2012;23:381–399.

10 Hopfer H, Holzer J, Hunemorder S, Paust HJ, Sachs M, Meyer-Schwesinger C, Turner JE, Panzer U, Mittrucker HW: Characterization of the renal CD4+ T-cell response in experimental autoimmune glomerulonephritis. Kidney Int 2012;82:60–71.

11 Paust HJ, Turner JE, Riedel JH, Disteldorf E, Peters A, Schmidt T, Krebs C, Velden J, Mittrucker HW, Steinmetz OM, Stahl RA, Panzer U: Chemokines play a critical role in the cross-regulation of Th1 and Th17 immune responses in murine crescentic glomerulonephritis. Kidney Int 2012;82:72–83.

12 Paust HJ, Ostmann A, Erhardt A, Turner JE, Velden J, Mittrucker HW, Sparwasser T, Panzer U, Tiegs G: Regulatory T cells control the Th1 immune response in murine crescentic glomerulonephritis. Kidney Int 2011;80: 154–164.

13 Maloy KJ, Salaun L, Cahill R, Dougan G, Saunders NJ, Powrie F: CD4+CD25+ T(R) cells suppress innate immune pathology through cytokine-dependent mechanisms. J Exp Med 2003;197:111–119.

14 Nozaki Y, Nikolic-Paterson DJ, Snelgrove SL, Akiba H, Yagita H, Holdsworth SR, Kitching AR: Endogenous Tim-1 (Kim-1) promotes T-cell responses and cell-mediated injury in experimental crescentic glomerulonephritis. Kidney Int 2012;81:844–855.

15 Turner JE, Krebs C, Tittel AP, Paust HJ, Meyer-Schwesinger C, Bennstein SB, Steinmetz OM, Prinz I, Magnus T, Korn T, Stahl RA, Kurts C, Panzer U: IL-17A production by renal γδ T cells promotes kidney injury in crescentic GN. J Am Soc Nephrol 2012;23: 1486–1495.

16 Hochheiser K, Engel DR, Hammerich L, Heymann F, Knolle PA, Panzer U, Kurts C: Kidney dendritic cells become pathogenic during crescentic glomerulonephritis with proteinuria. J Am Soc Nephrol 2011;22:306–316.

17 Riedel JH, Paust HJ, Turner JE, Tittel AP, Krebs C, Disteldorf E, Wegscheid C, Tiegs G, Velden J, Mittrucker HW, Garbi N, Stahl RA, Steinmetz OM, Kurts C, Panzer U: Immature renal dendritic cells recruit regulatory CXCR6+ invariant natural killer T cells to attenuate crescentic GN. J Am Soc Nephrol 2012;23:1987–2000.

18 Bollee G, Flamant M, Schordan S, Fligny C, Rumpel E, Milon M, Schordan E, Sabaa N, Vandermeersch S, Galaup A, Rodenas A, Casal I, Sunnarborg SW, Salant DJ, Kopp JB, Threadgill DW, Quaggin SE, Dussaule JC, Germain S, Mesnard L, Endlich K, Boucheix C, Belenfant X, Callard P, Endlich N, Tharaux PL: Epidermal growth factor receptor promotes glomerular injury and renal failure in rapidly progressive crescentic glomerulonephritis. Nat Med 2011;17:1242–1250.
19 van Roeyen CR, Eitner F, Boor P, Moeller MJ, Raffetseder U, Hanssen L, Bucher E, Villa L, Banas MC, Hudkins KL, Alpers CE, Ostendorf T, Floege J: Induction of progressive glomerulonephritis by podocyte-specific overexpression of platelet-derived growth factor-D. Kidney Int 2011;80:1292–1305.
20 Iyoda M, Shibata T, Wada Y, Kuno Y, Shindo-Hirai Y, Matsumoto K, Akizawa T: Long- and short-term treatment with imatinib attenuates the development of chronic kidney disease in experimental anti-glomerular basement membrane nephritis. Nephrol Dial Transplant 2013 (E-pub ahead of print).
21 Smeets B, Uhlig S, Fuss A, Mooren F, Wetzels JF, Floege J, Moeller MJ: Tracing the origin of glomerular extracapillary lesions from parietal epithelial cells. J Am Soc Nephrol 2009;20:2604–2615.
22 Smeets B, Angelotti ML, Rizzo P, Dijkman H, Lazzeri E, Mooren F, Ballerini L, Parente E, Sagrinati C, Mazzinghi B, Ronconi E, Becherucci F, Benigni A, Steenbergen E, Lasagni L, Remuzzi G, Wetzels J, Romagnani P: Renal progenitor cells contribute to hyperplastic lesions of podocytopathies and crescentic glomerulonephritis. J Am Soc Nephrol 2009;20:2593–2603.
23 Le Hir M, Keller C, Eschmann V, Hahnel B, Hosser H, Kriz W: Podocyte bridges between the tuft and Bowman's capsule: an early event in experimental crescentic glomerulonephritis. J Am Soc Nephrol 2001;12:2060–2071.
24 Ohse T, Vaughan MR, Kopp JB, Krofft RD, Marshall CB, Chang AM, Hudkins KL, Alpers CE, Pippin JW, Shankland SJ: De novo expression of podocyte proteins in parietal epithelial cells during experimental glomerular disease. Am J Physiol 2010;298:F702–F711.
25 Pedchenko V, Bondar O, Fogo AB, Vanacore R, Voziyan P, Kitching AR, Wieslander J, Kashtan C, Borza DB, Neilson EG, Wilson CB, Hudson BG: Molecular architecture of the Goodpasture autoantigen in anti-GBM nephritis. N Engl J Med 2010;363:343–354.
26 Bosch X, Guilabert A, Font J: Antineutrophil cytoplasmic antibodies. Lancet 2006;368:404–418.
27 Chen YX, Yu HJ, Ni LY, Zhang W, Xu YW, Ren H, Chen XN, Wang XL, Li X, Pan XX, Wang WM, Chen N: Propylthiouracil-associated antineutrophil cytoplasmic autoantibody-positive vasculitis: retrospective study of 19 cases. J Rheumatol 2007;34:2451–2456.
28 Chen YX, Zhang W, Chen XN, Yu HJ, Ni LY, Xu J, Pan XX, Ren H, Chen N: Propylthiouracil-induced antineutrophil cytoplasmic antibody (ANCA)-associated renal vasculitis versus primary ANCA-associated renal vasculitis: a comparative study. J Rheumatol 2012;39:558–563.
29 Kain R, Exner M, Brandes R, Ziebermayr R, Cunningham D, Alderson CA, Davidovits A, Raab I, Jahn R, Ashour O, Spitzauer S, Sunder-Plassmann G, Fukuda M, Klemm P, Rees AJ, Kerjaschki D: Molecular mimicry in pauci-immune focal necrotizing glomerulonephritis. Nat Med 2008;14:1088–1096.
30 Roth AJ, Brown MC, Smith RN, Badhwar AK, Parente O, Chung H, Bunch DO, McGregor JG, Hogan SL, Hu Y, Yang JJ, Berg EA, Niles J, Jennette JC, Preston GA, Falk RJ: Anti-LAMP-2 antibodies are not prevalent in patients with antineutrophil cytoplasmic autoantibody glomerulonephritis. J Am Soc Nephrol 2012;23:545–555.
31 Kessenbrock K, Krumbholz M, Schonermarck U, Back W, Gross WL, Werb Z, Grone HJ, Brinkmann V, Jenne DE: Netting neutrophils in autoimmune small-vessel vasculitis. Nat Med 2009;15:623–625.

Prof. Nan Chen, MD
Department of Nephrology, Ruijin Hospital
School of Medicine, Shanghai Jiao Tong University
No. 197 Ruijin Er Road, Shanghai 200025 (PR China)
E-Mail chen-nan@medmail.com.cn

Chen N (ed): New Insights into Glomerulonephritis.
Contrib Nephrol. Basel, Karger, 2013, vol 181, pp 216–228 (DOI: 10.1159/000348634)

ANCA-Associated Renal Vasculitis – An Update

Vladimir Tesar · Zdenka Hruskova

Department of Nephrology, 1st Faculty of Medicine, Charles University and General University Hospital, Prague, Czech Republic

Abstract

ANCA-associated vasculitis (AAV) is a potentially life-threatening disease with frequent and often severe kidney involvement which may result in end-stage renal disease. Anti-PR3 and anti-MPO disease are genetically distinct diseases and may have a different pathogenesis. Recent discovery of new autoantibodies (anti-LAMP-2) and the role of complement activation in the pathogenesis of AAV could result in better monitoring of the activity of the disease and identification of new treatment targets. The outcome of patients with AAV has dramatically improved, but long-term mortality still remains relatively high partly due to effective but relatively toxic immunosuppressive treatment. Recent studies demonstrated that B-cell depletion with rituximab is comparable to cyclophosphamide as induction treatment in newly diagnosed AAV patients and better than cyclophosphamide in relapsing patients. Rituximab-based maintenance treatment is superior to standard treatment with azathioprine. The use of more targeted treatment will hopefully be translated into a better long-term outcome of AAV patients.

Copyright © 2013 S. Karger AG, Basel

ANCA-associated vasculitis (AAV) is a necrotizing small vessel vasculitis often affecting the kidney with or without granulomatous inflammation, most frequently, but not exclusively, of the upper and lower respiratory tract. AAV is a rare disease with an annual incidence in Europe and North America of about 20 patients/million inhabitants and with an overall prevalence of about 180 patients/million in UK inhabitants [1].

Table 1. Comparison of the old and new classification of AAV

New classification	Old classification	Clinical characteristics	ANCA
MPA	MPA	necrotizing vasculitis, necrotizing GN and pulmonary capillaritis are common	anti-MPO more common than anti-PR3 (58 vs. 26%)
GPA	Wegener's granulomatosis	necrotizing vasculitis, necrotizing granulomatous inflammation in the respiratory tract, no history of asthma, without eosinophilia	anti-PR3 more common than anti-MPO (66 vs. 24%)
EGPA	Churg-Strauss syndrome	necrotizing granulomatous inflammation of respiratory tract, necrotizing vasculitis, history of asthma, eosinophilia	mostly anti-MPO-positive (40%) or ANCA-negative, anti-PR3 <5%

The prevalence of ANCA in different clinicopathologic units is according to Kallenberg [13].

The aim of this brief update is to shortly review the changes in the nomenclature of AAV in light of recently published genome-wide association studies and the new histologic classification of renal AAV and its relation to the renal outcome, to focus on the news in the pathogenesis of AAV, and finally to discuss the relation of the new treatment options in improving the outcome of patients.

New Classification of AAV – Are Anti-PR3-Positive and Anti-MPO-Positive AAV Distinct Diseases?

AAV was historically subdivided into three clinicopathological syndromes: Wegener's granulomatosis (WG), microscopic polyangiitis (MPA) and Churg-Strauss syndrome (CSS). Antineutrophil cytoplasmic antibodies (ANCA) directed at proteinase-3 (anti-PR3) or myeloperoxidase (anti-MPO) are present in most patients with active disease, but the use of ANCA in monitoring the activity of the disease is still controversial [2].

According to the 2012 Revised International Chapel Hill Consensus Conference Nomenclature of Vasculitides [3] accepting the recommendations of ACR, ASN and EULAR with the aim of avoiding eponyms and preferring descriptive names related to the pathogenesis of the disease, WG has been renamed to granulomatosis with polyangiitis (GPA) and CSS to eosinophilic granulomatosis with polyangiitis (EGPA) while the name microscopic polyangiitis, MPA, has been retained (table 1).

The presence of anti-PR3 antibodies increases the risk of relapse almost twofold compared to anti-MPO and the type of ANCA has been repeatedly demonstrated to be a better predictor of the relapse risk than clinicopathological diagnosis of GPA or MPA [4]. The type of ANCA, however, seems to predict neither response to treatment nor end-stage renal disease and death [5].

A recent first genome-wide association study [6] which investigated the genetic basis of both GPA and MPA and anti-PR3 and anti-MPO disease in a large discovery and replication cohort of patients with AAV demonstrated a much tighter genetic association with autoantigen specificity than with the clinical syndrome. Anti-proteinase-3 ANCA was associated with HLA-DP and the genes encoding α_1-antitrypsin (SERPINA1) and proteinase-3. Anti-myeloperoxidase ANCA was associated with HLA-DQ. Anti-PR3 and anti-MPO AAV seem to be genetically distinct autoimmune diseases with a similar clinical picture and respective ANCA antigens are supposed to play a crucial role in determining the disease phenotype [7].

In clinical practice, classification and diagnostic systems should always specify the type of ANCA, i.e. anti-PR3 GPA or anti-MPO-positive MPA [3, 5].

New Classification of Renal Histology in AAV

The renal outcome in AAV patients seems to be unrelated to the type of ANCA and is clinically best predicted by age and renal function at presentation. Whereas glomerular filtration rate (GFR) at presentation and recovery of renal function are related to acute potentially reversible lesions in the kidney (e.g. fibrinoid necrosis and cellular crescents), long-term outcome (GFR at 12 or 18 months) is better predicted by the presence of normal glomeruli and chronic irreversible changes (e.g. fibrous crescents and tubular atrophy) [8, 9].

A new histopathological classification recently developed by the International Working Group of Renal Pathologists [10] divides ANCA-associated glomerulonephritis into four classes: focal, crescentic, mixed and sclerotic. Validation in 100 Caucasian patients demonstrated that the newly proposed classification is able to predict the 1- and 5-year renal survival of patients (table 2) and could have not only a prognostic impact, but could also help to guide the intensity of immunosuppressive treatment.

Recent verification of this classification demonstrated that in Japanese and Chinese patients [11] compared to European patients, anti-MPO AAV was (as expected) much more frequent (88 vs. 45%, respectively) and crescentic class was the most frequent one both in Europe and China (55 and 47%, respectively), but much less common in Japan (only 8%) with higher representation of focal

Table 2. Histopathological classification of ANCA-associated glomerulonephritis

Class	Definition	Renal survival, %		
		1 year	5 years	7 years
Focal	≥50% normal glomeruli	93	93	93
Crescentic	≥50% glomeruli with cellular crescents	84	76	76
Mixed	<50% normal, <50% crescentic, <50% globally sclerotic glomeruli	69	61	50
Sclerotic	≥50% globally sclerotic glomeruli	50	50	25

class. Although the sclerotic class was always associated with the highest risk of end-stage renal disease, there was no apparent difference in the risk between crescentic and mixed class among Chinese and even between the focal and mixed class in Japanese patients. Larger validation studies are undoubtedly necessary, but histologic classification should apparently always be evaluated in relation to the GPA/MPA and anti-PR3/anti-MPO ratio and especially in anti-MPO patients a variable degree of interstitial and vascular lesions should also be taken into consideration.

News in the Pathogenesis

ANCA is believed to play a direct role in the pathogenesis of AAV [12]. Neutrophils activated by ANCA through both Fab and Fc fragments adhere to endothelial cells and damage them [13]. Whereas animal models of MPO-ANCA-associated glomerulonephritis and/or vasculitis were long known, a model of anti-PR3 antibody-mediated vasculitis in mice with a humanized immune system has only recently been described [14, 15].

Several recent theories contributed to our understanding of the origin of ANCA in AAV. Production of ANCA (more frequently PR3-ANCA) by B lymphocytes may be triggered in patients with AAV by CpG motifs highly prevalent in bacterial DNA interacting with TLR9 receptors. This mechanism could explain the association of relapses of AAV with nasal carriage of *Staphylococcal aureus* [16].

Another type of autoantibody, anti-LAMP-2 (directed against lysosomal membrane protein-2), shown to be present in 80–90% of both anti-PR3 and anti-MPO-positive patients with active AAV [17], cross-reacts with bacterial adhesin FimH and may be formed as a result of the molecular mimicry between shared epitope of bacterial adhesin FimH and LAMP-2. Anti-LAMP-2 antibodies induce apoptosis of human microvascular endothelial cells and cause pauci-immune focal necrotizing glomerulonephritis in humans [18]. Recent failure to confirm the high prevalence of anti-LAMP-2 antibodies in AAV [19] may be caused by the rapid disappearance of these antibodies shortly after the initiation of immunosuppressive treatment, but also by the selection of patients and methods used for anti-LAMP-2 evaluation [20]. Only a more generally available assay for anti-LAMP-2 evaluation and prospective studies of larger cohorts of patients may finally define their putative role both in the pathogenesis and monitoring of activity of AAV.

Formation of neutrophil extracellular traps (NETs) was also recently suggested to play a role in the pathogenesis of AAV [21]. NETs may be released by ANCA-stimulated neutrophils and may contain the targeted autoantigens (PR3 and MPO). Deposition of NETs in the kidney suggests their putative role in the pathogenesis of kidney damage in AAV.

Another theory on the origin of ANCA includes the autoantigen complementarity. Autoimmunity may be in anti-PR3-positive patients primarily directed at the peptide translated from the antisense DNA strand of PR3 (complementary PR3 (cPR3) [22]) which may share at least a part of the epitope with some bacterial mimicry of this peptide. Both mouse and human antibodies to PR3 and cPR3 bound each other forming an idiotypic network. Autoimmune response in AAV primarily raised against complementary peptide may thereafter induce the anti-idiotypic antibodies reacting with the target autoantigen.

Interestingly, antibodies to cPR3 may also cross-react with plasminogen (but not plasmin), probably due to same amino acid motif found in both plasminogen and part of cPR3 [23]. Anti-plasminogen antibodies are related to the high incidence of venous thrombotic events in GPA and were found to be present in 25% of patients with AAV (both anti-PR3- and anti-MPO ANCA) compared to 2% in controls [24]. They are able to retard fibrinolysis and are associated with a higher percentage of glomeruli with fibrinoid necrosis and cellular crescents and more severely reduced renal function. Therapies enhancing fibrinolysis might supposedly improve the outcome in a subset of AAV patients.

As the renal lesions in AAV are pauci-immune, the putative role of complement in the pathogenesis of AAV was underestimated [25]. Only recently, IgG isolated from patients with both MPO- and PR3-ANCA was shown to cause activation of complement in the neutrophils [26] and supernatants from ANCA-activated neutrophils were able to activate complement in normal serum and release C5a contributing to the neutrophil respiratory burst. Mice with C5 knockout do not develop crescentic nephritis in one experimental model of AAV [27] and neutrophil priming can be completely abrogated by anti-C5 receptor antibodies [27]. Plasma levels of C3a, C5a, soluble C5b-9 and Bb and urinary levels of C5a are significantly increased in patients with active AAV compared to patients with AAV in remission [28, 29] and plasma levels of Bb correlate in patients with active AAV with the proportion of total and cellular crescents in renal biopsy [28]. C5a receptor expressed by neutrophils and monocyte/macrophages (C5L2) was significantly more abundant in renal biopsies from patients with AAV compared to controls [29]. Considering the role of C5a in amplification of neutrophil activation and formation of glomerular crescents, C5a and/or C5aR could be a good therapeutic target in AAV.

T-helper cells undoubtedly play an important role in the pathogenesis of AAV [30]. The population of effector memory CD4+ T cells is expanded and may contribute to tissue injury and disease progression, sustaining chronic granulomatous inflammation with ectopic lymphoid-like tissue formation, which further promotes ANCA production. There is also a dysfunction of regulatory T cells which may even differentiate in a proinflammatory environment into Th17 cells producing IL-17 and thus contribute to the formation of granuloma [31].

Impaired function of the regulatory immune mechanisms fails to normally resolve vascular inflammation. Low intracellular production of IL-10 with regulatory potential was recently shown as a predictor of relapse in AAV patients in remission [32]. Decreased IL-10 production may be mediated by persistently elevated activity of protein tyrosine phosphatase non-receptor 22 (PTPN22) and suppression of ERK in leukocytes [33].

Long-Term Outcome of the Patients

Despite high remission rates (80–90%), 5-year survival for GPA and MPA is still rather unsatisfactory (74–91 and 45–76%, respectively [34]), partly due to a high relapse rate (much higher in GPA compared to MPA) requiring repeated courses of immunosuppressive treatment and irreversible organ damage often present at presentation (more often in MPA). The outcome of patients with AAV (remission, relapse, overall and renal survival) is namely influenced by older age, male gender, type of organ involvement, presence and type of ANCA and immunosuppressive treatment.

A recently reported outcome of 535 European patients (median age 60.7 years, 53% GPA, median creatinine level 203 μmol/l, median follow-up 5.2 years) [35] confirmed still a relatively high standardized mortality rate (compared to age- and sex-matched population) of 2.6 with infections (48%) and active vasculitis (19%) being the main causes of death in the first year, followed by cardiovascular disease (26%), malignancy (22%) and infection (20%) after the first year. Negative outcome was associated with advancing age, low GFR (<15 ml/min) and high BVAS (reflecting activity of the disease at presentation). The risk of relapse was significantly higher in patients with anti-PR3 antibodies (HR 1.62), but significantly lower in patients with decreased GFR (HR 0.39 when patients with a creatinine level of >200 μmol/l were compared with patients with a creatinine level of <100 μmol/l) [36]. So, except for the type of ANCA, a decreased GFR is another important predictor of reduced relapse rate. This observation may be of practical importance concerning monitoring and treatment of patients with AAV.

A recent retrospective analysis of 445 patients with GPA treated in one large German academic center over four decades clearly demonstrated improving the outcome of the patients [37] characterized by decreasing the standardized mortality ratio (from 2.1 before 1993 to 1.03 after 1999) with a decreasing proportion of deaths related to either the activity of the disease or its treatment (from 86.4 to 50%) and decreasing the relapse rate (from 63.9 to 35.3%) despite similar organ manifestations. Due to better awareness, the diagnosis is earlier (interval between first symptoms and diagnosis decreased from 8 to 4 months) and although still 80% of patients are treated with cyclophosphamide the cumulative dose of cyclophosphamide decreased from 67 to 24 g.

Despite improving outcome, mortality of especially older patients with renal involvement remains high, partly due to the high short- and long-term toxicity of currently available treatment.

Standard Treatment

Introduction of cyclophosphamide in the treatment of AAV (at the time mostly defined as WG) in the 1970s dramatically improved the outcome of patients and transformed AAV from the acute life-threatening disease with a mean survival of only about 6 months to a chronic disease with slowly accumulating organ damage both due to the activity of the disease and the adverse events of the rather toxic treatment.

At the beginning of the 1990s, most patients with AAV (even with relatively mild disease) were treated for at least 1 year with cyclophosphamide and then for an undefined period of time with azathioprine with different doses of corticosteroids. As the oral cyclophosphamide was used for a very long time and repeated with each relapse of the disease, the cumulative dose of cyclophosphamide was usually very high and the patients were in high long-term risk of myelodysplasia and secondary malignancies [38].

Table 3. EUVAS disease categorization of AAV according to Mukhtyar et al. [39]

Category	Definition
Localized	Upper and/or lower respiratory tract disease without any other systemic involvement or constitutional symptoms
Early systemic	Any, without organ-threatening of life-threatening disease
Generalized	Renal or other organ-threatening disease, serum creatinine <500 μmol/l
Severe	Renal or other vital organ failure, serum creatinine >500 μmol/l
Refractory	Progressive disease unresponsive to glucocorticoids and cyclophosphamide

In the last 20 years, studies organized by the European Vasculitis Study Group (EUVAS) were dedicated to the optimization of the available treatment options [39]. To pinpoint the treatment to the severity of the disease, patients were subdivided into five groups (localized, early systemic, generalized, severe and refractory disease (table 3)).

It has been demonstrated that in patients with generalized AAV (serum creatinine <500 μmol/l) an early compared to later (3 vs. 12 months) switch from cyclophosphamide to azathioprine does not increase the (early) relapse rate (within 18 months [40]) and that oral continuous cyclophosphamide can be in terms of time to remission and remission rate safely replaced by intravenous cyclophosphamide pulses with a much lower cumulative dose of cyclophosphamide (8.2 vs. 15.9 g) and a lower rate of leukopenia [41]. Long-term (4.3 years) survival and renal survival was also the same despite an almost twofold risk of relapses in patients treated initially with pulsed cyclophosphamide suggesting that relapses may not always be closely related to the long-term outcome of the patients [42]. Rather surprisingly, mycophenolate was inferior to azathioprine as a maintenance treatment after remission induced by cyclophosphamide in newly diagnosed patients with AAV [43] in terms of relapse rate (RR 1.69) with no difference in severe adverse events rate, organ damage, estimated GFR, and proteinuria.

Patients with severe renal disease (serum creatinine >500 μmol/l) benefited from plasma exchange (compared to methylprednisolone pulses) as add-on treatment to oral cyclophosphamide and corticosteroids in terms of renal recovery (dialysis-independent renal function) at 3 months (69 vs. 49%) and 1 year [44]. Plasma exchange had, however, no impact on patient survival and long-term renal survival and its place in the therapeutic armamentarium in AAV is currently being tested in a much larger randomized trial.

Rituximab

The last decade was characterized by accumulating evidence in support of the use of rituximab in AAV patients coming both from randomized controlled trials in new and relapsing patients and retrospective analysis of rituximab use often in patients refractory to other treatment. Retrospective analyses suggested remission rates as high as 80–90% among patients with refractory AAV not responding to the standard immunosuppressive treatment. Treatment was usually well tolerated with only rare infusion reactions, decrease of IgM and sometimes also IgG and rarely late-onset neutropenia.

Rituximab induced B-cell depletion in all patients with refractory AAV [45] and relapsing GPA [46] with a response in 98% of patients (75% complete remission, 23% partial remission [45]) usually in parallel with a decrease of ANCA titer, lowering of the dose of corticosteroids and withdrawal of further immunosuppression [45]. The relapse rate was high and it is currently unclear if B-cell recovery and increasing titer of ANCA is [46] or is not [45] useful in predicting imminent relapse. Most authors recommend preemptive administration of rituximab in 4- to 6-month intervals at a dose of 500–1,000 mg for 2 years. Repeated rituximab administration may dramatically decrease the relapse rate (almost no relapses during rituximab administration [47] and much lower relapse rate during further follow-up (e.g. 26 vs. 85% within 44 months [48]). Rituximab can also be used in AAV in combination with low-dose cyclophosphamide pulses as a cyclophosphamide-sparing regimen [49].

Rituximab may however be less effective in granulomatous manifestations (namely retro-orbital granuloma) of GPA [50, 51], compared to e.g. renal disease (remission/improvement 44 vs. 89%), although its efficacy in granulomatous limited GPA [52] and in refractory ophthalmic GPA [53] was demonstrated in other small series of patients.

Efficacy and safety of rituximab in AAV was confirmed in three randomized controlled trials. In 197 patients with either newly diagnosed or relapsing AAV (GPA or MPA) rituximab (375 mg/m^2 weekly for 4 weeks) was non-inferior to cyclophosphamide (2 mg/kg daily) in terms of remission without the use of prednisone in newly diagnosed patients with AAV, including patients with major renal disease and alveolar hemorrhage, but was more efficacious than cyclophosphamide in patients with relapsing disease (67 vs. 42% of patients achieved remission [54]).

Rituximab (375 mg/m^2 weekly for 4 weeks) with two intravenous cyclophosphamide pulses was similarly effective in terms of remission (76 vs. 82%) as intravenous cyclophosphamide for 3–6 months followed by azathioprine in 44 patients with newly diagnosed AAV (median age 68 years) with severe renal involvement (median GFR 18 ml/min/1.73 m^2 [55]). There was the same death rate (18% in both limbs) and a similar median increase in the GFR between the entry and end of the study, namely 19 ml/min in the rituximab group and 15 ml/min in the control group.

In a small study of renal biopsies from the previous trial [55], GFR at 1 year was independently associated with CD3+ T-cell tubulitis and tubular atrophy. Tubular atrophy remained the independent predictor of GFR even after 2 years [56]. Therapy directed at T cells could further improve renal outcomes in rituximab-treated patients with AAV, but possibly at the expense of a high risk of infectious complications.

A recently published French randomized controlled trial compared rituximab and azathioprine in 114 patients with newly diagnosed (2/3) and relapsing (1/3) AAV. Once remission was achieved with conventional treatment, patients were randomized either to repeated rituximab administration (500 mg every 6 months for 18 months) or oral azathioprine for 22 months. The major relapse rate at 28 months was 27.1% in the azathioprine limb and only 3.6% in the rituximab limb with a similar rate of infection and low serious adverse event rates [57].

In all three randomized controlled trials there was no difference between rituximab on the one hand and cyclophosphamide or azathioprine on the other in terms of severe adverse events rate.

What is the current place of rituximab in the treatment of AAV? Should the cyclophosphamide be already abandoned in AAV patients? Although cyclophosphamide dramatically improved the outcome of AAV patients, its (especially long-term) side effects always warranted search for an alternative similarly effective and less toxic treatment [58]. In our opinion, cyclophosphamide can still be used in newly diagnosed patients with AAV, but should be replaced as a first-choice treatment by rituximab in relapsing patients. To avoid long-term side effects (sterility, myelodysplasias, secondary malignancies) repeated courses of cyclophosphamide should be abandoned [59]. Rituximab is also probably the best currently available option to maintain remission in AAV. The long-term outcome of rituximab-treated patients remains, however, uncertain and must be further explored.

Acknowledment

The preparation of the paper was supported by the research project PRVOUK-P25/LF1/2 of Charles University in Prague.

References

1 Watts RA, Lane SE, Bentham G, et al: Epidemiology of systemic vasculitis: a ten-year study in the United Kingdom. Arthritis Rheum 2000;43:414–419.

2 Finkielman JD, Merkel PA, Schroeder D, et al: Antiproteinase-3 antineutrophil cytoplasmic antibodies and disease activity in Wegener granulomatosis. Ann Intern Med 2007; 147:611–619.

3 Jennette J, Falk R, Bacon P, et al: 2012 Revised International Chapel Hill Consensus Conference Nomenclature of Vasculitides. Arthritis Rheum 2013;65:1–11.

4 Pagnoux C, Hogan SL, Chin H, et al: Predictors of treatment resistance and relapse in antineutrophil cytoplasmic antibody-associated small-vessel vasculitis: comparison of two independent cohorts. Arthritis Rheum 2008;58:2908–2918.

5 Lionaki S, Blyth ER, Hogan SL, et al: Classification of antineutrophil cytoplasmic autoantibody vasculitides: the role of antineutrophil cytoplasmic autoantibody specificity for myeloperoxidase or proteinase-3 in disease recognition and prognosis. Arthritis Rheum 2012;64:3452–3462.

6 Lyons PA, Rayner TF, Trivedi S, et al: Genetically distinct subsets within ANCA-associated vasculitis. N Engl J Med 2012;367:214–223.

7 Watts RA, Scott DG: ANCA vasculitis: to lump or split? Rheumatology (Oxford) 2012; 51:2115–2117.

8 Hauer HA, Bajema IM, Van Houwelingen HC, et al: Determinants of outcome in ANCA-associated glomerulonephritis: a prospective clinico-histopathological analysis of 96 patients. Kidney Int 2002;62:1732–1742.

9 Lind van Wijngaarden RA, Hauer HA, Wolterbeek R, et al: Clinical and histologic determinants of renal outcome in ANCA-associated vasculitis: a prospective analysis of 100 patients with severe renal involvement. J Am Soc Nephrol 2006;17:2264–2274.

10 Berden AE, Ferrario F, Hagen EC, et al: Histopathological classification of ANCA-associated glomerulonephritis. J Am Soc Nephrol 2010;21:1628–1636.

11 Muso E, Endo T, Itabashi M, et al: Evaluation of the newly proposed simplified histological classification in Japanese cohorts of myeloperoxidase-anti-neutrophil cytoplasmic antibody-associated glomerulonephritis in comparison with other Asian and European cohorts. Clin Exp Nephrol 2012 (E-pub ahead of print).

12 Falk RJ, Jennette JC: ANCA disease: where is this field heading? J Am Soc Nephrol 2010; 21:745–752.

13 Kallenberg CGM: Pathogenesis of ANCA-associated vasculitides. Ann Rheum Dis 2011; 70(suppl 1):i59–i63.

14 Xiao H, Schreiber A, Heeringa P, et al: Antineutrophil cytoplasmic autoantibodies specific for myeloperoxidase cause glomerulonephritis and vasculitis in mice. J Clin Invest 2002;110:955–963.

15 Little MA, Al-Ani B, Ren S, et al: Anti-proteinase-3 anti-neutrophil cytoplasm autoantibodies recapitulate systemic vasculitis in mice with a humanized immune system. PLoS One 2012;7:e28626.

16 Tadema H, Abdulahad WH, Lepse N, et al: Bacterial DNA motifs trigger ANCA production in ANCA-associated vasculitis in remission. Rheumatology (Oxford) 2011;50:689–696.

17 Kain R, Tadema H, McKinney EF, et al: High prevalence of autoantibodies to hLAMP-2 in anti-neutrophil cytoplasmic antibody-associated vasculitis. J Am Soc Nephrol 2012;23: 556–566.

18 Kain R, Exner M, Brandes R, et al: Molecular mimicry in pauci-immune focal necrotizing glomerulonephritis. Nat Med 2008;14:1088–1096.

19 Roth AJ, Brown MC, Smith RN, et al: Anti-LAMP-2 antibodies are not prevalent in patients with antineutrophil cytoplasmic autoantibody glomerulonephritis. J Am Soc Nephrol 2012;23:545–555.

20 Kain R, Rees AJ: What is the evidence for antibodies to LAMP-2 in the pathogenesis of ANCA-associated small vessel vasculitis? Curr Opin Rheumatol 2013;25:26–34.

21 Kessenbrock K, Krubholz M, Schönermarck U, et al: Netting neutrophils in autoimmune small-vessel vasculitis. Nat Med 2009;15:623–625.

22 Pendergraft WF, Preston GA, Shah RR, et al: Autoimmunity is triggered by cPR3 (105–201), a protein complementary to human autoantigen proteinase-3. Nat Med 2004;10:72–79.
23 Bautz DJ, Preston GA, Lionaki S, et al: Antibodies with dual reactivity to plasminogen and complementary PR3 in PR3-ANCA vasculitis. J Am Soc Nephrol 2008;19:2421–2429.
24 Berden AE, Nolan SL, Morris HL, et al: Antiplasminogen antibodies compromise fibrinolysis and associate with renal histology in ANCA-associated vasculitis. J Am Soc Nephrol 2010;21:2169–2179.
25 Kallenberg CG, Heeringa P: Complement is crucial in the pathogenesis of ANCA-associated vasculitis. Kidney Int 2013;83:16–18.
26 Xiao H, Schreiber A, Heeringa P, et al: Alternative complement pathway in the pathogenesis of disease mediated by antineutrophil cytoplasmic autoantibodies. Am J Pathol 2007;170:52–64.
27 Schreiber A, Xiao H, Jennette JC, et al: C5a receptor mediates neutrophil activation and ANCA-induced glomerulonephritis. J Am Soc Nephrol 2009;20:289–298.
28 Guo SJ, Yuan J, Chen M, et al: Circulating complement activation in patients with antineutrophil cytoplasmic antibody-associated vasculitis. Kidney Int 2013;83:129–137.
29 Yuan J, Gou SJ, Huang J, et al: C5a and its receptors in human anti-neutrophil cytoplasmic antibody-associated vasculitis. Arthritis Res Ther 2012;14:R140.
30 Abdulahad WH, Lamprecht P, Kallenberg CG: T-helper cells as new players in ANCA-associated vasculitides. Arthritis Res Ther 2011;13:236.
31 Lamprecht P, Wieczorek S, Epplen JT, et al: Granuloma formation in ANCA-associated vasculitides. APMIS Suppl 2009;127:32–36.
32 Hruskova Z, Rihova Z, Mareckova H, et al: Intracelllular cytokine production in ANCA-associated vasculitis: low levels of interleukin-10 in remission are associated with a higher relapse rate in the long-term follow-up. Arch Med Res 2009;40:276–284.
33 Cao Y, Yang J, Colby K, et al: High basal activity of the PTPN22 gain-of-function variant blunts leukocyte responsiveness negatively affecting IL-10 production in ANCA vasculitis. PLoS One 2012;7:e42783.
34 Mukhtyar C, Flossmann O, Hellmich B, et al: Outcomes from studies of antineutrophil cytoplasm antibody associated vasculitis: a systematic review by the European League Against Rheumatism systemic vasculitis task force. Ann Rheum Dis 2008;67:1004–1010.
35 Flossmann O, Berden A, de Groot K, et al: Long-term patient survival in ANCA-associated vasculitis. Ann Rheum Dis 2011;70:488–494.
36 Walsh M, Flossmann O, Berden A, et al: Risk factors for relapse of antineutrophil cytoplasmic antibody-associated vasculitis. Arthritis Rheum 2012;64:542–548.
37 Holle JU, Gross WL, Latza U, et al: Improved outcome in 445 patients with Wegener's granulomatosis in a German vasculitis center over four decades. Arthritis Rheum 2011;63:257–266.
38 Faurschou M, Sorensen IJ, Mellemkjaer L, et al: Malignancies in Wegener's granulomatosis: incidence and relation to cyclophosphamide therapy in a cohort of 293 patients. J Rheumatol 2008;35:100–105.
39 Mukhtyar C, Guillevin L, Cid MC, et al: EULAR recommendations for the management of primary small and medium vessel vasculitis. Ann Rheum Dias 2009;68:310–317.
40 Jayne D, Rasmussen N, Andrassy K, et al: A randomized trial of maintenance therapy for vasculitis associated with antineutrophil cytoplasmic autoantibodies. N Engl J Med 2003;349:36–44.
41 de Groot K, Harper L, Jayne DR, et al: Pulse versus daily oral cyclophosphamide for induction of remission in antineutrophil cytoplasmic antibody-associated vasculitis: a randomized trial. Ann Intern Med 2009;150:670–680.
42 Harper L, Morgan MD, Walsh M, et al: Pulse versus daily oral cyclophosphamide for induction of remission in ANCA-associated vasculitis: long-term follow-up. Ann Rheum Dis 2012;71:955–960.
43 Hiemstra TF, Walsh M, Mahr A, et al: Mycophenolate mofetil vs. azathioprine for remission maintenance in antineutrophil cytoplasmic antibody-associated vasculitis: a randomized controlled trial. JAMA 2010;304:2381–2388.
44 Jayne DR: Randomized trial of plasma exchange of high-dose methylprednisolone as adjunctive therapy for severe renal vasculitis. J Am Soc Nephrol 2007;18:2180–2188.

45 Jones RB, Ferraro AJ, Chaudhry AN, et al: A multicenter survey of rituximab therapy for refractory antineutrophil cytoplasmic antibody-associated vasculitis. Arthritis Rheum 2009;60:2156–2168.

46 Cartin-Ceba R, Golbin JM, Keogh KA, et al: Rituximab for remission induction and maintenance in refractory granulomatosis with polyangiitis (Wegener's): ten-year experience at a single centre. Arthritis Rheum 2012;64:3770–3778.

47 Roubaud-Baudron C, Pagnoux C, Méaux-Ruault N, et al: Rituximab maintenance therapy for granulomatosis with polyangiitis and microscopic polyangiitis. J Rheumatol 2012;39:125–130.

48 Smith RM, Jones RB, Guerry MJ, et al: Rituximab for remission maintenance in relapsing antineutrophil cytoplasmic antibody-associated vasculitis. Arthritis Rheum 2012;64:3760–3769.

49 Mansfield N, Hamour S, Habib AM, et al: Prolonged disease-free remission following rituximab and low-dose cyclophosphamide therapy for renal ANCA-associated vasculitis. Nephrol Dial Transplant 2011;26:3280–3286.

50 Aries PM, Hellmich B, Voswinkel J, et al: Lack of efficacy of rituximab in Wegener's granulomatosis with refractory granulomatous manifestations. Ann Rheum Dis 2006;65:853–858.

51 Holle JU, Dubrau C, Herlyn K, et al: Rituximab for refractory granulomatosis with polyangiitis (Wegener's granulomatosis): comparison of efficacy in granulomatous versus vasculitic manifestations. Ann Rheum Dis 2012;71:327–333.

52 Seo P, Specks U, Keogh KA: Efficacy of rituximab in limited Wegener's granulomatosis with refractory granulomatous manifestations. J Rheumatol 2008;35:2017–2023.

53 Taylor SR, Salama AD, Joshi L, et al: Rituximab is effective in the treatment of refractory ophthalmic Wegener's granulomatosis. Arthritis Rheum 2009;60:1540–1547.

54 Stone JH, Merke PA, Spiera R, et al: Rituximab versus cyclophosphamide for ANCA-associated vasculitis. N Engl J Med 2010;363:221–232.

55 Jones RB, Tervaert JW, Hauser T, et al: Rituximab versus cyclophosphamide in ANCA-associated renal vasculitis. N Engl J Med 2010;363:211–220.

56 Berden AE, Jones RB, Erasmus DD, et al: Tubular lesions predict renal outcome in antineutrophil cytoplasmic antibody-associated glomerulonephritis after rituximab therapy. J Am Soc Nephrol 2012;23:313–321.

57 Guillevin L, Pagnoux C, Karras A, et al: Rituximab versus azathioprine for maintenance in antineutrophil cytoplasmic antibodies (ANCA)-associated vasculitis. Arthritis Rheum 2012;64:S706.

58 Kallenberg CG: Could we abandon cyclophosphamide in systemic vasculitis and lupus nephritis? Ann Rheum Dis 2012 (E-pub ahead of print).

59 Guerry MJ, Brogan P, Bruce IN, et al: Recommendations for the use of rituximab in antineutrophil cytoplasm antibody-associated vasculitis. Rheumatology (Oxford) 2012;51:634–643.

Vladimir Tesar, MD, PhD
Department of Nephrology, General University Hospital
U nemocnice 2
CZ–128 08 Prague 2 (Czech Republic)
E-Mail vladimir.tesar@vfn.cz

Chen N (ed): New Insights into Glomerulonephritis.
Contrib Nephrol. Basel, Karger, 2013, vol 181, pp 229–239 (DOI: 10.1159/000348479)

Management of Crescentic Glomerulonephritis: What Are the Recent Advances?

Xiao Li · Nan Chen

Department of Nephrology, Ruijin Hospital, School of Medicine, Shanghai Jiao Tong University, Shanghai, PR China

Abstract

Crescentic glomerulonephritis (GN) is a life-threatening clinical syndrome characterized by crescents which are morphological manifestations of severe glomerular injury. Immunologically, crescentic GN are classified as anti-glomerular basement membrane nephritis, immune complex-mediated GN and pauci-immune GN. Conventional treatment for crescentic GN mainly consists of corticosteroids, immunosuppressants and plasma exchange. Plasmapheresis is beneficial for patients with Goodpasture's syndrome and those with severe pauci-immune GN. Recently, new therapeutic agents have emerged, such as monoclonal antibodies to T cells, B cells and cytokines (e.g. anti-CD20 antibodies and TNF-α inhibitors) and signal transduction inhibitors, which may provide satisfactory alternatives. However, most of these treatments have only been established in experimental crescentic GN, described in single cases or reported in open-label trials; thus, the safety and efficacy of these agents remain to be investigated via controlled clinical trials.

Copyright © 2013 S. Karger AG, Basel

Crescentic glomerulonephritis (GN) often presents with a rapid decline of renal function and pathology exhibiting extensive crescent formation. The disease is also called rapid progressive glomerulonephritis (RPGN) because if untreated, patients develop renal failure within days or weeks of diagnosis [1]. RPGN is usually caused by one of three mechanisms: anti-glomerular basement membrane (anti-GBM) antibody disease with or without pulmonary hemorrhage, severe immune complex GN or pauci-immune GN, which is often associated with anti-neutrophil cytoplasmic autoantibodies (ANCA) [2].

Early aggressive treatment is recommended. Current therapeutic regimens consist of corticosteroids, immunosuppressive drugs and plasma exchange. Recent investigation has focused on other immunomodulatory methods involving monoclonal antibodies to T cells, B cells and cytokines or pharmacological inhibitors of signal transduction [3]. However, the efficacy and safety of these agents remain to be clarified by randomized clinical trials (RCTs). The main aims of therapy are to reduce the risk of death and end-stage renal disease, but minimizing drug-related toxicity and preventing relapse should also be considered.

Current Therapies for Crescentic Glomerulonephritis

Treatment of Anti-GBM Antibody-Induced Crescentic GN

Anti-GBM GN is caused by autoantibodies to the noncollagenous domain of the α3 chain of type IV collagen. Anti-GBM GN is responsible for 10–20% of patients presenting with RPGN, although the estimated annual incidence is only 0.5–1 per million population [4]. Approximately one half of patients with anti-GBM GN experience pulmonary hemorrhage, which is called Goodpasture's syndrome. The strategy for treating anti-GBM GN is to remove the pathogenic autoantibodies from the circulation and simultaneously prevent further autoantibody production while attenuating the existing glomerular inflammation and injury [5].

A large retrospective study conducted at the Hammersmith Hospital [6] revealed the long-term outcomes of 71 patients with anti-GBM GN. The patients were treated with high-dose prednisolone (1 mg/kg/day to a maximum of 60 mg/day) tapered over 6–9 months, oral cyclophosphamide (2–3 mg/kg/day, with reduced dosages in patients >55 years of age) for 2–3 months and a daily plasma exchange (50 ml/kg to a maximum of 4 liters) for 14 days or until the anti-GBM antibody was no longer detectable. A Chinese retrospective survey [7] of 221 anti-GBM GN patients demonstrated that combination therapy using plasmapheresis plus corticosteroids and cyclophosphamide had an overall beneficial effect on both patient survival (HR for patient mortality, 0.31; $p = 0.001$) and renal survival (HR for renal failure, 0.60; $p = 0.032$), particularly, patient survival for those with Goodpasture's syndrome (HR for patient mortality, 0.29; $p = 0.004$) and renal survival for those with anti-GBM nephritis and an initial serum creatinine (SCr) >6.8 mg/dl (HR for renal failure, 0.52; $p = 0.014$).

The Kidney Disease: Improving Global Outcomes (KDIGO) guidelines [5] (table 1) recommend initiating immunosuppression with cyclophosphamide and corticosteroids plus plasmapheresis in all patients with anti-GBM GN except

Table 1. Therapy for anti-GBM GN [data modified from 5]

	Administration dosage and route	Duration
Corticosteroids		
Methylprednisolone	500–1,000 mg/day i.v.	3 days
Prednisone or prednisolone	1 mg/kg/day ideal body weight (max. 80 mg/day) and gradually taper until discontinuation	6 months
Cyclophosphamide	2 mg/kg/day orally or 0.5–0.75 g/m^2 body surface area i.v. monthly	3 months
Plasmapheresis	One 4-liter exchange per day with 5% albumin. Add 150–300 ml fresh frozen plasma at the end of each pheresis session if the patients exhibit pulmonary hemorrhage or have had recent surgery, including a kidney biopsy	14 days or until anti-GBM antibodies are no longer detectable

those who are dialysis-dependent at presentation, exhibit 100% crescents in an adequate biopsy sample and do not have pulmonary hemorrhage (1B). Treatment should be started immediately once the diagnosis of anti-GBM GN is confirmed. If the diagnosis is highly suspected, then it is appropriate to begin high-dose corticosteroids and plasma exchange while waiting for confirmation (no grade).

The long-term outcomes of patients depend on the severity of renal disease at presentation [6]. Patients presenting an initial SCr <504 μmol/l have a 1-year overall survival rate of 100% and a kidney survival rate of 95%; at 5 years, the patient and kidney survival rates are both 94%. However, among patients who require dialysis at presentation, the patient and kidney survival rates are reduced to 65 and 8%, respectively, at 1 year and 44 and 13%, respectively, at 5 years. In other series of anti-GBM GN, not all necessarily using the same treatment regimens, the patient survival rate at 6–12 months was approximately 67–94%, and the kidney survival rate was approximately 15–58% [8, 9].

The predictors of kidney survival in anti-GBM GN are SCr levels at presentation, the requirement for dialysis at presentation and the percentage of glomerular crescents [6, 7]. The requirement for hemodialysis on initial presentation, a SCr >5 mg/dl and 50–100% crescents on renal biopsy suggest the necessity for long-term hemodialysis [10]. It should be considered whether the risk of aggressive immunosuppression exceeds the small potential benefit in patients who are unlikely to recover their renal function. Maintenance immunosuppressive therapy is not recommended because of the self-limited nature of anti-GBM antibody disease.

Approximately 20–30% of patients with anti-GBM disease will also have ANCA, usually with anti-myeloperoxidase specificity. The initial case series suggested that these double positive patients respond to immunosuppressive therapy with a better prognosis than that of classical anti-GBM GN [11]. However, results from other studies [7, 12] have suggested that the double-antibody-positive patients do not appear to have a different prognosis or disease course compared with classical anti-GBM GN.

Treatment of Immune Complex-Mediated Crescentic GN

Patients with immune complex-induced RPGN should be treated according to their specific underlying etiologies, such as lupus nephritis, IgA nephropathy, postinfectious GN, membranoproliferative GN, Henoch-Schönlein purpura, cryoglobulinemia, etc.

Crescentic IgA Nephropathy

Crescentic IgA nephropathy has a poor prognosis [13]. Cumulative published cases suggest that renal survival in crescentic IgA nephropathy is only 50% at 1 year and 20% at 5 years [14]. It has been reported that patients with >50% crescents develop end-stage renal disease in 75% of cases after 10 years of follow-up. Compared with patients with general IgA nephropathy, most patients with crescentic IgA nephropathy exhibit a rapidly progressive GN that is associated with more severe pathological changes, including glomerular, tubular, interstitial and vascular lesions.

Large observational studies [14, 15] have all indicated that immunosuppression is potentially useful. The recommended therapeutic regimens in these reports vary, but initial therapy usually includes high-dose oral or intravenous corticosteroids plus oral or intravenous cyclophosphamide. Recently published KDIGO guidelines recommend the use of steroids and cyclophosphamide in patients with rapidly progressive crescentic IgA nephropathy, analogous to the treatment of ANCA vasculitis (2D) [5]. RCTs are necessary for investigating the exact benefits of corticosteroids, cyclophosphamide, MMF and azathioprine in crescentic IgA nephropathy. KDIGO guidelines recommend that children presenting crescentic Henoch-Schönlein purpura with nephrotic syndrome and/or deteriorating kidney function should be treated the same as for crescentic IgA nephropathy (2D) [5].

Postinfectious Crescentic GN

The postinfectious form of rapidly progressive GN with crescents is a form of immune-complex GN that appears to have a better prognosis. Low serum complement levels and glomerular C3 hump-like deposits indicate a complement-mediated injury in postinfectious crescentic GN. 23 cases of postinfectious crescentic GN were followed up for a mean period of 40.1 ± 28.9 months; the persistence of hypertension and nephrotic-range proteinuria were determined to be major, negative prognostic predictors for renal dysfunction [16].

The optimal treatment for severe poststreptococcal crescentic GN or crescentic GN induced by *Staphylococcus aureus* infection remains uncertain. A complete recovery may be achieved in most patients with appropriate supportive treatment (including vigorous antibiotic therapy). The use of combined immunosuppressants for these patients yielded no advantages over supportive therapy alone [17].

Crescentic Lupus Nephritis

The treatment of severe lupus nephritis has been extensively reviewed recently [18]. Cyclophosphamide is reserved for high-risk patients with proliferative lupus nephritis, such as crescentic lupus nephritis. Both mycophenolate mofetil and tacrolimus are possible alternatives to intravenous cyclophosphamide as induction therapies for active lupus nephritis [19]. In brief, initial-pulse prednisolone and cyclophosphamide are preferred for the treatment of severe or crescentic lupus nephritis. Oral prednisolone and monthly pulse cyclophosphamide are administered for 6–12 months until the patient has attained complete remission [5].

Treatment of Pauci-Immune Crescentic GN

The ANCA-associated small-vessel vasculitides include granulomatosis with polyangiitis (Wegener's), microscopic polyangiitis, Churg-Strauss syndrome and the renal-limited form of microscopic polyangiitis, also known as pauci-immune or idiopathic crescentic GN. Cyclophosphamide and corticosteroids are well-established initial therapies for crescentic GN caused by renal vasculitis (1A) [5]. KDIGO guidelines recommend that rituximab and corticosteroids be used as an alternative initial treatment in patients without severe disease or in whom cyclophosphamide is contraindicated (1B). The guidelines also recommend the addition of plasmapheresis for patients who require dialysis or exhibit rapidly increasing SCr levels (1C), patients with diffuse pulmonary hemorrhage (2C) and patients with an overlapping syndrome of ANCA vasculitis and anti-GBM GN (2D) [5]. The detailed treatment update is discussed elsewhere in this issue.

Novel Therapeutic Strategies

Because toxicities are associated with the long-term use of nonspecific immunosuppressive agents, newer and alternative therapies are emerging (table 2).

Table 2. Novel therapeutic strategies for crescentic glomerulonephritis (CGN)

	Anti-GBM GN	Immune-complex CGN	Pauci-immune CGN
MMF	Refractory cases with Goodpasture's syndrome [19, 20]	Controversial in IgA nephropathy [21–23]	Only 1 RCT suggested that MMF is similar to cyclophosphamide in inducing the remission of AVV
Rituximab	Successfully used for a relapse case [27]	An option for refractory lupus nephritis [28]	Combined with glucocorticoids as a front-line therapy [26]
TNF-α antagonists	No data	No data	A possible role for infliximab in vasculitis etanercept does not appear to have a role in Wegener's granulomatosis [31]
Cyclosporin, leflunomide, IVIG, anti-T-cell mAb, CTLA4-Ig, etc.	Case report form and open-label trials	Case report form and open-label trials	Case report form and open-label trials
Imatinib, small novel peptides, EGFR inhibitors, regulatory cytokines, etc.	Experimental evidence [33]	Experimental evidence	Experimental evidence [32]

Mycophenolate Mofetil

Mycophenolate mofetil (MMF) is effective in the prevention of organ transplant rejection. This drug suppresses the immune response by inhibiting activated lymphocytes, blocks purine synthesis and inhibits the specific proliferative responses of both T and B cells to antigens.

Two patients with Goodpasture's syndrome refractory to conventional therapy responded well to treatment with MMF [21]. Both patients remained dialysis-dependent throughout the treatment. The initial dose of MMF was 1 g every 12 h. After 6 or 8 months, MMF was reduced to 500 mg every 12 h. As circulating anti-GBM antibodies remained undetectable, 1 patient underwent successful renal transplantation after 12 months on dialysis, and the other patient was placed on the active waiting list for renal transplantation.

At present, the role of MMF in the treatment of crescentic IgA nephropathy is controversial. Two studies reported no benefit from MMF (2.0 g/day) in patients either at risk of progression [22] or with more advanced disease [23]. In contrast, a Chinese study in 40 patients with a mean initial GFR of 72 ml/min per 1.73 m^2 and mean proteinuria of 1.8 g/day revealed a significant reduction in proteinuria at 18 months after administration of MMF (1.5–2 g/day) for 6 months compared with the control subjects [24]. A 6-year follow-up of the same cohort demonstrated a kidney survival benefit. The reasons for heterogeneity of these outcomes require further investigation, but the various ethnicities or differences in the achieved drug levels might have been contributory factors.

There are insufficient data to support the use of MMF for induction therapy in ANCA-associated vasculitis (AAV). There is only one published controlled study of MMF (1.5–2 g/day) versus cyclophosphamide (monthly intravenous pulse of 0.75–1 g/m^2); this investigation of 35 patients from China [25] demonstrated similar remission rates between the two groups. In summary, the evidence to support the use of MMF in crescentic GN is lacking and larger randomized controlled trials are necessary.

Rituximab

Rituximab is a chimeric monoclonal antibody that targets the CD20 antigen on the surface of B cells and clears circulating B cells from the circulation. In April 2011, the US Food and Drug Administration (FDA) approved rituximab combined with glucocorticoids as a front-line therapy for adult patients suffering from granulomatosis with polyangiitis (Wegener's granulomatosis) or microscopic polyangiitis based on the promising results of the RAVE trial [26]. This new indication for rituximab provided the first ever FDA-approved therapy for these two diseases and the first alternative to cyclophosphamide for the treatment of severe disease in almost 40 years. The RAVE trial also suggests that rituximab is superior to cyclophosphamide in treating relapsing ANCA-associated GN.

Rituximab was successfully used for a relapse patient with anti-GBM disease [27]. The patient had persistent hemoptysis, hematuria and positive circulating anti-GBM antibodies despite treatment with corticosteroids, cyclophosphamide and plasma exchange. The patient was treated with weekly infusions of rituximab 375 mg/m^2. After the sixth cycle, the patient experienced complete remission, and the anti-GBM antibodies were undetectable. An RCT is necessary to compare rituximab with cyclophosphamide, both drugs combined with prednisone and plasmapheresis for anti-GBM GN.

Two RCTs investigating rituximab (the EXPLORER and LUNAR studies) failed to show efficacy of this drug on systemic lupus erythematosus. However, data from open-source and observational studies disagree with these clinical trial results. As supported by off-label use experience, rituximab appears to be a therapeutic option for cases of refractory lupus nephritis or recurrent flares that persist despite the use of standard therapy [28].

Tumor Necrosis Factor-α Antagonists
Tumor necrosis factor-α (TNF-α) is a proinflammatory cytokine produced by macrophages, renal mesangial cells and tubular epithelial cells. In experimental crescentic GN, anti-TNF-α treatment significantly reduces glomerular inflammation, crescent formation and tubulointerstitial scarring, with preservation of renal function [29].

Currently, three TNF-α antagonists are marketed worldwide for use in patients: adalimumab (Abbott Laboratories, Abbott Park, Ill., USA), etanercept (Amgen, Inc., Thousand Oaks, Calif., USA) and infliximab (Centocor, Inc., Horsham, Pa., USA). The various anti-TNF-blocking reagents have different biological actions.

There are several published trials of anti-TNF therapy in patients with systemic vasculitis but other types of crescentic GN. The safety and efficacy of infliximab was studied in 32 patients in the ACTIVE (Anti-Cytokine Therapy with Infliximab for Vasculitis) trial. This open-label, multicenter, prospective trial supported a possible role for infliximab in the treatment of patients with vasculitis [30]. Wegener's Granulomatous Etanercept Trial (WGET) is a multicenter, randomized, placebo-controlled trial that enrolled 174 patients with Wegener's granulomatous [31] and the data suggest that etanercept does not play a role in the induction or maintenance of remission in Wegener's granulomatosis. Larger randomized controlled trials that provide long-term follow-up data will be necessary prior to recommending anti-TNF therapy for patients with crescentic GN.

Other Treatment Approaches
Therapies for crescentic GN that include cyclosporin, leflunomide, high-dose intravenous immunoglobulin (IVIG), anti-T-cell monoclonal antibody (mAb), humanized anti-CD4 antibodies, anti-thymocyte globulin, cytotoxic T-lymphocyte-associated antigen 4 (CTLA4)-Ig and recombinant IL-1 receptor antagonists have been described in single cases or open-label trials. Because of limited information, no recommendations have been made on these therapy regimens.

AAV is triggered by interaction between FcγRIIa of neutrophils and ANCA. Based on computer-aided drug design technology, three small novel peptides were developed that block the interaction between IgG and FcγRIIa and might be potential targets of AAV [32]. Strong experimental evidence exists for the beneficial effects of imatinib (a selective tyrosine kinase inhibitor that can block platelet-derived growth factor receptor activity) [33], epidermal growth factor receptor (EGFR) inhibitors, regulatory cytokines (recombinant IL-4 or IL-10), p38 MAPK inhibitor, phosphodiesterase IV inhibitor, P2X7 receptor inhibitor and low-level laser therapy for suppressing the disease activity and providing therapeutic options for RPGN.

Conclusion

Crescentic GN is caused by various etiologies and pathogenic mechanisms that determine the natural outcome and appropriate treatment. Therapy for vasculitis, anti-GBM disease or connective tissue disease-related RPGN (such as crescentic lupus nephritis) includes high-dose methylprednisolone and cyclophosphamide. In addition, patients with anti-GBM GN and severe vasculitis require plasma exchanges. Conventional immunosuppressive treatment has reduced the rate of fatal outcomes but is stymied by cytotoxic side effects and frequent relapses. Recent advances in biological agents, such as anti-CD20 monoclonal antibodies and TNF-α antagonists, have provided new options for patients who are resistant to conventional treatment. The efficacy and safety of these new agents must be assessed in large controlled clinical trials.

Acknowledgements

This work was supported by grants from the National Basic Research Program of China 973, Program No. 2012CB517600 (No. 2012CB517604); the National Key Technology R&D Program (12–5)2011BAI10B00 (No. 2011BAI10B06); the National Nature Science Foundation of China (No. 81170671), and the Key Program of Shanghai Science and Technology Commission (No. 08dz1900502).

References

1 Couser WG: Rapidly progressive glomerulonephritis: classification, pathogeneticl mechanisms, and therapy. Am J Kidney Dis 1988; 11:449–464.
2 Jennette JC, Thomas DB: Crescentic glomerulonephritis. Nephrol Dial Transplant 2001; 16(suppl 6):80–82.
3 Tam FW: Current pharmacotherapy for the treatment of crescentic glomerulonephritis. Expert Opin Investig Drugs 2006;15:1353–1369.
4 Bolton WK: Goodpasture's syndrome. Kidney Int 1996;50:1753–1766.
5 KDIGO clinical practice guideline for glomerulonephritis. Kidney Int Suppl 2012; 2:200–242.
6 Levy JB, Turner AN, Rees AJ, Pusey CD: Long-term outcome of anti-glomerular basement membrane antibody disease treated with plasma exchange and immunosuppression. Ann Intern Med 2001;134:1033–1042.
7 Cui Z, Zhao J, Jia XY, Zhu SN, Jin QZ, Cheng XY, Zhao MH: Anti-glomerular basement membrane disease: outcomes of different therapeutic regimens in a large single-center Chinese cohort study. Medicine (Baltimore) 2011;90:303–311.
8 Li FK, Tse KC, Lam MF, Yip TP, Lui SL, Chan GS, Chan KW, Chan EY, Choy BY, Lo WK, Chan TM, Lai KN: Incidence and outcome of antiglomerular basement membrane disease in Chinese. Nephrology (Carlton) 2004;9:100–104.
9 Hirayama K, Yamagata K, Kobayashi M, Koyama A: Anti-glomerular basement membrane antibody disease in Japan: part of the nationwide rapidly progressive glomerulonephritis survey in Japan. Clin Exp Nephrol 2008;12:339–347.

10 Chan AL, Louie S, Leslie KO, Juarez MM, Albertson TE: Cutting edge issues in Goodpasture's disease. Clin Rev Allergy Immunol 2011;41:151–162.
11 Bosch X, Mirapeix E, Font J, Borrellas X, Rodríguez R, López-Soto A, Ingelmo M, Revert L: Prognostic implication of anti-neutrophil cytoplasmic autoantibodies with myeloperoxidase specificity in anti-glomerular basement membrane disease. Clin Nephrol 1991; 36:107–113.
12 Levy JB, Hammad T, Coulthart A, Dougan T, Pusey CD: Clinical features and outcome of patients with both ANCA and anti-GBM antibodies. Kidney Int 2004;66:1535–1540.
13 Xie J, Kiryluk K, Wang W, Wang Z, Guo S, Shen P, Ren H, Pan X, Chen X, Zhang W, Li X, Shi H, Li Y, Gharavi AG, Chen N: Predicting progression of IgA nephropathy: new clinical progression risk score. PLoS One 2012;7:e38904.
14 Barratt J, Feehally J: Treatment of IgA nephropathy. Kidney Int 2006;69:1934–1938.
15 Pankhurst T, Lepenies J, Nightingale P, Howie AJ, Adu D, Harper L: Vasculitic IgA nephropathy: prognosis and outcome. Nephron Clin Pract 2009;112:c16–c24.
16 El-Husseini AA, Sheashaa HA, Sabry AA, Moustafa FE, Sobh MA: Acute postinfectious crescentic glomerulonephritis: clinicopathologic presentation and risk factors. Int Urol Nephrol 2005;37:603–609.
17 Zaffanello M, Cataldi L, Franchini M, Fanos V: Evidence-based treatment limitations prevent any therapeutic recommendation for acute poststreptococcal glomerulonephritis in children. Med Sci Monit 2010;16:RA79–RA84.
18 Borchers AT, Leibushor N, Naguwa SM, Cheema GS, Shoenfeld Y, Gershwin ME: Lupus nephritis: a critical review. Autoimmun Rev 2012;12:174–194.
19 Li X, Ren H, Zhang QY, Zhang W, Wu XJ, Xu YW, Shen PY, Chen N: Mycophenolate mofetil, tacrolimus and intravenous cyclophosphamide for the induction treatment of active lupus nephritis. Nephrol Dial Transplant 2012;27:1467–1472.
20 Malho A, Santos V, Cabrita A, Silva AP, Pinto I, Bernardo I, Neves PL: Severe relapsing Goodpasture's disease successfully treated with mycophenolate mofetil. Int J Nephrol 2012;2010:383548.
21 García-Cantón C, Toledo A, Palomar R, Fernandez F, Lopez J, Moreno A, Esparza N, Suria S, Rossique P, Diaz JM, Checa D: Goodpasture's syndrome treated with mycophenolate mofetil. Nephrol Dial Transplant 2000;15:920–922.
22 Maes BD, Oyen R, Claes K, Evenepoel P, Kuypers D, Vanwalleghem J, Van Damme B, Vanrenterghem YF: Mycophenolate mofetil in IgA nephropathy: results of a 3-year prospective placebo-controlled randomized study. Kidney Int 2004;65:1842–1849.
23 Frisch G, Lin J, Rosenstock J, Markowitz G, D'Agati V, Radhakrishnan J, Preddie D, Crew J, Valeri A, Appel G: Mycophenolate mofetil (MMF) vs. placebo in patients with moderately advanced IgA nephropathy: a double-blind randomized controlled trial. Nephrol Dial Transplant 2005;20:2139–2145.
24 Tang S, Leung JC, Chan LY, Lui YH, Tang CS, Kan CH, Ho YW, Lai KN: Mycophenolate mofetil alleviates persistent proteinuria in IgA nephropathy. Kidney Int 2005;68:802–812.
25 Hu W, Liu C, Xie H, Chen H, Liu Z, Li L: Mycophenolate mofetil versus cyclophosphamide for inducing remission of ANCA vasculitis with moderate renal involvement. Nephrol Dial Transplant 2008;23:1307–1312.
26 Stone JH, Merkel PA, Spiera R, Seo P, Langford CA, Hoffman GS, Kallenberg CG, St Clair EW, Turkiewicz A, Tchao NK, Webber L, Ding L, Sejismundo LP, Mieras K, Weitzenkamp D, Ikle D, Seyfert-Margolis V, Mueller M, Brunetta P, Allen NB, Fervenza FC, Geetha D, Keogh KA, Kissin EY, Monach PA, Peikert T, Stegeman C, Ytterberg SR, Specks U; RAVE-ITN Research Group: Rituximab versus cyclophosphamide for ANCA-associated vasculitis. N Engl J Med 2010; 363:221–232.
27 Arzoo K, Sadeghi S, Liebman HA: Treatment of refractory antibody-mediated autoimmune disorders with an anti-CD20 monoclonal antibody (rituximab). Ann Rheum Dis 2002;61: 922–924.
28 Wiesik-Szewczyk E, Olesinska M: B-cell targeted therapy in systemic lupus erythematosus: potential of rituximab. Biologics 2012;6: 347–354.
29 Khan SB, Cook HT, Bhangal G, Smith J, Tam FW, Pusey CD: Antibody blockade of TNF-α reduces inflammation and scarring in experimental crescentic glomerulonephritis. Kidney Int 2005;67:1812–1820.

30 Booth A, Harper L, Hammad T, Bacon P, Griffith M, Levy J, Savage C, Pusey C, Jayne D: Prospective study of TNF-α blockade with infliximab in anti-neutrophil cytoplasmic antibody-associated systemic vasculitis. J Am Soc Nephrol 2004;15:717–721.
31 The Wegener's Granulomatosis Etanercept Trial (WGET) Research Group: Etanercept plus standard therapy for Wegener's granulomatosis. N Engl J Med 2005;352:351–361.
32 Chen N, Zhang J, Yu HJ, Zhang W, Chen YX, Wang WM: Computer-aided drug design of a potential and selective small peptide inhibiting interaction of FcγIIa and ANCA. Clin Exp Immunol 2011;164(suppl 1):74–74.
33 Iyoda M, Shibata T, Kawaguchi M, Yamaoka T, Akizawa T: Preventive and therapeutic effects of imatinib in Wistar-Kyoto rats with anti-glomerular basement membrane glomerulonephritis. Kidney Int 2009;75:1060–1070.

Prof. Nan Chen, MD
Department of Nephrology, Ruijin Hospital
School of Medicine, Shanghai Jiao Tong University
No. 197 Ruijin Er Road, Shanghai 200025 (PR China)
E-Mail chen-nan@medmail.com.cn

Chen N (ed): New Insights into Glomerulonephritis.
Contrib Nephrol. Basel, Karger, 2013, vol 181, pp 240–247 (DOI: 10.1159/000348632)

Plasma Exchange in the Treatment of Rapidly Progressive Glomerulonephritis

Xiaonong Chen · Nan Chen

Department of Nephrology, Ruijin Hospital, School of Medicine, Shanghai Jiao Tong University, Shanghai, PR China

Abstract

Plasma exchange has been used for the treatment of rapidly progressive glomerulonephritis (RPGN) for more than 40 years. Its main application has been the rapid removal of special pathogenic factors that are of great importance for treatment of antiglomerular basement membrane disease, antineutrophil cytoplasmic antibody-associated vasculitis, and immune complex-mediated crescentic nephritis. Although the number of controlled trials is increasing, evidence for the benefit of plasma exchange in RPGN has varied in quality and the results are controversial. Moreover, its efficacy appears to depend on the type and duration of crescentic nephritis. Patients with pulmonary hemorrhage need plasma exchange to alleviate their symptoms and improve their chance of survival. Further trials are needed to ascertain the optimal dose, frequency and number of plasma exchange sessions, long-term outcomes, as well as the role of plasma exchange in the treatment of RPGN.
Copyright © 2013 S. Karger AG, Basel

Rapidly progressive glomerulonephritis (RPGN) is defined as a clinical syndrome involving hematuria, proteinuria, anemia, and rapidly progressing renal failure. As the most severe type of glomerulonephritis, renal biopsy usually reveals crescentic glomerulonephritis [1]. Due to rapid onset and progression of the disease, if left untreated more than 90% of patients die within 6 months or become dependent on dialysis. Plasmapheresis, also known as therapeutic plasma exchange (TPE), was introduced in the 1970s to treat a patient with Goodpasture's syndrome and its application was readily extended to patients with type I and type III RPGN [2, 3]; since then, the treatment has been widely applied to many autoimmune renal diseases. Furthermore, plasmapheresis has emerged as a useful therapy for multiple myeloma, scleroderma, multiple sclerosis, Guillain-Barré syndrome, and various other diseases. Since the first introduction of TPE in clinical practice four decades ago, significant advances have been made to this technology, especially related to its application in autoimmune diseases. In this review, we focus on recent advances of plasmapheresis in treatment of RPGN.

Modalities of Plasmapheresis

Intensified Plasma Exchange

Intensified plasma exchange uses a plasma filter or centrifugal plasma cell separator to separate plasma from blood cells and then replace them with normal plasma or a plasma component, such as albumin. Plasmapheresis is carried out daily or every other day, with each replacement consisting of 2–4 liters. This method is effective to clear pathogenic antibodies and circulating immune complexes, and has been widely used in clinical practice [4].

Double-Filtration Plasmapheresis

Double-filtration plasmapheresis was developed on the basis of intensified plasmapheresis. Separated plasma is passed through the second membrane filter that blocks globulin and macromolecular proteins due to the small pore size of the filter membrane. Plasma without the above-mentioned components is then reinfused. Double-filtration plasmapheresis cannot only clear pathogenic antibodies and immune complexes from the blood, but also avoids abundant plasma infusion and eventually save costs. However, the clinical application of double-filtration plasmapheresis in the treatment of the RPGN is scarcely reported, and its efficacy compared to intensified plasma exchange has yet to be verified [4].

Immunoadsorption

Immunoadsorption is a therapeutic procedure in which plasma of the patient, after separation from the blood, is passed through the immunochromatographic absorption column (e.g. specific absorption of antiglomerular basement membrane (GBM) antibody, or protein A adsorption column). These columns have the capacity to remove immunoglobulins by specifically binding them to the active component (e.g. staphylococcal protein A) of the device. After the pathogenic components are cleared, the plasma can then be reinfused into the patient. Immunoadsorption is effective for clearing pathogenic antibodies and/or circulating immune complexes, but is limited by high cost [4].

Application of Plasmapheresis in RPGN

Anti-GBM Nephritis

The standard treatment of anti-GBM nephritis is a combination of TPE, cyclophosphamide, and corticosteroids [5]. The American Society for Apheresis (ASFA) has indicated the use of TPE for the management of anti-GBM disease as the first-line of therapy when the patient is in the earlier stages of the disease and not on renal replacement therapy [4]. If the patient has developed diffuse alveolar hemorrhage (DAH) from the anti-GBM antibody disease, plasma exchange would be considered as first-line therapy. Several case studies and case reports, along with a randomized controlled trial (RCT), have demonstrated that these treatments provide improved survival of both patients and their kidneys [6].

It is critical that TPE is implemented early in the course of anti-GBM disease. Several studies have demonstrated that most patients with creatinine levels <6.6 mg/dl are able to recover their renal function [7]. However, those with an initial creatinine level >6.6 mg/dl or who are on dialysis at the time of TPE initiation are not able to recover their renal function due to irreversible glomerular injury. Such patients do not benefit from TPE, and it should not be performed unless the patient has DAH. DAH may have relatively mild manifestations, but can be rapidly fatal and responds to TPE in 90% of affected patients [8]. Therefore, a low threshold for initiating TPE is warranted in the presence of DAH. When applied, plasma should be used for the last portion of the replacement fluid. Of note, some studies have found that patients with DAH but with no renal involvement have a good outcome irrespective of the use of TPE.

Johnson et al. [6] performed the only RCT evaluating the role of TPE reported in the publicly available literature. This trial compared immunosuppression alone to immunosuppression plus plasma exchange in 17 patients, with 9 in one arm and 8 in the other. The clinical course was followed in these patients by monitoring the rate of disappearance of the anti-GBM antibody from circulation. The authors found that the rate of disappearance of pathogenic anti-GBM antibody was more rapid in the test group, which received both immunosuppressant and TPE. Moreover, the creatinine level in this group at the end of treatment was half of that of the group receiving immunosuppressant without TPE, and the use of TPE was associated with a decrease in mortality from 85 to 40% [6].

On the other hand, Cui et al. [9] reported the largest retrospective cohort of 221 patients, who had been treated from 1998 to 2008 in China. The authors compared the effects of three different treatment regimens: (1) combination therapy of plasmapheresis and immunosuppression, (2) steroids and cytotoxic agents, and (3) steroids alone. The patient and kidney survival rates were 72.7 and 25.0%, respectively, at 1 year after disease presentation. The serum levels of anti-GBM antibodies and the presentation of positive antineutrophil cytoplasmic antibodies (ANCA) were independent predictors of patient death. The serum creatinine (SCrl) at presentation was an independent predictor for renal failure. The combination therapy of plasmapheresis plus corticosteroids and cyclophosphamide had an overall beneficial effect on both patient survival and kidney survival. Specifically, the survival for patients with Goodpasture's syndrome and the kidney survival for patients with anti-GBM nephritis with initial SCr >6.8 mg/dl were both increased. The treatment with corticosteroids plus cyclophosphamide did not improve the renal outcome of disease.

Overall, TPE was shown to cause rapid lowering of anti-GBM antibody titers that was associated with lower posttreatment SCr and decreased incidence of end-stage renal disease (ESRD) and mortality. Therefore, it is recommended that patients without dialysis dependency should receive TPE initially for 14 days with each replacement consisting of 2–4 liters every day or every other day, or until anti-GBM antibodies detection is negative, using human serum albumin but adding fresh frozen plasma for patients who have undergone recent biopsy or who have DAH [10]. However, if the patient has anti-GBM disease with SCr levels >600 μmol/l, >85% glomeruli with crescent, or is already on dialysis with no alveolar hemorrhage, then TPE would be of no benefit [4, 10].

Immune Complex Crescentic Glomerulonephritis

Patients with immune complex crescentic glomerulonephritis should be treated according to their specific underlying etiology. Since TPE is weak in some forms of immune complex crescentic glomerulonephritis, such as crescentic IgA nephropathy or postinfectious crescentic glomerulonephritis, we have introduced its application in crescentic lupus nephritis patients. In patients with lupus nephritis, the standard treatment includes steroids and immunosuppressants. Our data, however, has suggested that TPE may be beneficial to improve the prognosis in patients with life-threatening lupus, severe DAH, crescentic glomerulonephritis, neurological involvement, or hemolytic uremic syndrome/thrombotic thrombocytopenic purpura (HUS/TTP) [11–13].

ANCA-Associated Glomerulonephritis

The standard approach to management of ANCA associated small vessel vasculitis is combination of corticosteroids and cytotoxic immunosuppressive drugs according to each patient's clinical situation. In patients with severe renal involvement and pulmonary hemorrhage, TPE might be beneficial as it can remove ANCAs [14]. The European League Against Rheumatism (EULAR) recommends TPE as an adjunct for selected patients with rapidly progressive severe small and medium vessel vasculitis of the kidney.

TPE is of significant value in the management of Wegener's granulomatosis with RPGN and evidence of pulmonary or DAH [15]. The strong supporting evidence for that indication, including eight RCTs along with several case studies, has led to its designation as an ASFA Category I (first-line therapy) indication for TPE. The evidence came from the European Vasculitis Study Group [16]. The study group was created in the early 1990s to investigate the diagnostic role of ANCA, to standardize ANCA testing, and to facilitate therapeutic clinical trials. This was followed by first wave testing aiming to optimize the existing therapeutic approaches and second wave testing to develop newer therapeutic approaches. Among the first-wave studies was the MEPEX study, which investigated the role of TPE in comparison to methylprednisolone as additional therapy for severe ANCA-positive glomerulonephritis [4, 16]. This RCT included all patients with ANCA-associated glomerulonephritis and SCr >6 mg/dl. The patients were randomized to receive standard drug treatment and one of the additional following treatments: (1) TPE (60 ml/kg) for seven treatments of plasma or plasma volume to a maximum of 4 liters exchange, or (2) 3 days of intravenous Solu-Medrol (15 mg/kg) for three doses, with a maximum pulse dose of 1 g. 151 patients, from nine European countries, with a minimum age of 64 years, entered the study between 1995 and 2001. Two thirds of the patients were on dialysis, and all were oliguric [17]. 14 patients withdrew, and the remaining 137 patients were randomized to either receive TPE or Solu-Medrol. After 3 months, 80% of surviving patients who received TPE had become dialysis independent, compared with the 59% of surviving patients who received steroids. While these differences were statistically significant, the difference became statistically insignificant when both therapeutic modalities were compared in all patients (alive or dead); patients that were alive and dialysis-dependent at the end of 3 months accounted for 66% of the TPE group and for 48% of the Solu-Medrol group [17]. In the long-term follow-up of the MEPEX trail, although plasma exchange decreased the risk of progression to end-stage renal failure by almost 25% at 1 year, it had no effect on longer-term renal function or survival [18]. However, another single-center study found that TPE was beneficial in treating Wegener granulomatosis patients with creatinine levels >250 μmol/l (2.85 mg/dl). The long-term prognosis was better in the TPE-treated group [19]. Although there is still uncertainty regarding the long-term effects of TPE, a recent meta-analysis demonstrated that the effect of plasma exchange did not differ significantly across the range of baseline SCr values or number of plasma exchange treatments. Though plasma exchange may decrease the composite endpoint of ESRD or death in patients with renal vasculitis, the influence on the composite endpoint of death or ESRD after 1 year was not convincing [20].

Patients with a severe form of Churg-Strauss syndrome did not display a change in 5-year cumulative survival rates following the addition of plasma ex-

Table 1. Studies showing the efficacy of plasmapheresis therapy in ANCA-associated glomerulonephritis

Study (first author)	Year	Design	TPE/no TPE	Result
Glockner [27]	1988	RCT	14/12	no benefit
Pusey [28]	1991	RCT	25/23	benefit in dialysis-dependent cases
Zauner [29]	2002	RCT	18/15	no benefit
Walsh [20]	2011	meta-analysis (nine trials)	387	may decrease the composite endpoint of ESRD or death
Gregersen [30]	2012	retrospective study	NA	benefit in improving renal function among surviving plasma exchange-treated patients not on dialysis

change to the combined treatment of corticosteroids and pulse cyclophosphamide [21]. As ANCA-associated glomerulonephritis is a disease with poor prognosis and there is still uncertainty regarding long-term effects of TPE and its application in different subgroups of the disease, more studies are needed to fully understand its benefits [22, 23].

In summary, ANCA vasculitis is a disease that presents with a varying degree of severity; thus, no standard therapy is suggested for all patients. As summarized in table 1, TPE may benefit certain patients, but the long-term effects need further study.

Potential Complications

Though complications from plasmapheresis are uncommon and minor [24, 25], they do include vascular access complication, allergic reactions, hypotension, and bleeding. Therefore, plasmapheresis in critically ill patients, especially those with cardiovascular complications, should be performed with careful monitoring [26]. Furthermore, as fresh frozen plasma and human albumin are used during the procedure, the risk of blood-borne viral infections, such as hepatitis C virus and human immunodeficiency virus, should not be neglected.

Conclusions

RPGN has traditionally been an entity of kidney disease with very poor prognosis. However, in recent years, increased awareness of RPGN and more widespread use of methylprednisolone, cyclophosphamide pulse therapy, plasmapheresis and specific immune adsorption therapy have led to significant improvements in the prognosis of RPGN. Plasmapheresis has proved to be beneficial in treating patients with RPGN and other diseases by effectively removing pathogenic molecules, however there remains much uncertainty and many studies are still needed to validate the long-term benefits and efficacy of this therapy for RPGN.

Acknowledgments

This work was supported by grants from the National Basic Research Program of China 973, Program No. 2012CB517600 (No.2012CB517604), the Leading Program of Shanghai Science and Technology Commission (114119b2700), the National Key Technology R&D Program (12-5)2011BAI10B00 (2011BAI10B06).

References

1 Jennette JC: Rapidly progressive crescentic glomerulonephritis. Kidney Int 2003;63:1164–1177.

2 Lockwood CM, Rees AJ, Pearson TA, Evans DJ, Peters DK, Wilson CB: Immunosuppression and plasma exchange in the treatment of Goodpasture's syndrome. Lancet 1976;1:711–715.

3 Lockwood CM, Boulton-Jones JM, Lowenthal RM, Simpson IJ, Peters DK: Recovery from Goodpasture's syndrome after immunosuppressive treatment and plasmapheresis. Br Med J 1975;2:252–254.

4 Szczepiorkowski ZM, Winters JL, Bandarenko N, Kim HC, Linenberger ML, Marques MB, Sarode R, Schwartz J, Weinstein R, Shaz BH: Guidelines on the use of therapeutic apheresis in clinical practice – evidence-based approach from the Apheresis Applications Committee of the American Society for Apheresis. J Clin Apher 2010;25:83–177.

5 Pusey CD: Anti-glomerular basement membrane disease. Kidney Int 2003;64:1535–1550.

6 Johnson JP, Moore J Jr, Austin HA 3rd, Balow JE, Antonovych TT, Wilson CB: Therapy of anti-glomerular basement membrane antibody disease: analysis of prognostic significance of clinical, pathologic and treatment factors. Medicine (Baltimore) 1985;64:219–227.

7 Levy JB, Turner AN, Rees AJ, Pusey CD: Long-term outcome of anti-glomerular basement membrane antibody disease treated with plasma exchange and immunosuppression. Ann Intern Med 2001;134:1033–1042.

8 Simpson IJ, Doak PB, Williams LC, Blacklock HA, Hill RS, Teague CA, Herdson PB, Wilson CB: Plasma exchange in Goodpasture's syndrome. Am J Nephrol 1982;2:301–311.

9 Cui Z, Zhao J, Jia XY, Zhu SN, Jin QZ, Cheng XY, Zhao MH: Anti-glomerular basement membrane disease: outcomes of different therapeutic regimens in a large single-center Chinese cohort study. Medicine (Baltimore) 2011;90:303–311.

10 Pusey CD, Levy JB: Plasmapheresis in immunologic renal disease. Blood Purif 2012;33:190–198.

11 Li X, Zhang W, Chen XN, Chen N: Plasma exchange and immunoadsorption in the treatment of severe lupus nephritis. Lupus 2007;16:228.

12 Li X, Ren H, Zhang Q, Zhang W, Wu X, Xu Y, Shen P, Chen N: Mycophenolate mofetil or tacrolimus compared with intravenous cyclophosphamide in the induction treatment for active lupus nephritis. Nephrol Dial Transplant 2012;27:1467–1472.

13 Zhang W, Shi H, Ren H, Shen PY, Pan XX, Li X, Chen YX, Xu YW, Chen XN, Zhu P, Chen N: Clinicopathological characteristics and outcome of Chinese patients with thrombotic thrombocytopenic purpura-hemolytic uremic syndrome: a 9-year retrospective study. Nephron Clin Pract 2009;112:c177–c183.

14 Bosch X, Guilabert A, Espinosa G, Mirapeix E: Treatment of antineutrophil cytoplasmic antibody-associated vasculitis: a systematic review. JAMA 2007;298:655–669.

15 Klemmer PJ, Chalermskulrat W, Reif MS, Hogan SL, Henke DC, Falk RJ: Plasmapheresis therapy for diffuse alveolar hemorrhage in patients with small-vessel vasculitis. Am J Kidney Dis 2003;42:1149–1153.
16 de Lind van Wijngaarden RA, Hauer HA, Wolterbeek R, Jayne DR, Gaskin G, Rasmussen N, Noel LH, Ferrario F, Waldherr R, Bruijn JA, Bajema IM, Hagen EC, Pusey CD: Chances of renal recovery for dialysis-dependent ANCA-associated glomerulonephritis. J Am Soc Nephrol 2007;18:2189–2197.
17 Jayne DR, Gaskin G, Rasmussen N, Abramowicz D, Ferrario F, Guillevin L, Mirapeix E, Savage CO, Sinico RA, Stegeman CA, Westman KW, van der Woude FJ, de Lind van Wijngaarden RA, Pusey CD: Randomized trial of plasma exchange or high-dosage methylprednisolone as adjunctive therapy for severe renal vasculitis. J Am Soc Nephrol 2007;18:2180–2188.
18 Smith RM, Jones RB, Jayne DR: Progress in treatment of ANCA-associated vasculitis. Arthritis Res Ther 2012;14:210.
19 Szpirt WM, Heaf JG, Petersen J: Plasma exchange for induction and cyclosporine A for maintenance of remission in Wegener's granulomatosis – a clinical randomized controlled trial. Nephrol Dial Transplant 2011;26:206–213.
20 Walsh M, Catapano F, Szpirt W, Thorlund K, Bruchfeld A, Guillevin L, Haubitz M, Merkel PA, Peh CA, Pusey C, Jayne D: Plasma exchange for renal vasculitis and idiopathic rapidly progressive glomerulonephritis: a meta-analysis. Am J Kidney Dis 2011;57:566–574.
21 Guillevin L, Lhote F, Cohen P, Jarrousse B, Lortholary O, Genereau T, Leon A, Bussel A: Corticosteroids plus pulse cyclophosphamide and plasma exchanges versus corticosteroids plus pulse cyclophosphamide alone in the treatment of polyarteritis nodosa and Churg-Strauss syndrome patients with factors predicting poor prognosis. A prospective, randomized trial in sixty-two patients. Arthritis Rheum 1995;38:1638–1645.
22 Chen YX, Yu HJ, Zhang W, Ren H, Chen XN, Shen PY, Xu YW, Li X, Pan XX, Ni LY, Wang WM, Chen N: Analyzing fatal cases of Chinese patients with primary antineutrophil cytoplasmic antibodies-associated renal vasculitis: a 10-year retrospective study. Kidney Blood Press Res 2008;31:343–349.
23 Chen YX, Zhang W, Chen XN, Shen PY, Shi H, Xu YW, Ren H, Chen N: Application of rifle criteria in Chinese patients with ANCA-associated renal vasculitis. Clin Exp Rheumatol 2011;29:951–957.
24 Pusey C, Dash C, Garrett M, Gascoigne E, Gesinde M, Gillanders K, Wallington T: Experience of using human albumin solution 4.5% in 1,195 therapeutic plasma exchange procedures. Transfus Med 2010;20:244–249.
25 Stegmayr B, Ptak J, Wikstrom B, Berlin G, Axelsson CG, Griskevicius A, Centoni P, Liumbruno G, Molfettini P, Audzijoniene J, Mokvist K, Sojka BN, Norda R, Knutson F, Ramlow W, Blaha M, Witt V, Evergren M, Tomaz J: World apheresis registry 2003–2007 data. Transfus Apher Sci 2008;39:247–254.
26 Chen XN, Pan XX, Yu HJ, Shen PY, Zhang QY, Zhang W, Ren H, Qian Y, Zhu P, Chen N: Analysis of cardiovascular disease in Chinese inpatients with chronic kidney disease. Intern Med 2011;50:1797–1801.
27 Glockner WM, Sieberth HG, Wichmann HE, Backes E, Bambauer R, Boesken WH, Bohle A, Daul A, Graben N, Keller F, et al: Plasma exchange and immunosuppression in rapidly progressive glomerulonephritis: a controlled, multi-center study. Clin Nephrol 1988;29:1–8.
28 Pusey CD, Rees AJ, Evans DJ, Peters DK, Lockwood CM: Plasma exchange in focal necrotizing glomerulonephritis without anti-GBM antibodies. Kidney Int 1991;40:757–763.
29 Zauner I, Bach D, Braun N, Kramer BK, Funfstuck R, Helmchen U, Schollmeyer P, Bohler J: Predictive value of initial histology and effect of plasmapheresis on long-term prognosis of rapidly progressive glomerulonephritis. Am J Kidney Dis 2002;39:28–35.
30 Gregersen JW, Kristensen T, Krag SR, Birn H, Ivarsen P: Early plasma exchange improves outcome in PR3-ANCA-positive renal vasculitis. Clin Exp Rheumatol 2012;30:S39–S47.

Prof. Nan Chen, MD
Department of Nephrology, Ruijin Hospital
School of Medicine, Shanghai Jiao Tong University
No. 197 Ruijin Er Road, Shanghai 200025 (PR China)
E-Mail chen-nan@medmail.com.cn

Author Index

Appel, G.B. 163

Barbour, S. 143

Cattran, D. 143
Chen, N. IX, 1, 22, 31, 75, 91, 101, 109, 152, 175, 207, 229, 240
Chen, X. 22, 31, 175, 240
Chen, Y.-X. 207
Chuang, P.Y. 41
Coppo, R. 84

Feng, X. 22, 109
Frémeaux-Bacchi, V. 185

Gharavi, A. 52
Glassock, R.J. 131

Hao, C. 22
Hao, X. 91, 101
He, J.C. 41
Hruskova, Z. 216

Lai, K.N. 194
Lesavre, P. 185
Li, X. 22, 31, 109, 152, 175, 229
Li, Y. 31
Lu, Y. 175

Ma, J. 91, 101
Menon, M.C. 41

Noël, L.-H. 185

Pan, X. 22, 101, 109, 152, 175

Reich, H. 143
Remuzzi, G. 12
Ren, H. 22, 31, 91, 101, 109, 152, 175
Ronco, P. 12

Servais, A. 185
Shapiro, S. 52
Shen, P. 22, 31, 109, 152, 175
Shi, H. 31

Tang, S.C.W. 194
Tesar, V. 216
Tomino, Y. 65

Wang, W. 22, 31, 75, 152, 175
Wang, X. 119
Wang, Z. 91, 101
Weening, J.J. 12
Wu, P. 22
Wu, X. 31

Xie, J. 1, 31, 52, 91, 101
Xu, H. 119
Xu, J. 22, 152
Xu, Y. 22, 152, 175

Yang, L. 91, 101

Zhang, Q. 91, 109
Zhang, W. 22, 31, 109, 152, 175
Zhou, Q. 101
Zhu, B. 91

Subject Index

ACTN4, mutation analysis in focal segmental glomerulosclerosis 94–98
Aldose reductase, autoantibodies in membranous nephropathy 138
ANCA-associated glomerulonephritis
  endothelial cell injury 17
  lysosome-associated membrane protein-2 in pathogenesis 212
  neutrophil extracellular traps 212
  plasmapheresis 243–245
ANCA-associated vasculitis
  classification
    overview 217, 218
    renal histology 218, 219
  epidemiology 216
  outcomes 221, 222
  pathogenesis 219–221
  treatment
    overview 222, 223
    rituximab 224, 225
Anemia, glomerulonephritis studies
  etiology of kidney disease impact 34
  glomerular filtration rate 34, 35, 37
  overview 31–33
  study design 33, 34
Angiotensin-converting enzyme inhibitors
  focal segmental glomerulosclerosis management 112, 113
  idiopathic membranous nephropathy management 7
  immunoglobulin A nephropathy management 6, 78, 79
Angiotensin receptor blockers
  focal segmental glomerulosclerosis management 112, 113
  idiopathic membranous nephropathy management 7
  immunoglobulin A nephropathy management 6, 78, 79
Antiglomerular basement membrane glomerulonephritis
  dendritic cells in pathogenesis 210
  Goodpasture autoantigen 211
  growth factor receptors 210
  parietal epithelial cells in pathogenesis 211
  plasmapheresis 242, 243
  podocyte activation 211
  T cells in pathogenesis 209, 210
  Toll-like receptors in pathogenesis 208, 209
  treatment 230–232
Azathioprine
  ANCA-associated vasculitis management 222, 225
  immunoglobulin A nephropathy management 7
  nephrotic syndrome management 127

Bovine serum albumin, antibodies in membranous nephropathy 138

C3 glomerulopathy
  classification 186
  clinical features 190, 191

complement activation alternative pathway 188–190
deposits and composition 188
histology 186–188
treatment 191, 192
Calcium, *see* Hypocalcemia
CD68, immunoglobulin A nephropathy susceptibility 60
Chlorambucil, nephrotic syndrome management 125
COL11A2, immunoglobulin A nephropathy susceptibility 55, 56
Complement, *see also* C3 glomerulopathy
C3 glomerulopathy
complement activation alternative pathway 188–190
deposits and composition 188
hemolytic uremic syndrome endothelial cell injury 18, 19
immunoglobulin A nephropathy
activation 44
prognostic value 69, 70
membranoproliferative glomerulonephritis, complement activation pathway 168–170
overview of pathways 17, 18
Complement factor H
C3 glomeropathy management 191
CFH and *CFHR* in immunoglobulin A nephropathy susceptibility 54, 58, 59
deficiency 189, 190
Corticosteroids, *see* Glucocorticoids
Crescentic glomerulonephritis, *see also* Antiglomerular basement membrane glomerulonephritis, Rapidly progressive glomerulonephritis
renal biopsy findings in China 26, 27
treatment
antiglomerular basement membrane glomerulonephritis 230–232
immunoglobulin A nephropathy 232
lupus nephritis 233
mycophenolate mofetil 234, 235
novel therapies 236
pauci-immune disease 233
postinfectious disease 232, 233
rituximab 235
tumor necrosis factor-α antagonism 236
Cyclophosphamide
ANCA-associated vasculitis management 222, 223
antiglomerular basement membrane glomerulonephritis management 230, 231
C3 glomerulopathy management 191
focal segmental glomerulosclerosis management 115
idiopathic membranous nephropathy management 7, 8
immunoglobulin A nephropathy management 7, 80
membranoproliferative glomerulonephritis management 8, 178
membranous nephropathy idiopathic disease treatment with steroid combination therapy 153–161
minimal change disease management 3, 4
nephrotic syndrome management 124–126
Cyclosporin A
focal segmental glomerulosclerosis management 5, 115
minimal change disease management 3
nephrotic syndrome management 123, 124

Deep vein thrombosis, *see* Venous thromboembolism
DEFA, immunoglobulin A nephropathy susceptibility 59, 60
Dendrin, injured podocyte marker 71, 72
Dendritic cell, antiglomerular basement membrane glomerulonephritis pathogenesis 210
Dyslipidemia, membranous nephropathy association 144, 145

Eculizumab
C3 glomerulopathy management 191, 192
membranoproliferative glomerulonephritis management 172
Edema, management in membranous nephropathy 145, 146
EIF4A1, immunoglobulin A nephropathy susceptibility 60
Endocapillary proliferative glomerulonephritis, renal biopsy findings in China 26, 27
α-Enolase, autoantibodies in membranous nephropathy 138

FK506, *see* Tacrolimus
Focal segmental glomerulosclerosis, *see also* Minimal change disease
clinicopathological characteristics 105
epidemiology in China 4, 5, 92
familial versus sporadic disease prognosis 102–107
gene mutation studies in familial disease
ACTN4 94–98
overview 102
TRPC6 94–98
podocyte injury 13, 14
prognosis by treatment type
angiotensin-converting enzyme inhibitors 112, 113
baseline characteristics 111, 112
glucocorticoid
alone 112–114
immunosuppressant combination therapy 114–116
study design 110, 111
renal biopsy findings in China 26–28
spectrum of disease 13
therapeutic targets 85–88

GATSL3, immunoglobulin A nephropathy susceptibility 59
Glomerular filtration rate, anemia and glomerulonephritis 34, 35, 37
Glomerular visceral epithelial cell, membranous glomerulonephropathy pathology 14–16
Glomerulonephritis, *see also* specific diseases
anemia studies 31–39
epidemiology in China 1–3, 23–26
hepatitis C virus-associated glomerulonephritis, *see* Hepatitis C virus
mineral metabolism studies 31–39
Glucocorticoids
ANCA-associated vasculitis management 223
antiglomerular basement membrane glomerulonephritis management 230, 231
focal segmental glomerulosclerosis management
alone 112–114
immunosuppressant combination therapy 4, 5, 114–116
immunoglobulin A nephropathy management 79–81
membranoproliferative glomerulonephritis management 178
membranous nephropathy idiopathic disease treatment with steroid combination therapy
cyclophosphamide 153–161
tacrolimus 153–161
minimal change disease management 3
nephrotic syndrome management with methylprednisolone 123
Goodpasture autoantigen, antiglomerular basement membrane glomerulonephritis 211

Hemolytic uremic syndrome, complement and endothelial cell injury 18, 19
Hepatitis B virus, membranoproliferative glomerulonephritis association 180, 181
Hepatitis C virus
discovery 194, 195

epidemiology 195, 196
extrahepatic manifestations 196, 197
glomerulonephritis
clinical findings 198
cryoglobulinemic
glomerulonephritis 198, 199
laboratory findings 199–201
pathogenesis 197, 198
treatment 201–204
membranoproliferative
glomerulonephritis
association 180, 181, 198
natural history of infection 196
virology and replication 195
HLA-DPA1, immunoglobulin A
nephropathy susceptibility 55
HLA-DPB1, immunoglobulin A
nephropathy susceptibility 55, 56
HLA-DQA1, immunoglobulin A
nephropathy susceptibility 55
HLA-DQB1, immunoglobulin A
nephropathy susceptibility 55
HLA-DQB2, immunoglobulin A
nephropathy susceptibility 55
HLA-DRB1, immunoglobulin A
nephropathy susceptibility 55
HORMAD2, immunoglobulin A
nephropathy susceptibility 59
Hyperphosphatemia, glomerulonephritis
studies 35, 36, 38, 39
Hypertension, membranous nephropathy
association 144
Hypocalcemia, glomerulonephritis
studies 35, 38, 39

Idiopathic membranous nephropathy, *see*
Membranous nephropathy
Immunoglobulin A nephropathy
epidemiology in China 5–7
genetics
candidate gene studies 53, 54
genome-wide association
studies 54–60
linkage analysis 53, 54
risk score 60, 61
lesions and frequency 41, 42
management
clinical factors in treatment
scheme 77
cyclophosphamide 80
glucocorticoids 79–81
mycophenolate mofetil 80, 81
supportive therapy
angiotensin blockade 78, 79
antiplatelet/anticoagulant
therapy 79
tonsillectomy 78
tacrolimus 91
overview 75, 76
pathogenesis
complement activation 44
immunoglobulin A1 complexes 43,
44
modeling 61
overview 53
podocyte injury
clinicopathological
correlation 47, 48
mesangio-podocyte
crosstalk 45–47
prognosis
clinical predictors 66, 67
complement activation 69, 70
glomerular changes 67–69
histological changes 67–69
podocyte injury 70–72
tubulointerstitial changes 69
renal biopsy findings in China 26–28
Intercellular adhesion molecule-1,
immunoglobulin A nephropathy
expression 67
Interferon, *see* Pegylated interferon

LIF, immunoglobulin A nephropathy
susceptibility 59
Lupus nephritis
chronic lesions 19
crescentic glomerulonephritis
management 233
renal biopsy findings in China 26
Lysosome-associated membrane
protein-2
ANCA-associated glomerulonephritis
pathogenesis 212

ANCA-associated vasculitis
autoantibodies 219

Mast cell, immunoglobulin A nephropathy prognostic use 69
Membranoproliferative glomerulonephritis
classification 163–167, 176, 180–182, 186
clinicopathological correlates 177, 178
complement activation 168–170
complement and endothelial cell injury 18
epidemiology in China 8, 9
evaluation of patients 170, 171
pathology 14–17, 167, 168
renal biopsy findings in China 26, 27
treatment
angiotensin blockade 171
cyclophosphamide 178
eculizumab 172
glucocorticoids 178
tacrolimus 178
Membranous nephropathy
autoantibodies in primary disease
aldose reductase 138
α-enolase 138
neutral endopeptidase 137
phospholipase A_2 receptor-1 132–136
superoxide dismutase 138
bovine serum albumin antibodies 138
complications
dyslipidemia 144, 145
edema management 145, 146
hypertension 144
infection 149
proteinuria 145
venous thromboembolism
anticoagulation therapy 148, 149
hypoalbuminuria in prediction 147
pathophysiology 148
risks 146, 147
history of study 131, 132
idiopathic disease
epidemiology in China 7, 8
treatment with steroid combination therapy
cyclophosphamide 153–161
tacrolimus 153–161
pathology 13–17
recurrence in renal allografts 136, 137
renal biopsy findings in China 26
secondary disease and antigen-antibody systems 138, 139
Mesangioproliferative glomerulonephritis, renal biopsy findings in China 26, 28
Methylprednisolone, *see* Glucocorticoids
Minimal change disease
epidemiology in China 3, 4
podocyte injury 13, 14
progression, *see* Focal segmental glomerulosclerosis
renal biopsy findings in China 27
Mixed cryoglobulinemia, hepatitis C virus association 197
Mizoribine, nephrotic syndrome management 127
MPDU1, immunoglobulin A nephropathy susceptibility 60
MTMR3, immunoglobulin A nephropathy susceptibility 59
Mycophenolate mofetil
C3 glomerulopathy management 191
crescentic glomerulonephritis management 234, 235
immunoglobulin A nephropathy management 7, 80, 81
membranoproliferative glomerulonephritis management 8
minimal change disease management 3
nephrotic syndrome management 123, 124, 126

Nephrotic syndrome
classification 120
overview of primary disease 119, 120
pathology 121, 122

relapse 120
remission 121
renal biopsy indications 122
treatment
initial episode 122
monitoring 127, 128
relapse
azathioprine 127
chlorambucil 125
cyclophosphamide 124–126
cyclosporine 123, 124
methylprednisolone 123
mizoribine 127
mycophenolate mofetil 123, 124, 126
rituximab 126, 127
tacrolimus 123, 124
Neutral endopeptidase, autoantibodies in membranous nephropathy 137
Neutrophil extracellular traps, ANCA-associated glomerulonephritis 212
Nuclear factor-κB, focal segmental glomerulosclerosis therapeutic targeting 87, 88

OSM, immunoglobulin A nephropathy susceptibility 59

Parathyroid hormone, glomerulonephritis serum levels 36, 37
Parietal epithelial cell, antiglomerular basement membrane glomerulonephritis pathogenesis 211
Pegylated interferon, hepatitis C virus-associated glomerulonephritis management 202–204
Phospholipase A_2 receptor-1, autoantibodies in membranous nephropathy 132–136
Phosphorus, *see* Hyperphosphatemia
Plasmapheresis
antiglomerular basement membrane glomerulonephritis management 231
overview 240, 241
rapidly progressive glomerulonephritis management
ANCA-associated glomerulonephritis 243–245
antiglomerular basement membrane glomerulonephritis 242, 243
complications 245
immune complex crescentic glomerulonephritis 243
modes
double-filtration plasmapheresis 241
immunoadsorption 241
intensified plasma exchange 241
Podocalyxin, injured podocyte marker 71
Podocyte
antiglomerular basement membrane glomerulonephritis activation 211
focal segmental glomerulosclerosis
pathogenesis 13, 14
therapeutic targeting 85, 86
immunoglobulin A nephropathy
injury
clinicopathological correlation 47, 48
mesangio-podocyte crosstalk 45–47
prognostic value 70–72
Prednisolone, *see* Glucocorticoids
Prednisone, *see* Glucocorticoids
Proteinuria, membranous nephropathy association 145
PSMB8, immunoglobulin A nephropathy susceptibility 57, 58
PSMB9, immunoglobulin A nephropathy susceptibility 57, 58
Pulmonary embolism, *see* Venous thromboembolism

Rapidly progressive glomerulonephritis, *see also* Crescentic glomerulonephritis
ANCA-associated glomerulonephritis
lysosome-associated membrane protein-2 in pathogenesis 212
neutrophil extracellular traps 212

antiglomerular basement membrane glomerulonephritis
dendritic cells in pathogenesis 210
Goodpasture autoantigen 211
growth factor receptors 210
parietal epithelial cells in pathogenesis 211
podocyte activation 211
T cells in pathogenesis 209, 210
Toll-like receptors in pathogenesis 208, 209
classification 207, 208
plasmapheresis
ANCA-associated glomerulonephritis 243–245
antiglomerular basement membrane glomerulonephritis 242, 243
complications 245
immune complex crescentic glomerulonephritis 243
modes
double-filtration plasmapheresis 241
immunoadsorption 241
intensified plasma exchange 241
overview 240, 241
Renal vein thrombosis, *see* Venous thromboembolism
Ribavirin, hepatitis C virus-associated glomerulonephritis management 201–204
Rituximab
ANCA-associated vasculitis management 224, 225
crescentic glomerulonephritis management 235
nephrotic syndrome management 126, 127

Saquinavir, focal segmental glomerulosclerosis studies 87, 88
SF3A1, immunoglobulin A nephropathy susceptibility 59
SOX15, immunoglobulin A nephropathy susceptibility 60
Superoxide dismutase, autoantibodies in membranous nephropathy 138

Tacrolimus (FK506)
focal segmental glomerulosclerosis management 5, 115
immunoglobulin A nephropathy management 91
membranoproliferative glomerulonephritis management 178
membranous nephropathy idiopathic disease treatment with steroid combination therapy 153–161
minimal change disease management 3
nephrotic syndrome management 123, 124
TAP1, immunoglobulin A nephropathy susceptibility 57, 58
TAP2, immunoglobulin A nephropathy susceptibility 57, 58
T cell
antiglomerular basement membrane glomerulonephritis pathogenesis 209, 210
focal segmental glomerulosclerosis therapeutic targeting 85, 86
TNSF13, immunoglobulin A nephropathy susceptibility 60
Toll-like receptors, antiglomerular basement membrane glomerulonephritis pathogenesis 208, 209
Tonsillectomy, immunoglobulin A nephropathy management 78
TP53, immunoglobulin A nephropathy susceptibility 60
Transforming growth factor-β, immunoglobulin A nephropathy role 45, 46
TRPC6, mutation analysis in focal segmental glomerulosclerosis 94–98
Tumor necrosis factor-α
crescentic glomerulonephritis therapeutic targeting 236

immunoglobulin A nephropathy
role 45, 46

Venous thromboembolism, membranous nephropathy
anticoagulation therapy 148, 149
hypoalbuminuria in prediction 147
pathophysiology 148
risks 146, 147
Vitamin D, metabolism in glomerulonephritis 38, 39